航空类专业职业教育系列"十三五"规划教材

HANGKONG DIMIAN SHEBEI ZHUANYE YINGYU

航空地面设备专业英语

主　编　殷向东　徐红波
副主编　杨　林　李家宇

西北工业大学出版社
西安

【内容简介】　本书结合车辆专业英语和航空地面设备维修手册(英文版),主要介绍汽车发动机、汽车底盘的构造和工作原理等有关车辆的专业知识,学习有关车辆的英语词汇;介绍飞机牵引车、飞机气源车、飞机地面电源、货运升降平台和飞机加油车等航空地面设备构造、操作和维护的专业知识,学习航空地面设备的英语词汇。

本书可作为高等学校航空地面设备维修专业教材,也可作为航空地面设备维修相关从业人员的参考用书。

图书在版编目（CIP）数据

航空地面设备专业英语/殷向东,徐红波主编 . —西安：
西北工业大学出版社,2018.4
航空类专业职业教育系列"十三五"规划教材
ISBN 978 - 7 - 5612 - 5900 - 9

Ⅰ.①航… 　Ⅱ.①殷… ②徐… 　Ⅲ.①航空设备—
地面设备—英语—职业教育—教材 　Ⅳ.①V24

中国版本图书馆 CIP 数据核字(2018)第 057730 号

策划编辑:华一瑾
责任编辑:李阿盟　朱辰浩

出版发行　西北工业大学出版社
通信地址　西安市友谊西路 127 号　　邮编:710072
电　　话　(029)88493844　88491757
网　　址　www.nwpup.com
印 刷 者　兴平市博闻印务有限公司
开　　本　787 mm×1 092 mm　　1/16
印　　张　23.75
字　　数　580 千字
版　　次　2018 年 4 月第 1 版　　2018 年 4 月第 1 次印刷
定　　价　58.00 元

前　言

随着中国民航业的快速发展,飞机数量不断增加,机队规模日益增大,国内各大型机场的运输周转量也在持续增加。民航运输业的高速发展使航空地面设备的需求不断增加。航空地面设备的增加,提高了对航空地面设备人才的需求量。在各类航空地面设备中,价格昂贵、操作与维修复杂的型号往往属于进口产品,其操作与维护说明等技术资料采用英文撰写,这对航空地面设备维修人员的英语水平提出了较高的要求。因此,掌握一定的航空地面设备专业英语就成为航空地面设备维修人员的必备技能。

本书根据民航行业标准对航空地面设备维修专业人才的要求,从实际需要出发,围绕航空地面设备维修高级技术人员工作所需的基本知识和技能而选材、编写。为便于读者阅读和理解,本书采用图文并茂的形式,对航空地面设备进行简要介绍,使专业内容更加直观、具体、形象、生动,从而巩固、加深对航空地面设备专业术语的理解。本书可以作为民航高等职业教育航空地面设备维修专业的专业英语课程教材,也可供民航航空地面设备维修企业职业教育培训使用。

本书共分14章,内容覆盖航空地面设备基础知识、主要航空地面设备型号的结构和维护知识、航空地面设备的历史与未来,具体包括汽车基础知识、液压与液力传动基础知识、清水车、污水车、食品配餐车、升降平台车、行李传送车、电源车、气源车、无杆式飞机牵引车、有杆式飞机牵引车、飞机除冰车、航空地面设备发展史以及航空地面设备减排战略。

本书由上海民航职业技术学院殷向东和广州民航职业技术学院徐红波担任主编并统稿,广州民航职业技术学院杨林、李家宇担任副主编。具体编写分工如下:殷向东编写第4,5,7,11,12和13章,徐红波编写第2,6,8和9章,杨林编写第1和10章,李家宇编写第3和14章。

本书参考了大量的国内外技术资料,也得到了许多同行的大力支持,在此谨向所有参考资料的作者及关心支持本书编写的同仁们表示感谢。

由于水平有限,书中不妥之处,垦请读者批评指正。

编　者

2018 年 1 月

目 录

Chapter 1　General of an Automobile

Section 1. 1　General Information

1. 1. 1　Basic Components of an Automobile

An automobile probably has about 7,000 different parts in it. Some of them make it more comfortable or better looking, but most of them are to make it run. The three basic components of the automobile are the engine, chassis and body (see Figure 1.1).

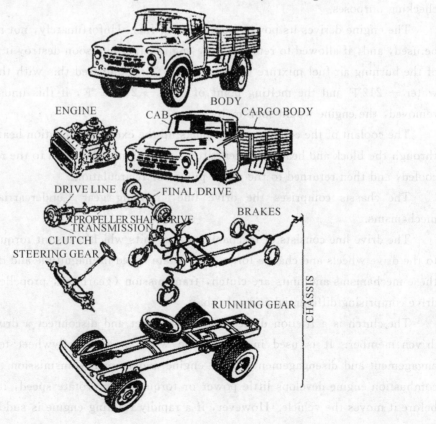

Figure 1. 1　Basic components of the automobile

The engine converts the fuel energy into mechanical power. An internal combustion engine powers our modern automobile. The engine burns its fuel within the engine proper, as compared to a steam engine where the fuel is burned externally. The gasoline and air mixture of the internal combustion engine is compressed by a piston inside an airtight cylinder and ignited by a spark. The trapped air-fuel mixture burns fiercely, causing tremendous heat which expands the trapped gases and pushes the piston down. This is the motive power of the automobile. The automobile engine is essentially a heat engine. It requires fuel to burn, a spark to ignite, lubrication to minimize friction, and a cooling system to dissipate unwanted heat.

The fuel system takes a correctly proportioned mixture of gasoline and air to burn and develops the power needed to push the piston down the cylinder. To store, mix, and deliver this air-fuel mixture is the duty of the fuel system.

When the explosive mixture of air and gasoline is compressed tightly, a spark is sent through the mixture, setting it on fire, which results in the heat and expansion used to push the piston down the cylinder. The ignition system furnishes a spark to each spark plug when its cylinder is full of the compressed air-fuel mixture.

The lubrication system provides a constant flow of filtered oil to all moving parts of the engine. The system consists of an oil pan to store the oil, a pump to circulate it, a filter to remove solid abrasive particles, and an oil gauge or light in the driver's compartment for checking purposes.

The engine derives its power from burning fuel. Unfortunately, not all of this heat can be used, and, if allowed to remain in the engine, it would soon destroy it. The temperature of the burning air-fuel mixture is about 4,500 °F. Compared this with the boiling point of water — 212°F and the melting point of iron — 2,500 °F, if this unused heat were not removed, the engine would soon melt.

The coolant of the cooling system picks up the excess combustion heat as it is circulated through the block and heads by a centrifugal-type pump, delivered to the radiator where it is cooled, and then returned to the water pump for recirculation.

The chassis comprises the drive line, running gear (undercarriage) and control mechanisms.

The drive line consists of mechanisms and units which transmit torque from the engine to the drive wheels and change torque and rotate speed in magnitude and directions. Among these mechanisms and units are clutch, transmission (gearbox), propeller shaft, and final drive comprising differential and axle shaft.

The clutch is a friction device used to connect and disconnect a driving force from a driven member. It is used in conjunction with an engine flywheel to provide smooth engagement and disengagement of the engine and manual transmission. Since an internal combustion engine develops little power or torque at low rotate speed, it must gain speed before it moves the vehicle. However, if a rapidly rotating engine is suddenly connected to the drive line of a stationary vehicle, a violent shock will result.

A transmission is a speed and power changing device installed at some point between the engine and driving wheels of the vehicle. It provides a means for changing the ratio between engine rotate speed and driving wheel rotate speed to best meet each particular driving situation. It converts torque in magnitude and the direction, allows the automobile to move forth and back and the engine to be disconnected from the drive line for a longer period of time.

The propeller shaft is used to transmit torque from the transmission to the final drive at varying angles. The universal joints serve to compensate for changes in the line of drive by transmitting power from a driving shaft through an angle to a driven shaft. Most cars use two or three universal joints in the drive line between the transmission and differential.

The final drive changes torque and transmits it from the propeller shaft through the differential to the axle shafts at a constant angle.

The differential is a gear system that transfers power from the drive shaft to the driving axles. It also permits one driving wheel to turn faster than the other to prevent skidding and scuffing of tires on turns.

The running gear is the backbone of the automobile; it includes the frame, front and rear axles, springs, shock absorbers, wheels and tires.

The control mechanism consists of the steering system for changing the direction of movement and the brakes for decelerating and stopping the automobile.

The body of the truck comprises a cargo body and a driver's cab. The fenders, radiator grille, hood, and mudguards also belong to the body.

1.1.2　Layout of an Automobile

The layout of different types of vehicles is different. A private car which is to carry up to eight persons has generally four sectors.

The layout of a car is shown in Figure 1.2. It shows the position of the main parts of an automobile. It consists of engine located at the front of the vehicle and followed by a clutch, gearbox, propeller shaft, universal joint, differential, rear axle etc. The radiator is located in front of the engine. Other various parts of the vehicle shown in the layout are dynamo, horn, steering box, fan, timing gear, carburetor, air filter, gear control, steering wheel, cylinder, petrol tank, rear axle and front axle. The drive from the gear box is conveyed through a short shaft to the front universal joint of the propeller shaft. From the propeller shaft it is conveyed to the rear universal joint through a sliding splined type of joint. The bevel gear of the short shaft is driven by the rear universal joint. This bevel gear meshes with a larger bevel gear which drives the two rear axle shafts through a differential gear.

The layout also consists of independent front-wheel springing with quarter-elliptic leaf springs, steering column bevel-gear control and hydraulic braking system.

The wheels which are four in number are fitted below the car chassis to support the load of the vehicle and passengers as well as to run the car. They are fitted with hollow rubber tyres filled with air in rubber tubes under sufficient pressure necessary for carrying the load. The shocks caused by road irregularities are absorbed by them. By fitting springs between the wheels and the vehicle to allow the vertical movement of wheels in relation to vehicle, greater part of unevenness of road surfaces is taken care of.

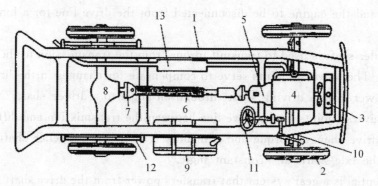

Figure 1.2　Layout of a car

1—frame；　2—wheels；　3—radiator；　4—engine；　5—clutch and gearbox；　6—propeller shaft；　7—universal joint；
8—differential, rear axle；　9—fuel tank；　10—front axle；　11—steering wheel；　12—road springs；　13—exhaust/silencer

Front axle is used for steering front wheels carried on stub axles swiveling upon king pins at the axle extremities.

Steering arms and a track rod link the two stub axles together for swiveling them by a steering wheel above the king pins. The steering wheel linked to one of the stub axles by a shaft, a gear box and, a suitable linkage is operated by the driver's hand wheel. Previously the axle — a one-piece beam was used to support the vehicle through springs. An arrangement known as independent front suspension has replaced the axle and spring arrangement. Under the control of springs, the wheels are free to rise and fall vertically independently of each other.

For fixing rear wheels, a tube like shaft enclosing driving shafts with suitable bearings for rotating the wheels is used. It is enlarged at the center for enclosing the final-drive gears used for providing main speed reduction between the engine and the driving wheels. The change of direction of the drive from the fore and aft line of the propeller shaft to the transverse line of the axle shafts is also provided by this tube known as rear axle.

When going round a curve, the inner wheel has to travel a smaller distance in comparison to the outer wheel. But both the rear wheels would rotate at the same speed if they are connected by a shaft. This rotation of both the wheels would result in slipping of one or both of them on the road surface causing excessive tyre wear as well as severe twisting loads on the shaft. Moreover, the two wheels of the exactly similar diameter (which is not usually so) can only turn at the same speed without slipping on the straight road. Tyres fitted on the opposite sides may be of different states of wear and even tyre of same nominal diameter made by different or same manufacturer may differ in actual dimensions or may not be exactly similar. Due to change of rolling radius (the distance from the wheel center to the ground), the effective size of the tyre may be altered by different inflation pressure also.

Each wheel is provided with its own separate half-shaft connected by a differential gear and meeting at about the center of the axle. The wheels are free to rotate at different speeds although they are provided with equal drive by the differential gear.

For preventing the transmission of shock from uneven road surfaces to the vehicle, springs are used to support the vehicle on the axle.

In order to allow for the vertical movements of the wheels relative to the frame as well as to allow the parts of the shaft to operate at different angle, another increasing arrangement is used. It consists of mounting the final-drive gears and the differential gear in a casing attached to the frame with independently sprung wheels attached to them by means of shafts through devices called universal joints.

Power unit consists of an internal combustion engine. It is usually mounted at the end point of the car. The clutch and the gear box are placed immediately behind it. The three components — engine, clutch and gear box — are assembled into a single unit.

For connecting the output shaft of the gear box to the rear axle, a long shaft known as propeller shaft is used. This shaft is either enclosed in a tubular casing or kept exposed or opened with a universal joint fitted at each end for allowing the changes in the shaft alignment with the rise and fall of the rear axle due to road surface variations. Universal joints cannot be eliminated even if the final drive gears are fixed to the frame with the wheel springing independently. Neither the misalignment resulting from the flexing of the vehicle structure over bumpy road surfaces can be avoided nor the precise alignment of shaft can be ensured without them.

For controlling the movement of the vehicles or to stop them, efficient braking system is a necessity for a vehicle. Brakes attached to each of the four wheels are of two types. In the initial type, a pair of shoes carried on a stationary plate is expanded in contact with a rotating drum mounted on the wheels to arrest the motion of the drum. In the modern type of brakes, one or more pairs of pads are carried in a caliper attached to the axle or wheel supporting linkage. The sides of the disk mounted on the wheel are griped by these pads. By applying pressure on a pedal, the brakes are applied. A hand lever acting through a separate linkage and locked in the on-position is used.

For operating the brake, either mechanical or hydraulic system is used. Mechanical system requiring gearing system for mechanical and hydraulic fluid for the hydraulic brakes are used.

New Words

1. automobile [ɔːtəməbiːl] *n.* 汽车，车辆
2. mechanical [miˈkænikəl] *adj.* 机械的
3. chassis [ʃæsi] *n.* 底盘，底架

4. airtight ['eətait] *adj.* 不漏(透)气的

5. ignite [ig'nait] *vt.* 点火(燃)

6. cylinder ['silində(r)] *n.* 汽缸

7. coolant [ku:lənt] *n.* 冷却器

8. radiator ['reidieitə] *n.* 冷却器

9. clutch [klʌtʃ] *n.* 离合器

10. skidding ['skidiŋ] *n.* 打滑,滑动

11. brake [breik] *n.* 制动器

12. fender [fendə(r)] *n.* 挡泥板

13. vehicle ['vi:əkl] *n.* 车辆

14. dynamo [dainəməʊ] *n.* 发电机

15. carburetor ['ka:bjʊretə] *n.* 化油器

16. mesh[meʃ] *n.* *v.* 啮合

17. drum[drʌm] *n.* 制动鼓

18. caliper['kælipə] *n.* 制动钳

Phrases and Expressions

1. an internal combustion engine 内燃机

2. engine proper 发动机机体

3. air-fuel mixture 空气-燃油混合气

4. motive power 驱动功率,推进力

5. spark plug 火花塞

6. oil pan 油底壳

7. drive line 动力传动系统(路线)

8. universal joint 万向节,万向接头

9. propeller shaft 传动轴

10. the bevel gear 锥齿轮,伞齿轮,斜形齿轮

11. quarter-elliptic leaf springs 1/4 椭圆形钢板弹簧

12. stub axle 转向轴,丁字轴

13. swiveling upon king pins 在转向节主销上转动

14. independent front suspension 前独立悬架

Notes to the Text

1. The engine burns its fuel within the engine proper, as compared to a steam engine where the fuel is burned externally.

与燃料在其外部燃烧的蒸汽机相比,内燃机的燃料燃烧在其机体内部进行。

2. The drive line consists of mechanisms and units which transmit torque from the engine to the drive wheels and change torque and rpm in magnitude and directions.

行驶(或传动)系统由将发动机的转矩传递到车轮的机构和元件构成,并控制转矩和转速

的大小及行驶的方向。

3. When going round a curve, the inner wheel has to travel a smaller distance in comparison to the outer wheel. But both the rear wheels would rotate at the same speed if they are connected by a shaft. This rotation of both the wheels would result in slipping of one or both of them on the road surface causing excessive tyre wear as well as severe twisting loads on the shaft.

曲线行驶时,里面车轮行驶的距离比外面车轮行驶的距离小。但如果后轮用同一根轴相连,后轮则必须以相同的速度转动。这将使部分或全部车轮与地面之间产生滑动摩擦,加速轮胎的磨损,并对轴产生附加的扭矩。

4. Neither the misalignment resulting from the flexing of the vehicle structure over bumpy road surfaces can be avoided nor the precise alignment of shaft can be ensured without them.

如果没有它们(万向节),既不能避免在坎坷不平的路上由车辆结构挠曲引起的不对中,也不能保证轴的准确对中。

5. In the initial type, a pair of shoes carried on a stationary plate is expanded in contact with a rotating drum mounted on the wheels to arrest the motion of the drum. In the modern type of brakes, one or more pairs of pads are carried in a caliper attached to the axle or wheel supporting linkage.

传统的制动系统,固定在定子上的一对制动蹄张开与同车轮一起转动的制动鼓产生摩擦力,使车轮制动;现代制动系统中,则由安装在车桥轴上或车轮支撑连接处的制动钳钳住旋转的制动盘而使车轮制动。

Section 1. 2　Engine Operating Principles and Engine Construction

1. 2. 1　Engine Operating Principles

Most automobile engines are internal combustion, reciprocating 4-stroke gasoline engines, but other types have been used, including the diesel, the rotary (Wankel), the 2-stroke, and the stratified charge.

Reciprocating means "up and down" or "back and forth". It is the up and down action of a piston in the cylinder that produces power in a reciprocating engine. Almost all engines of this type are built upon a cylinder block. The block is an iron or aluminum casting that contains engine cylinders and passages called water jackets for coolant circulation. The top of the block is covered with the cylinder head, which forms the combustion chamber. The bottom of the block is covered with an oil pan or oil sump, as shown in Figure 1. 3.

Power is produced by the linear motion of a piston in a cylinder. However, this linear motion must be changed into rotary motion to turn the wheels of cars or trucks. The piston is attached to the top of a connecting rod by a pin, called a piston pin or wrist pin. The bottom of the connecting rod is attached to the crankshaft. The connecting rod transmits the up-and-down motion of the piston to the-crankshaft, which changes it into rotary motion. The connecting rod is mounted on the crankshaft with large bearings called rod bearings. Similar bearings, called main bearings, are used to mount the crankshaft in the block.

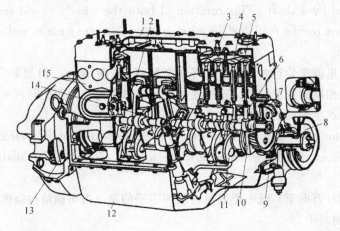

Figure 1.3　The engine construction

1—cylinder head;　2—piston;　3—spark plug;　4—intake valve;　5—exhaust valve;　6—lifter;
7—camshaft gear-wheel;　8—crankshaft pulley;　9—crankshaft timing gear;　10—camshaft;
11—crankshaft;　12—oil pan;　13—flywheel;　14—connecting rod;　15—cylinder block

The diameter of the cylinder is called the engine bore. Displacement and compression ratio are two frequently used engine specifications. Displacement indicates engine size, and compression ratio compares the total cylinder volume to compression chamber volume.

The term "stroke" is used to describe the movement of the piston within the cylinder, as well as the distance of piston travel. Depending on the type of engine the operating cycle may require either two or four strokes to complete. The 4-stroke engine is also called Otto cycle engine, in honor of the German engineer — Dr. Nikolaus Otto, who first applied the principle in 1876. In the 4-stroke engine, four strokes of the piston in the cylinder are required to complete one full operating cycle. Each stroke is named after the action it performs — intake, compression, power, and exhaust in that order, as shown in Figure 1.4.

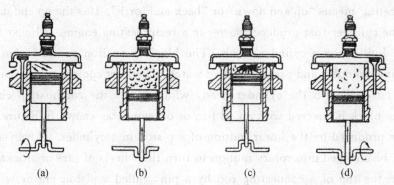

Figure 1.4　The four strokes cycle

(a) intake stroke;　(b) compression stroke;　(c) power stroke;　(d) exhaust stroke

1.2.1.1   Intake Stroke

As the piston moves down, the vaporized mixture of fuel and air enters the cylinder through the open intake valve. To obtain the maximum filling of the cylinder, the intake valve opens about 10° before t. d. c. (top dead center), giving 20° overlap. The inlet valve remains open until some 50° after b. d. c. (bottom dead center) to take advantage of incoming mixture.

1.2.1.2   Compression Stroke

The piston turns up, the intake valve closes, the mixture is compressed within the combustion chamber, while the pressure rise to about 1 MPa, depending on various factors including the compression ratio, throttle opening and engine speed. Near the top of the stroke the mixture is ignited by a spark which bridges the gap of the spark plug.

1.2.1.3   Power Stroke

The expanding gases of combustion produces a rise in pressure of the gas to some 3.5 MPa, and the piston is forced down in the cylinder. The exhaust valve opens near the bottom of the stroke.

1.2.1.4   Exhaust Stroke

The piston moves back up with the exhaust valve open some 50° before b. d. c., allowing the pressure within the cylinder to fall and to reduce "back pressure" on the piston during the exhaust stroke, and the burned gases are pushed out to prepare for the next intake stroke. The intake valve usually opens just before the exhaust stroke.

This 4-stroke cycle is continuously repeated in every cylinder as long as the engine remains running.

A "2-stroke" engine also goes through four actions to complete one operating cycle. However, the intake and the compression actions are combined in one stroke, and the power and exhaust actions are combined in the other stroke. The term 2-stroke cycle or 2-stroke is preferred to the term 2-cycle, which is really not accurate.

In automobile engines, all pistons are attached to a single crankshaft. The more cylinders an engine has, the more power strokes produced for each revolution. This means that an 8-cylinder engine runs more smoothly because the power strokes are closer together in time and in degrees of engine rotation.

The cylinders of multi-cylinder automotive engines are arranged in one of three ways.

Inline engines use a single block of cylinder. Most 4-cylinder and some 6-cylinder engines are of this design. The cylinders are not vertical. They can be inclined to either side.

V-type engines use two equal banks of cylinders, usually inclined 60 degrees or 90 degrees from each other. Most V-type engines have 6 or 8 cylinders, although V-4 and V-12 engines have been built.

Horizontally opposed or "pancake" engines have two equal banks of cylinders inclined 180 degrees apart. These space saving engine designs are often air-cooled, and are found in the Chevrolet Carvairs, Porsches, Subarus and Volkswagens. Subarus' design is liquid cooled. Late-model Volkswagen vans use a liquid-cooled version of the air cooled VW horizontally opposed engine.

1.2.2 Engine Construction

The automobile engine is essentially a heat engine. The heat engines used in modern automobile are internal combustion engines. Each of these engines has a few main working parts; the auxiliary parts are necessary to hold the working parts together or to assist the main working parts in their performance. The main parts are: (a) the engine block; (b) the cylinder; (c) cylinder head, usually holding inlet and exhaust valves; (d) the piston; (e) the connecting rod; (f) the crankshaft; (g) the crankshaft or main and connecting-rod bearing; and (h) the fuel pump and fuel nozzle.

1.2.2.1 Cylinder Block

The cylinder block is cast in one piece. Usually, this is the largest and the most complicated single piece of metal in the automobile.

The cylinder block is a complicated casting made of gray iron (cast iron) or aluminum. It contains the cylinders and the water jackets that surround them. To make the cylinder block, a sand form called a mold is made. Then molten metal is poured into the mold. When the metal has cooled, the sand mold is broken up and removed. This leaves the rough cylinder-block casting. The casting is then cleaned and machined to make the finished block. Figure 1.5 shows a finished cylinder block.

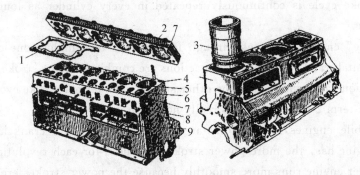

Figure 1.5   A cylinder block

1—cylinder head gasket; 2—piston head; 3—cylinder liner; 4—cylinder block; 5—cylinder;
6—valve seat; 7—water cooling line; 8—intake and exhaust valves; 9—valve chamber

Cylinder blocks for diesel engines are very similar to those for spark-ignition engines. The basic difference is that the diesel-engine cylinder block is heavier and stronger. This is because of the higher pressures developed in the diesel-engine cylinders.

Several engines have aluminum cylinder blocks. Aluminum is a relatively light metal, weighing much less than cast iron. Also, aluminum conducts heat more rapidly than cast iron. This means there is less chance for hot sports to develop. However, aluminum is too soft to use as cylinder-wall material. It wears too rapidly. Therefore, aluminum cylinder blocks must have cast-iron cylinder liners or be cast from an aluminum alloy that has silicon particles in it.

Some manufactures make an aluminum cylinder block that does not have cylinder liners, or sleeves. Instead, the aluminum is loaded with silicon particles. Silicon is a very hard material. After the cylinder-block is cast and the cylinders are honed. Then they are treated with a chemical that etches to eat away the surface aluminum. This leaves only the silicon particles exposed. The piston and rings slide on the silicon with minimum wear.

1.2.2.2　Piston

The piston converts the potential energy of the fuel into the kinetic energy that turns the crankshaft. The piston is a cylindrical shaped hollow part that moves up and down inside the engines cylinder. It has grooves around its perimeter near the top where the rings are placed. The piston fits snugly in the cylinder. The pistons are used to ensure a snug "air tight" fit (see Figure 1.6).

The pistons in your engine's cylinder are similar to your legs when you ride a bicycle. Thinking of your legs as pistons; they go up and down on the pedals, providing power. Pedals are like the connecting rods; they are "attached" to your legs. The pedals are attached to the bicycle crank which is like the crank shaft, because it turns the wheels.

To reverse this, the pistons (legs) are attached to the connecting rods (pedals) which are attached to the crankshaft (the bicycle rank). The power from the combustion in the cylinders powers the piston to push the connecting rods to turn the crankshaft.

1.2.2.3　Connecting-rod

The connecting rod shown in Figure 1.6 is made of forged high-strength steel. It transmits force and motion from the piston pin, or "wrist pin", and the rod to the piston. The pin usually is pressed into the small end of the connecting-rod. Some rods have a lock bolt in the small end. As the piston moves up and down in the cylinder, the pin rocks back and forth in the hole, or bore, in the piston. The big end of the connecting rod is attached to a crank-pin by a rod bearing cap.

Connecting rod and rod-bearing caps are assembled during manufacture. Then the hole for the bearing is bored with the cap in place. This is called line-boring. It makes each rod and its cap a matched set. Usually, the same number is stamped on the rod and cap. This prevents the caps setting mixed during engine service. If the caps are mixed, the bearing bore will not be round. An engine assembled with the rod-bearing caps switched will probably lock the crankshaft. If the crankshaft turns, the bearing will probably have improper clearance and early bearing failure will result.

Another reason for keeping the cap and rod matched is to prevent engine unbalance and unwanted vibration. All connecting rods in an engine must be as light as possible. But they must all weight the same. If one rod is heavier than the others, the engine will vibrate. This could damage the engine.

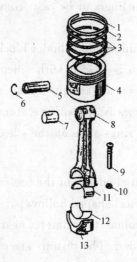

Figure 1. 6　A piston and a connecting-rod

1, 2—compression rings；　3—oil rings；　4—piston；　5—piston pin；　6—piston-pin ring；

7—connecting rod bushing；　8—connecting rod；　9—connecting rod bolt；　10—connecting rod nut；

11,12—crank bearing half shells；　13—connecting rod cap

1. 2. 2. 4　Crankshaft

The crankshaft shown in Figure 1. 7 is the main rotating member, or shaft, in the engine. It has crank-pins, to which the connecting rods from the pistons attached. When the power strokes, the connecting rods force the crank-pins and therefore the crankshaft to rotate. The reciprocating motion is transmitted through the power train to the car wheels.

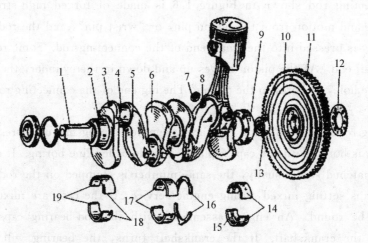

Figure 1. 7　The crankshaft

1—crankshaft front end；　2—front main journal；　3—oil passage hole；　4—crank pin；　5—crank web；

6—counter weight；　7—oil passage plug；　8—oil passage；　9—crankshaft collar；　10—flywheel；

11—flywheel gear ring；　12—flywheel lock plate；　13—clutch shaft bearing；　14—rear main bearing half shell；

15—oil groove；　16—crankshaft thrust；　17—central main bearing half shell；　18—bearing half shell；

19—front main bearing half shell

The crankshaft is a strong, one-piece casting, or forging, of heat-treated alloy steel. It must be strong to take the downward force of the power strokes without excessive bending. It must be balanced so the engine will run without excessive vibration.

New Words

1. combustion [kəm'bʌstʃən] n. 燃烧
2. reciprocating [ri'siprəkeitiŋ] n. 往复移动
3. stratified ['strætifaid] adj. 复叠式的
4. casting [kɑːstiŋ] n. 铸造
5. pin [pin] n. 轴销
6. crankshaft ['kræŋkʃɑːft] n. 曲轴
7. bore [bɔː(r)] n. 内径
8. displacement [dis'pleismənt] n. 排气量
9. stroke [strəuk] n. 行程, 冲程
10. intake ['inteik] n. 进气
11. compression [kəm'preʃn] n. 压缩
12. exhaust [ig'zɔːst] v. 排气
13. valve [vælv] n. 气门
14. overlap ['əuvəlæp] n. 进气门和排气门重叠的时间
15. throttle [θrɒtl] n. 节气门
16. revolution [revə'luːʃn] n. 旋转
17. horizontally [hɒri'zɒntəli] adv. 水平地
18. sleeve [sliːv] n. 衬套

Phrases and Expressions

1. stratified charge 分层进气
2. water jacket 水套
3. cylinder block 汽缸体
4. cylinder head 汽缸盖
5. linear motion 线性运动
6. combustion chamber 燃烧室
7. connecting rod 连杆
8. engine bore 发动机缸径
9. compression ratio 压缩比
10. inlet valve 进气门
11. top dead center (t. d. c.) 上止点
12. bottom dead center (b. d. c.) 下止点
13. exhaust valve 排气门
14. diesel engine 柴油机
15. spark-ignition engine 汽油机

Notes to the Text

1. The block is an iron or aluminum easting that contains engine cylinders and passages called water jackets for coolant circulation.

汽缸体是由铁或铝铸造而成的,在缸体内嵌有汽缸和水套,冷却水在水套内循环。

2. To obtain the maximum filling of the cylinder, the intake valve opens about 10° before t. d. c. (top dead center), giving 20° overlap.

为了使进气充分,进气门在活塞到达上止点之前约10°打开,使进气门和排气门有20°的气门重叠角。

3. The more cylinders an engine has, the more power strokes produced for each revolution. This means that an 8-cylinder engine runs more smoothly because the power strokes are closer together in time and in degrees of engine rotation.

发动机的汽缸数越多,每转的做功行程就越多。例如8缸的发动机,因其每缸做功行程的时间间隔和曲轴转角间隔更加接近,其运转也更加平稳。

4. It has grooves around its perimeter near the top where the rings are placed.

活塞周围顶部附近有放置活塞环的凹槽。

5. The crankshaft is a strong, one-piece casting, or forging, of heat-treated alloy steel. It must be strong to take the downward force of the power strokes without excessive bending. It must be balanced so the engine will run without excessive vibration.

曲轴由强度高的合金钢,经整体铸造或锻造及热处理而制成。它必须能够承受做功行程向下的力而不弯曲,且保持动平衡以避免发动机产生额外的振动。

Chapter 2　Hydraulic and Hydraulic Transmission Basics

Section 2. 1　Hydraulic Systems

Many different machines and processes use a fluid for developing a force to move or hold an object or for controlling an action. In automobiles, for example, hydraulic brakes are used to stop the car. In road construction and repair, another example, compressed air is used to operate chipping hammers. Machines and processed are becoming more and more automated in order to meet competition and to reduce human error.

Generally speaking, a number of fluids can be used in devices and systems. The term "hydraulic" refers to a liquid. For example, the term "hydraulic turbine" can be used to designate a turbine involving water flow. In a hydraulic system, oil, water, or other liquid can be used. Besides a liquid, either a gas or a compressible fluid can be used. In actual practice, two fluids most commonly used are oil and air. A fluid system that uses oil is called a "hydraulic system". A fluid system that uses compressed air is called a "pneumatic system". The "hydraulic system" is treated in this book.

Figure 2. 1 illustrates a typical hydraulic system. Oil from a tank or reservoir flows through a tube or pipe into a pump. The pump can be driven by an electric motor, air motor, gas turbine, or internal combustion engine. The pump increases the pressure of the oil; the oil pressure at the pump outlet may be 5 to 5,000 or more pounds per square inch. High-pressure oil flows in a tube or piping through a control valve; this valve can be used to change the oil flow. A relief valve is used to protect the system; the valve can be set at a desired maximum pressure. If the oil pressure in the system begins to rise above the maximum safe pressure, the relief valve opens to relieve the pressure and to prevent damage to either the system or the surroundings. The oil that enters the cylinder acts on the piston; this pressure action over the area of the piston develops a force on the piston rod. The force of the piston rod can be used to move a load or device. Oil from the cylinder returns to the reservoir. As the oil passes through the filter, dirt and foreign matter are removed from the oil. Each separate unit, such as the pump, the valve, the cylinder, or the filter is called a "component" of the system.

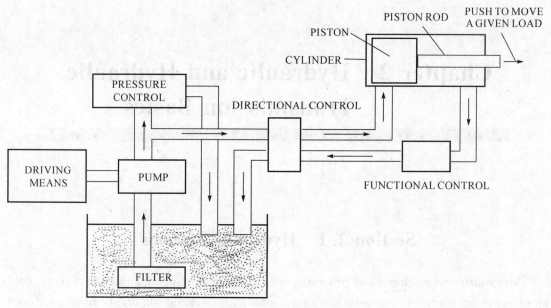

Figure 2. 1   A typical hydraulic system

There are certain advantages in the use of oil as the working fluid. Oil helps to lubricate the various sliding parts, such as piston elements, in the cylinder. Oil prevents rust and is readily available. For practical purposes, oil is a liquid that does not change its volume in the hydraulic system when the pressure is changed as the oil moves from one part of the system to another. If the oil fills the system completely, the movement of the piston can be controlled very closely by the oil flow.

For example, in the hydraulic system, shown in Figure 2.1, it can be assumed that the oil pressure at the cylinder is 1,500 pounds per square inch and the area of the piston over which the oil pressure acts is 2 square inches. Thus, the force of the oil on the piston is (2× 1,500), or 3,000 pounds. This indicates that a relatively large force can act on a load for a relatively small size of cylinder. This is one advantage of hydraulic devices.

A large number of circuits or systems can be devised, and a large number of different components can be used. Essentially, there are three main or basic features in the usual simple hydraulic system: (a) an oil pump; (b) a device with either a piston or a rotating member driven by the oil; and (c) piping and valve devices to control the flow of the oil. With these basic features of a hydraulic system as a beginning, one can imagine various combinations. Several simple systems can be combined. One or more oil pumps can be used to actuate one or more cylinders. More than one valve can be used.

The actuating piston can be given any movement that is required . Straight-line reciprocating motion is most often needed. Rotary motion can also be provided with various forms of hydraulic motors.

A hydraulic system has certain characteristics that are important in meeting certain service requirements. It is relatively easy to connect one component with another by tubing or piping. In some instances a flexible hose is used . A fluid can be used to cushion shocks. Many actions can be controlled by a simple manipulation of valves. The motion of an actuating piston can be changed quickly. A fluid system can give great flexibility in speed and motion control; it can give motion control in very small steps. Relief valves can easily be arranged to protect a system and to avoid damage. Control can be simple, efficient, and centralized. In general, hydraulic systems have relatively few moving mechanical parts; this means a high degree of reliability and a low maintenance cost.

Where can hydraulic systems be used? Today, nearly every industrial machine is somewhat dependent on the use of hydraulic systems equipment. From the high-speed printing presses which print our daily newspapers, magazines, and telephone directories to the large semi-trailer trucks which bring our foodstuffs to the supermarkets, applications for hydraulic devices may be found. A person can walk down the main street of any town or city and discover that practically all the products in each store are processed by some type of hydraulic equipment.

For example, automotive-power brakes, power steering, power windows, and powered seat adjustments are all typical hydraulic system devices. Automatic hydraulic transmissions rate as a high achievement of the art . At the production lines , hydraulic system is used to operate the presses which form body parts and fenders , to punch holes and heads of rivets which hold the frame together, to actuate heads , slides , and chucks of the machinery that is used to produce engine .

Figure 2. 2 shows an excavator, a hydraulic system automatically adjusts speed and power of the machine under all digging conditions. This is accomplished by an automatic regulating valve which varies the output of the pump-high speed in light digging conditions, high power output of the pump-high speed in light digging conditions, high power in heavy digging conditions.

Figure 2. 2   An excavator

New Words

1. hydraulic [haiˈdrɔːlik] *adj*. 液压的
2. pneumatic [njuːˈmætik] *adj*. 气动的
3. circuit [ˈsɜːkit] *n*. 环路
4. combinations [kɒmbiˈneiʃ(ə)ns] *n*. 组合
5. manipulation [mənipjuˈleiʃn] *n*. 操纵，操作
6. printing[ˈprintiŋ] *n*. 印刷
7. chucks[tʃʌks] *n*. 轻叩

Phrases and Expressions

1. chipping hammers 气锤
2. hydraulic turbine 水轮机
3. internal combustion engine 内燃机
4. straight-line reciprocating motion 直线往复运动

Notes to the Text

1. A relief valve is used to protect the system；the valve can be set at a desired maximum pressure. If the oil pressure in the system begins to rise above the maximum safe pressure，the relief valve opens to relieve the pressure and to prevent damage to either system or the surroundings.

安全阀用来保护系统；安全阀门可以设置为所需的最大压力。如果油压系统的压力开始超过最大安全压力，安全阀打开以减轻压力，从而防止破坏系统或环境。

2. For example，automotive-power brakes，power steering，power windows，and powered seat adjustments are all typical hydraulic system devices. Automatic hydraulic transmissions rate as a high achievement of the art .

例如，汽车动力制动器、动力转向、电动车窗和动力座椅调整都是典型的液压系统装置。自动液压传动率则是液压系统的最好应用，就像艺术的一个高成就一样。

Section 2.2　Hydraulic Power Units

2.2.1　Hydraulic Pump

The functions of a hydraulic power device are to provide hydraulic fluid under pressure to a hydraulic system and to provide a place for storing the oil that is not in use . The amount of fluid delivered depends on the capacity of the pump. The capacity of the pump depends on the fluid displacement per revolution of the pump multiplied by the speed at which the pump is to be operated. Pump capacity is usually measured in gallons per minute.

The heart of the power device is the hydraulic pump. Other components of the power unit include the oil reservoir, intake filter, pressure gauge, pressure relief valve, coupling for connecting the pump and the driving means, which may be an electric motor, an internal piping. Many power devices are equipped with coolers, heat exchangers, or heaters.

The pump creates a partial vacuum (pressure below atmospheric) on the intake side as the internal mechanism starts through its cycle; then the atmospheric pressure acting on the oil in the reservoir forces the oil into the pump, as the cycle progresses, traps this oil, and forces it through the outlet under pressure.

It should be remembered that the distance the fluid can be raised vertically depends on the atmospheric pressure acting on the surface of the fluid and the amount of vacuum created within the pump. The theoretical lift and the actual distance which a fluid can be raised may vary greatly, as mechanical imperfections in the pump, pipe friction, and wear of the pump parts should be taken into consideration. Although the pump can raise the fluid only a theoretical maximum of about 34 feet, it can force the liquid to much greater heights, depending on the force exerted on the fluid. Generally, the requirement for raising the liquid into the pump amounts to a maximum distance of a few feet. Most storage tanks (or reservoirs, as they are commonly termed in hydraulics) have the pump mounted on top of the tank, and the oil has to be raised only a short distance.

Pumps may be divided into two general classifications: the constant or positive-displacement type and the variable-displacement type. Since pump is the heart of hydraulic systems, as much as possible should be learned about them. Constant-delivery pumps have broad application in industry, and they are used in great numbers. Most constant-delivery pumps are of the rotary type, but some of the larger ones are of the reciprocating-piston type. The nomenclature for the constant-delivery pump is pump-fixed-delivery, and for the variable-delivery pumps, it is pump-variable-delivery.

2.2.2　Gear Pump

For simple systems with a relatively low level of pressure (about 140 to 180 bar or 14 to 18 MPa), the gear pump is the most used type of pump. The gear pump is a very simple, reliable, relatively cheap and less dirt sensitive hydraulic pump. Its parts are non-reciprocating, move at constant speed and experience a uniform force. Internal construction consists of just two close meshing gear wheels which rotate. Gear pumps are mainly divided into the external mesh gear pump and internal mesh gear pump.

2.2.2.1　External Mesh Gear Pump

External mesh gear pumps can come in single or double (two sets of gears) pump configurations with spur(see Figure 2.3), helical(see Figure 2.4), and herringbone gears (see Figure 2.5). Large-capacity external mesh gear pumps typically use helical or herringbone gears. Small external gear mesh pumps usually operate at 1,750 or 3,450 r/min and larger models operate at speeds up to 640 r/min. External gear mesh pumps have close tolerances and shaft support on both sides of the gears. This allows them to run to pressures beyond 3,000 psi/200 bar（psi：pound per square inch，1 pound＝0.454 kg，1 inch＝2.54 cm），making them well suited for use in hydraulics.

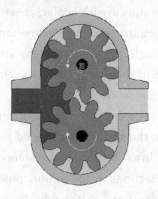

Figure 2.3　A liquid through an external mesh gear pump

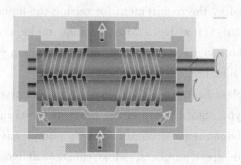

Figure 2.4　A helical gear hydraulic pump

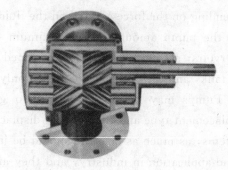

Figure 2.5　A herringbone gear hydraulic pump

External gear pumps are similar in pumping action to internal gear pumps, and in that two gears come into and out of mesh to produce flow（see Figure 2.6）. The external gear pump uses two identical gears rotating against each other — one gear is driven by a motor and it in turn drives the other gear. Each gear is supported by a shaft with bearings on both sides of the gear. There is a small clearance between the gear end face and the housing end cover. The clearance between the gear teeth top and the housing surface is also quite small and the gear housing can be separated as left and right oil tight chambers. As the gears rotate, a vacuum is formed as the teeth unmeshed, which causes liquid to be forced in through the inlet port. Fluid is displaced as the gear teeth meshed at the outlet side and is forced out of the pump into the hydraulic system. In this way, as the gears rotate continuously, the pump carries fluids through the housing and forces them out by a meshing action, this process repeats itself each revolution of the pump. There is no unique flow-deploying mechanism for the inlet and outlet ports separated from each other by the meshing gear teeth and the pump housing. To prevent cavitation, the pressure at the suction side of the pump should not exceed 0.1 to 0.2 bar（10 to 20 kPa）below atmospheric pressure（absolute pressure：0.8 bar or 80 kPa）.

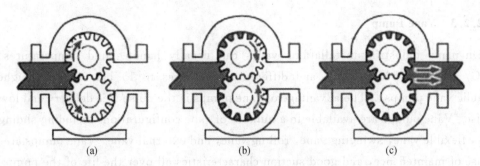

(a)   (b)   (c)

Figure 2. 6   Illustrating the movement of a liquid through a gear-type hydraulic pump
(a) liquid entering the pump;   (b) liquid being carried between the teeth of the gears;
(c) liquid being forced into the discharge line

2. 2. 2. 2   Inner Mesh Gear Pumps

The inner mesh gear pump is driven by a shaft, and the crescent shaped block acts as a division between suction and discharge(see Figure 2. 7). As the driving pinion meshes with and drives the internal gear, the size of the top oil tight chamber increases as the tops of teeth unmesh and begin to draw oil. The size of the bottom oil tight chamber decreases as the bottoms of teeth begin to mesh and then oil is forced out(see Figure 2. 8).

The greatest advantages of inner gear pumps are: no surrounded oil, low sound level and lower flow pulsation compared with external gear pumps. The rated pressure can reach 30 MPa with compensating measures taken towards the axial or radial clearance. Thus achieve higher volumetric and overall efficiencies.

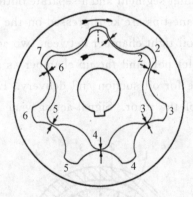

Figure 2. 7   Schematic diagram of the gyrator mechanism

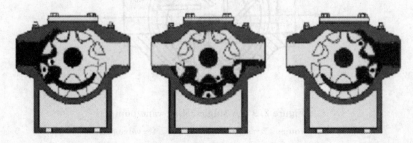

Figure 2. 8   Illustrating the movement of a liquid through an inner mesh gear pump

2.2.3 Vane Pump

On many industrial installations vanes can usually handle fluid temperatures from $-32℃/-25℉$ to $260℃/500℉$ and differential pressures to 15 bar/200 psi (higher for hydraulic vane pumps). The advantage of vane pumps is the pulse free delivery and low level of noise. Vane pumps are available in a number of vane configurations including sliding vane (left), flexible vane, swinging vane, rolling vane, and external vane. Vane pumps are noted for ease of maintenance, and good suction characteristics all over the life of the pump.

Each type of vane pump offers unique advantages. For example, external vane pumps can handle large solids. Flexible vane pumps, on the other hand, can only handle small solids but create good vacuum. Sliding vane pumps can run dry for short periods of time and handle small amounts of vapor. Vane pumps may be either of the single-acting or double-acting type. The former is used for variable displacement pumps, while the latter for fixed displacement pumps.

2.2.3.1 Single-acting Vane Pumps

The single-acting vane pump gets its name from the fact that there are only one inlet segment and a separate outlet segment during each revolution(see Figure 2.9).

A typical construction of signal-acting vane pump principal components are: a rotor, a stator, vane and a valve plate (ignored here), etc. Unlike double-acting vane pumps, the inner ring of the stator is circle. There exists an eccentricity e in the rotor with respect to the stator and there are only one inlet segment and a separate outlet segment on the valve plate. When the rotor rotates, the vane-tips are kept pressed on the inner ring of the stator under centrifugal force, forming an oil tight chamber between two adjacent vanes. Obviously, the down chamber serves as an inlet port and the up chamber as an outlet port while the rotor rotates. Their volumes change for oil suction and delivery. Each accomplishes suction and delivery once per revolution of the rotor. Signal-acting vane pumps have another name of non-unload type vane pumps.

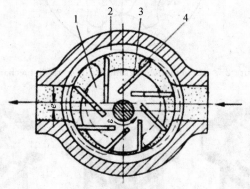

Figure 2.9 A single-acting vane pump

1—rotor; 2—stator; 3—blade; 4—valve plate

2.2.3.2 Double-acting Vane Pumps

The double-acting vane pump gets its name from the fact that there are two inlet segments and two outlet segments during each revolution(see Figure 2.10).

The principal parts of the pump include a transmission shaft, a rotor, a stator, left and right valve plates, front and rear pump housings, etc. The stator is anchored securely in the housing of the pump, which does not move. The rotor is slotted and is driven by a shaft. Each slot of the rotor serves the purpose of holding a fiat, rectangular vane. The vanes are free to move radially in the slot. As the rotor turns, centrifugal force ejects the vanes outward to contact and follow the inner wall of the stator. The inner ring of the stator is oval and the radii are different for each arc. Because of the existence of different radius, so as the rotor rotating clockwise, broken by the blades of airtight cubage reduces gradually and increases gradually. If each chamber is decreased in volume, the oil is pressurized and ejected from the pump through outlet port on the valve plate. And if each chamber is increased in volume, the oil is drawn into the pump under atmospheric pressure through inlet ports on the same valve plate.

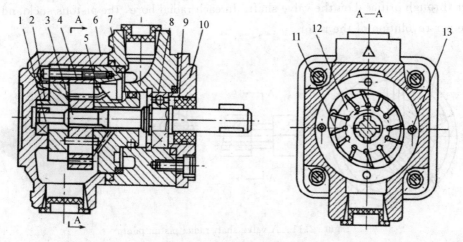

Figure 2.10 A double-acting vane pump

1—valve plate; 2—needle bearing; 3—transmission shaft; 4—stator; 5—valve plate;
6—after the pump; 7—before the pump; 8—ball bearings; 9—oil seal; 10—pump cover;
11—blade; 12—rotor; 13—locating pin

2.2.4 Plunger Pump

The oil suction and delivery actions of a piston pump depend upon the size change of the chamber produced by the reciprocating movement of the pistons within their cylinder bores. For piston pumps, the circular pistons and their corresponding cylinder bores can make high precise match-up and ensure good performance of seal. High volumetric and overall efficiencies can be acquired even in operation under high pressure. The piston pumps can be divided into radial and axial types according to their pistons' arrangements and their different moving directions with respect to the transmission shafts. However, signal piston pump cannot be used in industrial production directly due to its failure to provide oil continuously (a signal piston pump sucks oil in the first half circle and delivers oil in the second). A radial piston pump usually consists of three or more pistons for continuous suction and discharge.

2.2.4.1 Valve Shaft Radial Piston Pumps

As shown in Figure 2.11, the piston bores are arranged radially on rotor (cylinder) with an equal apart. Five pistons are set in the piston bores and can move freely within them. Bush is mounted in the rotor bores and rotates with the rotor. Valve shaft is stationary and there exists an eccentricity e between its center and that of the stator. The stator moves in the horizontal direction. As the rotor rotates in the clockwise direction, the pistons are pressed against the inner wall of stator under centrifugal force or low pressure oil and then forced outward in the upper semicircle, then a partial vacuum is formed, drawing the oil in the reservoir into chamber b through orifice a of the valve shaft; while the pistons in the next semicircle are pushed inward by the inner wall of the stator, which reduces the working volume of the oil tight chamber and forces a quantity of liquid in chamber c out of the cylinder through orifice d on the valve shaft. In each radial bore, the pistons suck and deliver oil once per revolution of the rotor.

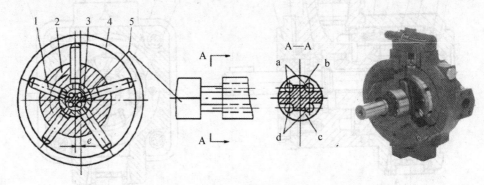

Figure 2.11   A valve shaft radial piston pump

1—piston; 2—rotator; 3—bush; 4—stator; 5—assignment shaft

2.2.4.2 Radial Piston Pumps with Load-sensitive Variable Displacement Capacity

This type of pump gets its name from the fact that the output pressure of the pump depends upon the load.

Recent improvements in the design of radial piston pumps have resulted in the widespread use of these pumps at a rated pressure high up to 35 MPa.

2.2.4.3 Swash Plate Axial Piston Pumps

The swash plate axial piston pump is also called straight shaft axial piston pump(see Figure 2.12, 2.13). The center lines of the pistons are parallel to the axial line of the cylinder. Several piston bores, each fitted with one piston, are arranged axially on the cylinder with an equal distance apart. The pistons can move freely within their corresponding piston bores. The swash plate is positioned at an angle relative to the piston axis to create a reciprocating motion. The swash plate and the valve plate themselves keep stationary and the pistons are pressed on the swash plate under low pressure oil or spring force. There are two waist shaping ports on the valve plate and are separated from each other by the transition region. The width of this transition is equal to or slight longer than the waist shaping ports at the bottom of the cylinder to prevent the connection of these two waist shaping ports (the inlet and outlet ports).

As the cylinder rotates with the transmission shaft in a direction, the pistons in the upper semicircle are forced outward gradually under low pressure oil. The oil tight working volume in the cylinder bores is on the increase and a vacuum is formed, drawing the oil into port a of the valve plate. When the pistons in the next semicircle are pushed inward by the swash plate gradually, they reduce the oil tight working volume and force the oil out through port b of the same valve plate. Each piston moves reciprocatingly once with one suction and delivery action per revolution of the cylinder.

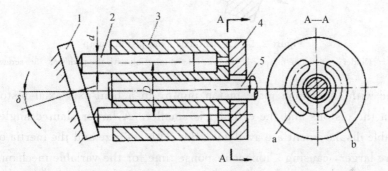

Figure 2.12　A swash plate axial piston pump

1—swash plate;　2—piston;　3—cylinder;　4—valve plate;　5—transmission shaft

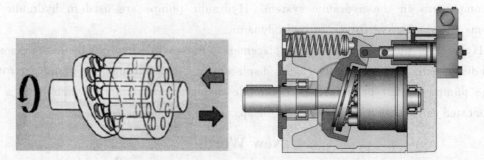

Figure 2.13　Illustrating the movement of a liquid through a swash plate axial piston pump

2.2.4.4　Bent-axis Axial Piston Pumps

As the bent-axis axial piston pump transmission shaft rotates with the electric motor, pistons are driven back and forth in their cylinder bores by connecting rod(see Figure 2.14). The side face of the connecting rod also brings the pistons and the cylinder together to revolve. Suction and delivery actions in the inlet and outlet ports respectively are accomplished through stationary valve plate. Similar to swash plate axial piston pumps, the delivery of a bent-axis axial piston pump can be changed by the change of cylinder slanted angle and the changing of slanted direction makes a double-action piston pump. The delivery formula of bent-axis axial piston pumps is the same as that of swash plate axial piston pumps.

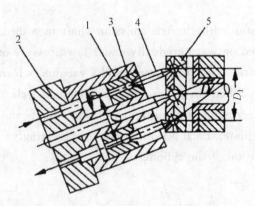

Figure 2. 14　A bent-axis axial piston pump

1—cylinder；　2—valve plate；　3—piston；　4—connecting with the plunger；　5—screws

Compared with the swash plate piston pump, in a bent-axis axial piston pump, the radial force on the pistons and the cylinder is smaller. A larger slanted angle and a larger scope of variable displacement are available. But both the size and the inertia of the variable mechanism are larger, causing a longer response time for the variable mechanism.

2.2.5　Short Passages for Reading

Hydraulic pumps convey mechanical energy into hydraulic energy. They supply fluid to the components in the hydraulic system. Hydraulic pumps are used in hydraulic drive systems and can be hydrostatic or hydrodynamic.

Hydrostatic pumps are positive displacement pumps while hydrodynamic pumps can be fixed displacement pumps, in which the displacement (flow through the pump per rotation of the pump) cannot be adjusted or variable displacement pumps, which have a more complicated construction that allows the displacement to be adjusted.

New Words

1. vacuum [ˈvækjʊəm] *n.* 真空
2. imperfections [impəˈfekʃən] *n.* 缺点
3. nomenclature [nə(ʊ)ˈmeŋklətʃə] *n.* 术语
4. spur [spɜː] *n.* 支柱
5. helical [ˈhelikl] *adj.* 螺旋形的
6. herringbone [ˈheriŋbəun] *n.* 人字形图案

7. cavitation [ˌkæviˈteiʃən] n. 气穴

8. pulsation [pʌlˈseiʃən] n. 脉动

9. volumetric [ˌvɔljuˈmetrik] adj. 体积的

10. segment [ˈsegmənt] n. 段,部分

11. eccentricity [ˌeksenˈtrisiti] n. 偏心距

12. anchored [ˈæŋkə] adj. 固定的

13. fiat [ˈfiːæt] n. 命令,许可

14. cubage [ˈkjuːbidʒ] n. 容积,体积

15. semicircle [ˈsemisɜːkl] n. 半圆,半圆形

16. hydrostatic [ˌhaidrəuˈstætik] adj. 流体静力学的

17. hydrodynamic [ˈhaidrəudaiˈnæmik] adj. 水力的

Phrases and Expressions

1. atmospheric pressure 大气压力,大气压强

2. variable-delivery pumps 变量输送泵

3. meshing gear teeth 轮齿啮合

4. absolute pressure 绝对压强

5. sliding vane 滑片,滑动叶片

6. swinging vane 摆动滑片

7. single-acting vane pumps 单作用叶片泵

8. centrifugal force 离心力

9. double-acting vane pumps 双作用叶片泵

10. transmission shaft n. 传动轴

11. plunger pump 柱塞泵

12. clockwise direction 顺时针方向

13. swash plate axial piston pump 旋转斜盘轴向活塞泵

14. bent-axis axial piston pumps 弯轴轴向活塞泵

15. slanted angle 倾斜角度

Notes to the Text

1. External mesh gear pumps can come in single or double (two sets of gears) pump configurations with spur(see Figure 2.3), helical(see Figure 2.4), and herringbone gears (see Figure 2.5).

外啮合齿轮泵有单齿轮泵或双齿轮泵(两套齿轮),可以分为直齿轮泵(见图 2.3)、螺旋齿轮泵(见图 2.4)和人字形齿轮泵(见图 2.5)。

2. As the gears rotate, a vacuum is formed as the teeth unmeshed, which causes liquid to be forced in through the inlet port. Fluid is displaced as the gear teeth meshed at the outlet side and is forced out of the pump into the hydraulic system.

当齿轮旋转时,右侧的齿轮逐渐脱离啮合,密闭工作容积逐渐增大,形成局部真空,油箱中

的油液在大气压的作用下,经吸油管进入吸油腔,将齿谷空间充满,并随着齿轮的转动,把油液带到左侧压油腔。左侧压油腔轮齿逐渐进入啮合,密封工作容积逐渐减小,把油液挤压出去,输送到压力管路中。

3. The inner mesh gear pump is driven by a shaft, and the crescent shaped block acts as a division between suction and discharge. As the driving pinion meshes with and drives the internal gear, the size of the top oil tight chamber increases as the tops of teeth unmesh and begin to draw oil.

内啮合齿轮泵是由一个轴驱动,月牙形板作为泵的吸油和放油之间的分隔物。当传动轴带动小齿轮旋转时,内齿轮同向旋转,其上半部齿轮脱开啮合,所在的密封容积增大,为吸油腔;下半部齿轮进入啮合,所在的密封容积减小,为压油腔。

4. Vane pumps are available in a number of vane configurations including sliding vane (left), flexible vane, swinging vane, rolling vane, and external vane. Vane pumps are noted for ease of maintenance, and good suction characteristics all over the life of the pump.

叶片泵包含许多配置,包括滑片泵(左),灵活的叶片,摆动的叶片,轧制叶片和外部泵。叶片泵因易于维护,以及有良好的吸入特点而闻名。

5. The swash plate axial piston pump is also called straight shaft axial piston pump. The center lines of the pistons are parallel to the axial line of the cylinder. Several piston bores, each fitted with one piston, are arranged axially on the cylinder with an equal distance apart. The pistons can move freely within their corresponding piston bores. The swash plate is positioned at an angle relative to the piston axis to create a reciprocating motion. The swash plate and the valve plate themselves keep stationary and the pistons are pressed on the swash plate under low pressure oil or spring force.

斜盘式轴向柱塞泵又称直轴式轴向柱塞泵,该液压泵的柱塞中心线平行于缸体的轴线。缸体上均匀分布着几个轴向排列的柱塞孔,柱塞可在孔内沿轴向滑动,斜盘的中心线与缸体中心线斜交成一个夹角,以产生往复运动。斜盘和配油盘固定不动。柱塞可在低油压或弹簧作用下压紧在斜盘上。

Section 2.3　Hydraulic Actuator — Hydraulic Cylinder and Hydraulic Motor

2.3.1　Hydraulic Cylinder

2.3.1.1　Pressure

The word pressure is defined as force per unit area. Although other units may be used, pressure is commonly expressed in such units a pound per square inch. The abbreviation psi is usually employed to indicate pounds per square inch.

Figure 2.15 shows an arrangement of two cylinders that are connected by a pipe or tube. A close-fitting piston is placed in each cylinder. In each cylinder (under the piston), the liquid and the connecting tube are shown. If it is assumed that there is no movement of each piston and that there is no leakage past each piston, the liquid and all the parts are at rest — a static condition. It is also assumed that a force F_1 of 100 pounds acts on piston NO. 1 and that there is no friction between each piston and its cylinder wall. If piston NO. 1 has a flat or face area of 2 square inches that is in direct contact with the liquid, the pressure in the liquid under piston NO. 1 is equal to the force divided by the area (100 divided by 2), or 50 pounds per square inch. Thus, the liquid pressure at the face of piston NO. 1 is 50 psi.

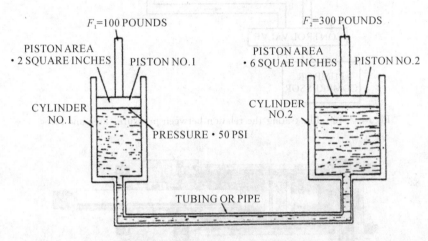

Figure 2. 15   Illustrating a hydraulic system having two cylinders and two pistons of different sizes

Assuming that piston NO. 2 is essentially at the same level as piston NO. 1, the liquid between the pistons serves as a medium to transmit the pressure from one piston face to the other piston face. Thus, the liquid pressure at the face of piston NO. 2 is 50 psi. If the area of piston NO. 2 is 6 square inches, the force F_2 on the face of piston NO. 2 is (6×50), or 300 pounds. Thus, a force of 100 pounds at piston NO. 1 develops a force of 300 pounds at piston NO. 2; this is accomplished by making the area of piston NO. 2 equal to three times the area of piston NO. 1. In a sense, the arrangement (see Figure 2. 15) is a fluid lever, similar to a mechanical lever using a metal bar and pivot.

Equal pressure at every point and in every direction in the body of a static liquid (a liquid at rest) is characteristic of all static fluids, liquids, or gases. This is called Pascal's law, after an early experimenter in this field of study. This law of pressure is very useful, and can be used to advantage in countless applications.

Figure 2. 16 and Figure 2. 17 are illustrations of a pump or compressor delivering fluid (either oil compressed air) to the left-hand side of a piston in a cylinder. Let P represent the fluid pressure, in psi, and A represent the piston area in sq. in. (abbreviation for square inches).

Then the force F acting on the left-hand face of the piston is PA. For a pressure P of 50 psi and an area A of 2 sq. in. , the force acting on the left-hand face of the piston is equal to (50×2), or 100 pounds. Assuming no friction due to the cylinder wall, the force F at the piston rod is equal to 100 pounds.

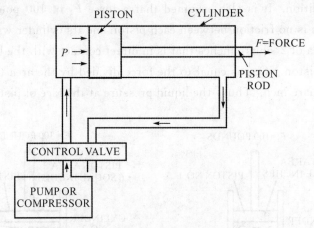

Figure 2.16   Illustrating the relation between pressure, area and force

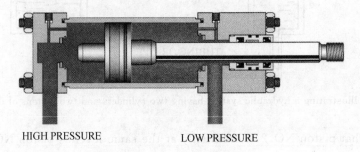

HIGH PRESSURE   LOW PRESSURE

Figure 2.17   Schematic diagram of the cylinder

2.3.1.2 Hydraulic Cylinder

Cylinders may be further divided into single-acting and double-acting according to function or piston type, plunger type and swing type based on their constructions. Some typical cylinders are described as follows:

1. Piston cylinders

There are double-rod type and single-rod type piston cylinders in construction, stationary cylinder body type and stationary piston rod type in installation.

(1) Single-rod piston cylinders.

The cylinder does not have a rod in the cap side of a single rod cylinder (see Figure 2.18, 2.19). Single-rod piston cylinders can also be divided into stationary cylinder body type and stationary piston rod type based on their installation. Unlike double-rod piston cylinders, the displacement range of the workbench is equal to double the piston effective stroke in both installation of this type.

The effective areas in the left and right chambers are different. When the pressures of the inlet and outlet ports are p_1 and p_2 respectively, and the flow rates q in the two ports are equal, then the thrust and speed in the opposite directions are different.

If both the left and the right chambers of the signal rod piston cylinder connect with pressurized oil at the same time, it forms differential connection. The signal rod piston cylinder in this form is called a differential cylinder. The pressures in the two chambers are equal, however, the effective area of the left chamber (non-rod chamber) is larger than that of the right chamber (rod chamber). The right direction force is larger than that in the left direction, causing the piston to move in the right, and the piston rod extends out. In addition, the flow rate q from the rod chamber converges with the pump's flow rate q and then enters into the left chamber of the cylinder together. The piston moves faster by this action. This type of cylinder can only move in one direction.

Figure 2.18   Structure of a single rod cylinder

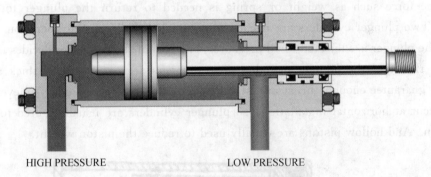

HIGH PRESSURE   LOW PRESSURE

Figure 2.19   A single rod cylinder

(2) Double-rod piston cylinders.

The double-rod piston cylinder of stationary cylinder body type consists of two rods with the same diameter at each side of the piston(see Figure 2.20). The inlet and outlet ports are located at the two ends of the cylinder respectively. The piston drives the workbench to move through the piston rod. The displacement range of the workbench equals treble effective strokes of the piston, so it takes up much space and is only suited for small size equipments. If the piston rods are fixed to machine tool with the help of a support and the cylinder connects with the workbench when in a stationary piston rod type, the displacement range of the workbench equals double effective strokes of the piston, thus it takes less space and is suited for large or medium size equipments.

The two piston rods in the two sides of the double rod piston cylinder are equal in diameters. When input flow rates and pressures are constant, the output thrust F and speed v are equal in the two opposite directions.

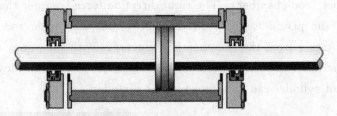

Figure 2.20　A double-rod piston cylinder

2. Plunger cylinders

The piston cylinders discussed above are used widely. But the cylinder bores request high precise-machining and are hard to be machined at long stroke occasion, and their manufacturing cost will be increased. However, the plunger cylinder does not meet such troubles and is widely used in practical production, specifically, when double direction control is not needed(see Figure 2.21, 2.22).

The pistons do not contact the cylinder barrel, thus free from precise machining, possessing a good craft and low cost. Note that plunger type cylinder is a single-acting type and outside force such as weight or spring is needed to return the plunger to its home position. Two plunger cylinders are needed to accomplish bidirectional operation. The end cover of the plunger is subjected to oil pressure. The area of the end cover decides the output speed and thrust of the plunger cylinder. The plungers are usually quite thick and more heavier to guarantee enough thrust and stability. For this type of cylinder, uneven wear is easy to occur at horizontal installation, so plunger cylinders are usually suited for vertical installation. And hollow pistons are usually used to reduce the piston weight.

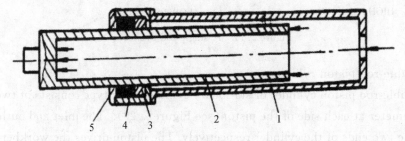

Figure 2.21　Structure of a plunger cylinder

1—cylinder body; 2—plunger; 3—guide sleeve; 4—V ring seal; 5—gland

Figure 2.22   A plunger cylinder

2.3.2   Hydraulic Motor

2.3.2.1   Gear Motor

To allow bi-rotational operation, a gear motor has two identical inlet and outlet ports which are set in the opposite direction and has a unique outside vent to guide the leakage in the bearings out of the motor(see Figure 2.23). Rolling bearings are used to reduce startup friction torque and the teeth of the motor are more than that of the pump to reduce torque pulsation.

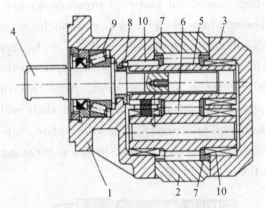

Figure 2.23   Structure of a gear motor

1—front cover;   2—shell;   3—back cover;   4—output shaft;   5—driving gear shaft;
6—driven gear shaft;   7—side panel;   8—shaft seal;   9—roller thrust bearing to the heart;   10—needle bearing

2.3.2.2   Sliding-vane Motor

Structure features: The vanes in a vane motor are positioned radially with a zero vane angle ($\theta=0$)(see Figure 2.24). A shuttle valve is set at the path between high/low pressure oil chambers and the vane bottom to ensure high pressure oil at the vane bottom. To ensure a normal startup, a spring with preload is mounted at the vane bottom to prevent the connection of high and low pressure chambers when high pressure oil enters into the vane motor. Vane type motors are small in size and thus the moment of inertia, making a short reaction time and applicable to frequent direction-changing occasions. But vane type motors find more applications in small torque, high speed, low mechanical performance request occasions due to its large leakage and inadequate stabilization at low speed.

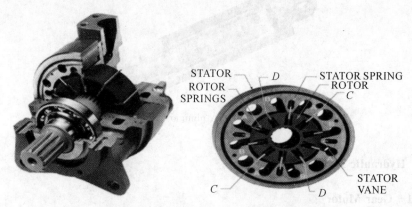

Figure 2.24　A sliding-vane motor

2.3.2.3　Axial Piston Motor

Valve plate and swash plate are stationary, while motor shaft and cylinder rotate together(see Figure 2.25). When pressurized oil enters into the piston bores of cylinder through valve plate, pistons extend out under oil pressure against the swash plate, which then produces a normal reaction force F on pistons, and this force can be further decomposed into axial branch force F_x and vertical branch force F_y. F_x is balanced out with the pressure acting on the pistons, while F_y develops a torque with respect to the cylinder center, which causes a counter clockwise rotation of motor shaft. Note that the instantaneous overall torque created by axial piston motor fluctuant and motor shaft will rotate clockwise if the pressurized oil is put from the other direction. The swash plate angle or the delivery not only decides the motor's torque, but also its rotating speed and direction. The larger the angle, the more the torque and the lower the speed.

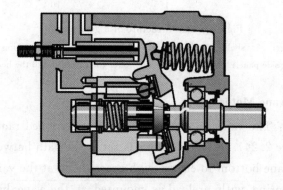

Figure 2.25　An axial piston motor

2.3.2.4　Radial Piston Motor

Low-speed motors are usually radial piston motors(see Figure 2.26). In order to meet the needs of low speed rotation and large torque, they possess high pressure and large delivery. Because of their large size and large moment of inertia, they are not suitable for the applications that require rapid response and frequent direction changing. Low-speed motors may be further classified based on how they function: single-acting motors which provide a push in one direction; multiple-acting motors which provide push in many directions.

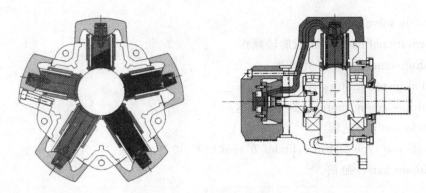

Figure 2.26　A radial piston motor

2.3.2.5　Short Passages for Reading

The popular concept of a hydraulic motor is that it is nothing but a pump runs backwards. This may be superficially true, but there are many differences in operating demands between pumps and motors. Because of this, a design that is completely acceptable as a pump may operate poorly as a motor in certain kinds of applications. Many motor designs have internal design features differing from those found in the corresponding types of pumps; in fact, some motors have no pump counterpart.

The function of hydraulic actuators is to translate hydraulic energy of fluid into energy of machine. They will drive the mechanism motion path like a straight, swing or rotation. The straight or swing is referred to as hydraulic actuator or swing motion actuator, and the rotational motion is called hydraulic motor. The output parameters are force and torque and rotating speed.

New Words

1. actuator ['æktjueitə] *n.* 执行机构
2. arrangement [ə'rein(d)ʒm(ə)nt] *n.* 布置,整理
3. pivot ['pivət] *n.* 枢轴
4. workbench ['wəːkbentʃ] *n.* 工作台
5. chamber ['tʃeimbə] *n.* 室,腔
6. converge [kən'vɜːdʒ] *vt.* 使汇聚,聚集
7. treble ['trebl] *adj.* 三倍的
8. stroke[strəʊk] *n.* 冲程
9. thrust [θrʌst] *n.* 推力

10. bidirectional [baidəˈrekʃənl] *adj.* 双向的,双向作用的

11. fluctuant [ˈflʌktʃuənt] *adj.* 变动的,波动的

12. superficially [suːpəˈfiʃ(ə)li] *adv.* 表面地

Phrases and Expressions

1. shuttle valve 换向阀
2. bi-rotational operation 双旋转操作
3. sliding-vane motor 叶片马达
4. radial piston motors 径向活塞式马达
5. multiple-acting motors 多功能马达
6. motor shaft 马达轴
7. single-rod type piston cylinder 单杆式活塞缸
8. cylinder barrel 缸筒

Notes to the Text

1. There are double-rod type and single-rod type piston cylinders in construction, stationary cylinder body type and stationary piston rod type in installation.

从结构上,活塞液压缸可以分为双杆式和单杆式;从安装形式上,又可以分为固定缸体类型和固定活塞杆式液压缸。

2. The piston drives the workbench to move through the piston rod. The displacement range of the workbench equals treble effective strokes of the piston, so it takes up much space and is only suited for small size equipments.

活塞驱动工作台移动活塞杆。工作台的位移范围等于活塞的三倍有效冲程,因此它占据了很大的空间,并且只适合小型的设备。

3. Two plunger cylinders are needed to accomplish bidirectional operation. The end cover of the plunger is subjected to oil pressure. The area of the end cover decides the output speed and thrust of the plunger cylinder. The plungers are usually quite thick and more heavier to guarantee enough thrust and stability. For this type of cylinder, uneven wear is easy to occur at horizontal installation, so plunger cylinders are usually suited for vertical installation.

完成双向操作需要两个柱塞汽缸。柱塞的顶盖受油压的影响。阀盖末端的面积决定了柱塞的输出速度和推力。一般来说,柱塞的厚度和质量都要大得多以保证足够的推力和稳定性。对于这种类型的液压缸,不均匀磨损在水平安装中很容易发生,因此柱塞汽缸通常适合垂直安装。

4. To ensure a normal startup, a spring with preload is mounted at the vane bottom to prevent the connection of high and low pressure chambers when high pressure oil enters into the vane motor.

为了确保正常启动,当高压油进入叶片马达时,在叶片底部安装预载弹簧,以防止高、低压室的连接。

5. The popular concept of a hydraulic motor is that it is nothing but a pump run backwards. This may be superficially true, but there are many differences in operating demands between pumps and motors. Because of this, a design that is completely acceptable as a pump may operate poorly as a motor in certain kinds of applications.

从工作原理上讲,液压传动中的泵和马达都是靠工作腔密闭容器的容积变化而工作的,所以说泵可以作马达用,反正也一样,即泵与马达有可逆性。实际上由于二者工作状况不一样,为了更好发挥各自工作性能,它们在结构上存在某些差别,使二者不能通用。

Section 2.4   Hydraulic Control Components

2.4.1   Hydraulic Valve

The function of pressure controls is to control pressures in the fluid power system. Pressure controls may be used to reduce, to relieve, to adjust pressure, or to begin another function.

Types of components that are considered to be pressure controls are:

(1)Relief valve. In hydraulics, the function of the pressure relief valve is to relieve the pump to protect it and the system from becoming overloaded. When the pressure reaches a certain point, the relief valve spills the oil back to the reservoir, and the pressure is relieved. There are many modifications of these valves, depending on the system in which they are used.

(2)Pressure reducing valve. The pressure reducing valve is used to reduce the pressure in part of the system to a lower pressure. In a hydraulic system, it is often desirable to reduce pressure in certain parts of the system.

The pressure regulating valve, as it is referred to in a pneumatic system, takes care of the pressure in the entire system. It smooths out the surges, and can be regulated in small amounts to give whatever reduced pressure is desired.

(3)Sequence valve. The function of a sequence valve is to set up the sequence of operations in either a pneumatic or a hydraulic circuit. Sequence valves are also used for other functions.

(4)Unloading valve. The function of an unloading valve, as used in hydraulic circuits, is to unload the pressure at the desired instant, in order to conserve horsepower and to afford protection to the system. This helps to reduce heat.

2.4.2   Pressure Valve

2.4.2.1   Relief Valve

The relief valve is installed in any system containing a confined liquid subject to pressure. Relief valves are devices installed in a circuit to make it certain that the system pressure does not exceed safety limits. Relief valves are intended to relieve occasional excess pressures arising during the course of normal operation.

The configurations for relief valves are either two-port or four-port. For safety, the relief valve is usually installed as close as possible to the pump, with no other valves between the relief valve and pump. Both types operate in the same way. The main reason for additional ports is convenience in connecting the plumbing.

The relief valve may be of the direct-acting type, the direct-operated pilot-type, or the remote-actuated pilot-type of valve. In the direct-acting valve, the fluid pressure acting on the piston must overcome the tension applied by a large spring in order to open the exhaust port. The direct-operated pilot-type is pilot-operated and used only a small spring. The remote-actuated pilot-type is controlled by a remote valve through a pilot connection. It protects the pump, electric motor, fluid lines, directional controls, other controls, cylinders, and fluid motors from being overloaded, or from having an operating pressure applied that is above the safe range of the components.

Although hydraulic relief valves are usually considered to be part of the power device (see Figure 2.27), they may be used in other places in the hydraulic system, such as to relieve pressure during a stand-by period. A stand-by period is referred to as the idle time when a hydraulic press machine is not doing work, such as during the loading period. The advantages of using an extra relief valve in the circuit to take care of the stand-by period are that it reduces heat, reduces the power consumption, and takes the full load off the system for fairly long intervals. Relief valves are known as a normally closed valve, because the exhaust passage is kept closed until the piston opens that passage to relieve fluid pressure.

Figure 2.27　The relief valve is a component of a power unit

Hydraulic pressure relief valves of the direct-acting type are usually built for pressures up to 3,000 psi; however, in some instances, they are designed for much higher hydraulic pressure. Usually, they do not cover the full range, but are found in ranges, such as: 50 to 750 psi, 700 to 1,500 psi, 1,500 to 2,500 psi, and 2,000 to 3,000 psi. To understand the principle on which the direct-acting relief valve functions, Figure 2.28 should be studied. The fluid flows unobstructed in one port and out the other port, until a resistance is met;

then the pressure inside the valve builds up to a point where the differential area between the top section and the bottom section of the valve piston, multiplied by the internal pressure in the valve, works against the spring. The piston rises to a point that allows the fluid to escape to the third port, relieving the pressure. The pressure against the piston depends on the tension placed against the spring by means of the adjusting screw.

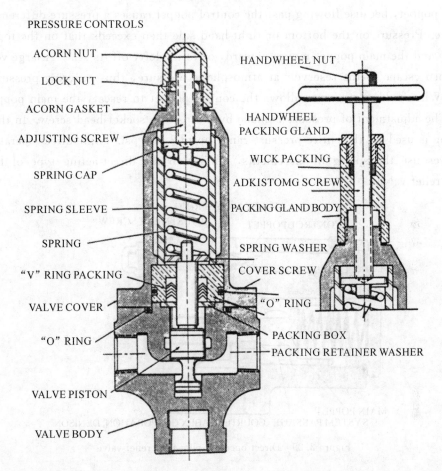

PRESSURE CONTROLS

ACORN NUT

LOCK NUT

ACORN GASKET

ADJUSTING SCREW

SPRING CAP

SPRING SLEEVE

SPRING

"V" RING PACKING

VALVE COVER

"O" RING

VALVE PISTON

VALVE BODY

HANDWHEEL NUT

HANDWHEEL

PACKING GLAND

WICK PACKING

ADKISTOMG SCREW

PACKING GLAND BODY

SPRING WASHER

COVER SCREW

"O" RING

PACKING BOX

PACKING RETAINER WASHER

Figure 2.28   Direct-acting relief valve

This type of valve is built in port sizes ranging from 1/4 inch to 2 inches, or more. The part names, as shown in Figure 2.28, should be studied. The piston is fitted closely to the valve body in order to reduce leakage to a minimum. The valve is simple in construction, having only two moving parts — the piston and the spring. The piston moves very rapidly.

The internal or external pilot-operated type of relief valve may be built for pressure up to 5,000 psi, in sizes that similar to the direct-acting type.

The direct-operated hydraulic relief valve is compact, since it does not require space for a large spring. Note that in the "M" series (see Figure 2.29) the spring is relatively small. The movable main poppet allows a large volume of oil to escape to the reservoir when the system pressure of the valve is reached. The action of the large main poppet is controlled by a much smaller poppet because of the small orifice. Since a greater area is exposed to system pressure on the top or left-hand side, the main poppet is held firmly on its seat, thus reducing leakage. System pressure also acts on the control poppet by the way of the orifice just mentioned. When pressure becomes great enough to overcome the adjustable spring pressure bearing on the control poppet, fluid flows to the reservoir. Forces are then upset on the main poppet, because flowing past the control poppet causes a pressure difference across the orifice. Pressure on the bottom or right-hand side then exceeds that on the top or left-hand side and the main poppet moves upward, or to the left, off its seat. A large volume of oil can then escape to the reservoir at atmospheric pressure, thus reducing pressure in the system. When reduced pressure allows the control poppet to reseat, the main poppet again closes. The adjustment of pressure is made by means of a socket-head screw. In the valve, one spring is used for the entire pressure range up to 2,000 psi. Some direct-operated pilot-type valves use the spring in increments, similar to the direct-acting type of hydraulic pressure relief valve.

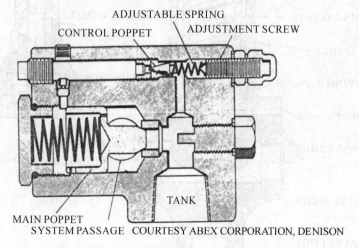

Figure 2.29 Direct-operated pilot-type relief valve

Large-capacity relief valves are often built with flange-type connections. These valves are generally used in large volumes concerned. Relief valves are also built for gasket mounting.

2.4.2.2 Pressure-reducing Valve

There are three kinds of pressure-reducing valves on request: (a) to maintain outlet pressure invariableness, called fixed value pressure-reducing valves; (b) to maintain a constant pressure differential between inlet and outlet ports, called fixed pressure differential pressure-reducing valves; (c) to ensure that the outlet pressure is proportional to that of the inlet, called proportion pressure-reducing valves. The fixed values are applied widely. They will be introduced here.

The pilot valve of the pressure-reducing valve is similar to that of a relief valve. But the leaking oil from the spring port should be induced back to the reservoir due to the pressure existence in the inlet and outlet ports. Unlike a relief valve for the main valve parts, a pressure-reducing valve is normally open and the main valve core stays at the bottom under the spring force. The orifice is the largest in size and performs no flow-restricting, so no pressure-reducing function occurs in this case. It is the valve outlet pressurized oil that is induced to the front port of the pilot valve. The purpose is to keep the outlet pressure constant and protects it from the influence of inlet pressure or flow rate.

The hydraulic pressure reducing valve may be either the direct-acting type or the direct-operated pilot-type. The direct-acting type is shown in Figure 2.30. The direct-operated pilot-type is somewhat more compact, since it does not use the large spring (see Figure 2.31).

In many hydraulic circuits or systems, more than one operating pressure are desirable in the system. The reducing valve may be the solution. In the hydraulic reducing valve shown in Figure 2.30, when the spring tension is relieved from the large spring, the pressure differential is highest between the inlet port and the outlet port. As the tension is increased on the spring, the difference is reduced. The pressure differential may be as high as 10 to 1. For example, the inlet pressure may be 500 psi, and the outlet pressure is 50 psi.

Reducing valves are built in pressure ranges similar to the pressure ranges of relief valves. They are usually not built in as many different sizes.

The parts which make up a hydraulic reducing valve (see Figure 2.30) should be studied. Note in Figure 2.30 that a check valve is built in, so that the oil can flow freely in the reverse direction, which eliminates the need for separate check valve. The piston is hardened, and is lapped into the valve body. The valve seat is also hardened.

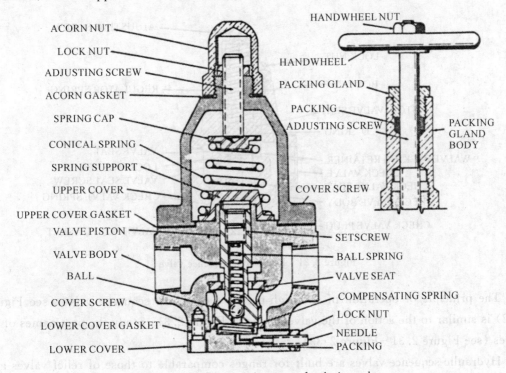

Figure 2.30 Direct-acting type of reducing valve

2.4.2.3　Sequence Valve

A sequence valve is used to direct the flow of fluid to more than one part of a circuit in sequence. The sequence valve is almost the same as a relief valve in construction. Their differences lie in: the outlet oil is directed to other pressure circuits of the system while in a relief valve, the outlet port is connected directly to the reservoir. In addition, discharged port in a sequence valve must be connected to reservoir separately for the pressure existence in both the inlet and outlet ports.

Sequence valves are widely used in both hydraulic and pneumatic systems. By using these valves, a second directional control valve can often be eliminated. Hydraulic sequence valves may be direct-acting, direct-operated pilot-type, or the remote-operated pilot-type, the same as relief valves. Pneumatic sequence valves are generally direct-acting. Figure 2.31 shows a pneumatic sequence valve, and Figure 2.32 shows a direct-acting hydraulic sequence valve. Figure 2.33 shows a direct-operated pilot-type valve that is designed for pipe line mounting. Such valves are also available for sub plate mounting. This valve has a built-in check. In Figure 2.31, when compressed air enters the port and when the pressure which acts on the piston overcomes the tension created by the spring, the orifice to the other ports is opened; this allows air pressure to flow out the second port. Air pressure can be returned through the valve when the check valve is depressed. The spring tension can be either increased or decreased by movement of the adjusting screw. Pneumatic sequence valves are often used for low-pressure oil service.

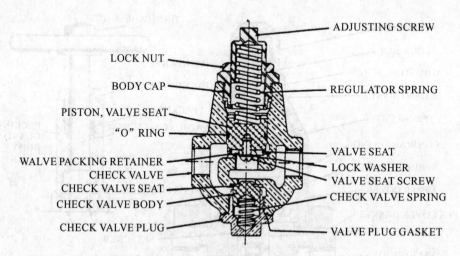

Figure 2.31　Pneumatic sequence valve

The principle of operation of the high-pressure hydraulic sequence valve (see Figure 2.32) is similar to the action of the valve shown in Figure 2.31. Study the part names of all valves (see Figure 2.31 – Figure 2.33).

Hydraulic sequence valves are built for ranges comparable to those of relief valves and reducing valves. Port sizes are also comparable. Pneumatic valves are generally built with threaded connections, while hydraulic valves may be threaded, flanged, or gasketed. Hydraulic sequence valves are sometimes used as spring-loaded check valves in complicated system where several directional control valves are used.

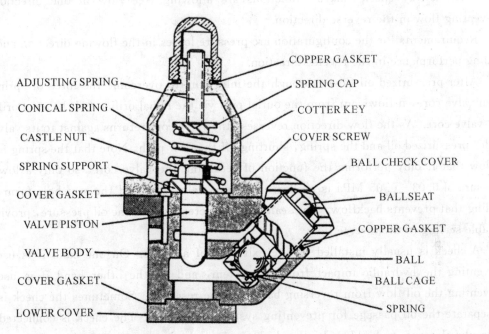

ADJUSTING SPRING
CONICAL SPRING
CASTLE NUT
SPRING SUPPORT
COVER GASKET
VALVE PISTON
VALVE BODY
COVER GASKET
LOWER COVER

COPPER GASKET
SPRING CAP
COTTER KEY
COVER SCREW
BALL CHECK COVER
BALLSEAT
COPPER GASKET
BALL
BALL CAGE
SPRING

Figure 2.32　Direct-acting type of hydraulic sequence valve

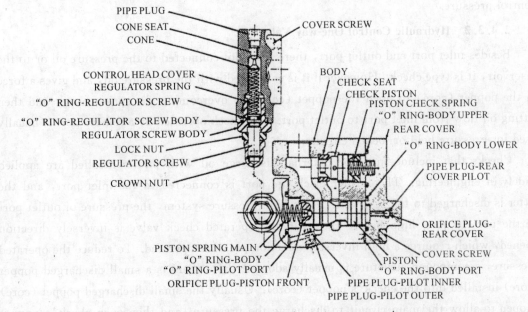

PIPE PLUG
CONE SEAT
CONE
CONTROL HEAD COVER
REGULATOR SPRING
"O" RING-REGULATOR SCREW
"O" RING-REGULATOR SCREW BODY
REGULATOR SCREW BODY
LOCK NUT
REGULATOR SCREW
CROWN NUT

COVER SCREW
BODY
CHECK SEAT
CHECK PISTON
PISTON CHECK SPRING
"O" RING-BODY UPPER
REAR COVER
"O" RING-BODY LOWER
PIPE PLUG-REAR
COVER PILOT
ORIFICE PLUG-
REAR COVER
COVER SCREW
PISTON
"O" RING-BODY PORT
PIPE PLUG-PILOT INNER
PIPE PLUG-PILOT OUTER

PISTON SPRING MAIN
"O" RING-BODY
"O" RING-PILOT PORT
ORIFICE PLUG-PISTON FRONT

Figure 2.33　Direct-operated pilot-type sequence valve

2.4.3 Direction Valve

2.4.3.1 One-way Valve

The one-way check valves' functions are allowing free flow in one direction and preventing flow in the reverse direction.

Requirements for the configuration are pressure losses in the flowing direction and good sealing performance in the reverse direction.

After pressurized oil flows through the inlet port to overcome the force of spring and open valve core, it flows out from the outlet port via the radial orifice and the axial orifice on the valve core. As the flow direction reverses, the valve core returns against to its valve seat under pressurized oil and the spring, shutting off the flow path. Note that the spring force is so low that it only performs the function of returning the valve core, so only a low open pressure of 0.03 – 0.05 MPa is needed. The cone valve core and the seat hole form a line sealing that prevents backflow. The sealing force increases with the oil pressure, providing a reliable sealing in the reverse flow.

A check is usually installed on the exit port of a pump. On one hand it is used for preventing the hydraulic impact from its system, and on the other hand it is used for preventing the oil flow from reversing back into the reservoir. Sometimes the check is used to separate the oil passage for preventing system turbulence. The check is connected with hydraulic elements to build multiple valves, such as check-pressure reducing valve or check-throttling valve. A high spring rate should be selected and the first open pressure is estimated by $p = 0.3 - 0.5$ MPa when it is installed in the oil circuit as a back check or control pressure.

2.4.3.2 Hydraulic Control One-way Valve

Besides inlet port and outlet port, there is a pilot connected to the pressure oil or to the reservoir, it is type check. However if it is connected with the pressure oil, and gives a force to the poppet (core) to open the poppet (core) by overcoming the spring force, and then letting oil flow from inlet port to outlet port. The check operated by a remote port is usually used for pressure held or locked circuits.

Checks that include inner or external discharge oil with pilot-controlled are applied widely in engineering. The former discharge port is connected to the inlet port, and the latter is discharged to the reservoir. For a high pressure system, the pressure of outlet port on the port is quite high before the hydraulic operated check valve is inversely direction opened, which requires a high inversely direction pressure operated. To reduce the operated pressure, a compound structure is usually adopted, here there is a small discharged poppet (core) installed into the check's poppet (core). Usually the small discharged poppet (core) is open to allow the main circuit to discharge the pressure, and then open check's poppet (core). In this way the control pressure is only designed at 4.5% of the working pressure.

Otherwise it is 40%–50% for a configuration without small discharged poppets(core).

Note that the port should be connected to the reservoir when it is not working, otherwise it cannot be returned.

2.4.3.3　Reversing Valve

1. Functions

According to the core motions relative to the body, a directional control valve in a hydraulic system is usually used to control the oil paths that can be connected, or the oil flow direction that can be changed, which allows the actuator to be started, stopped or the oil flow direction to be shifted.

2. Classfication

By configuration: rotary, sliding or ball.

By ports: two ports, three ports, four ports and so on.

By relative operating positions of spool in the valve body. There are two, three, and four positions.

By the operating means: these valve cores may be operated by a difference of oil pressure on the spool, or manually, mechanically, electrically, or by a combination of these means.

3. The construction of spool or directional control valve

Their differences lie in the ways and positions they possess. Two or three-position valves mean that their valve cores have two or three different working positions; similarly, two-way, three-way or four-way valves indicate that there are two, three, or four oil paths on the valve body which are separated from each other and connected with their independent oil lines. The functions of the direction valves depend upon their ways and working positions.

2.4.3.4　Magnetic Valve

A solenoid valve is an electromechanical valve for use with liquid or gas. The valve is controlled by an electric current through a solenoid: in the case of a two-port valve the flow is switched on or off; in the case of a three-port valve, the outflow is switched between the two outlet ports.

Solenoid valves are the most frequently used control elements in fluidics. Their tasks are to shut off, distribute or mix fluids. They are found in many application areas. Solenoids offer fast and safe switching, high reliability, long service life, good medium compatibility of the materials used, and compact design.

Besides the plunger-type actuator which is used most frequently, pivoted — armature actuators and rocker actuators are also used.

A solenoid valve has two main parts: the solenoid and the valve. The solenoid converts electrical energy into mechanical energy which, in turn, opens or closes the valve mechanically.

Solenoid valves may use metal seals or rubber seals, and may also have electrical interfaces to allow for easy control. A spring may be used to hold the valve opened or closed while the valve is not activated.

2.4.3.5 Hydraulic Controlled Reversing Valves and Electro-hydraulic Reversing Valve

1. Hydraulic reversing valves

Hydraulic reversing valves are using control liquid pressure promote valve core to change the direction of the flow direction control valves, starting force big, fluid accuse oil flow rate is greater, impulsion. In order to control movement speed of valve core, to reduce the impact, usually install one-way throttle device before hydraulic controlled pressure oil mouth. Hydraulic reversing valves according to which the access of the directional control valve control number: the two, the three links, four and five, etc.

2. Electro-hydraulic reversing valve

Electro-hydraulic reversing valve is combined by magnetic reversing valve (pilot valve) and hydraulic reversing valve.

2.4.4 Flow Valve

A throttle valve is a simple flow control valve. A throttle valve is virtually equal to available orifice. The orifice size is changed by the valve core movement, which is controlled by a certain mechanism. A throttle valve is quite widely applied in the ration pump and the throttling regulating speed circuit.

Throttle valves are usually combined with fixed delivery pumps and relief valves to constitute a throttle speed-regulating system. The speed can be adjusted by changing the throttle orifice size.

2.4.5 Proportional Valve

2.4.5.1 Proportional Relief Valve

A proportional relief servo valve by direct detection has a spool pilot-operated valve. The inlet oil pressure is directed to the left (action area A_0) of feedback rod and then into the left (action area A_1) of pilot-operated spool via fixed damping orifice. The oil in the right end of the pilot-operated spool is introduced to the spool valve orifice and the upper port of the main valve. The oil in the upper port is admitted to the right (action area A_2) of the pilot-operated spool. When the main valve core is balanced out, R_3 (a dynamic pressure feedback damping orifice located between the pilot-operated spool orifice and the upper port of the main valve) performs no restricting function and the equal forces act on two ends of it. Only $A_1 - A_0 = A_2$ is adopted for design can we keep a hydraulic pressure of $F = PA_0$. When hydraulic pressure F is equal to the electromagnetic force F_E, the pilot-operated spool valve core is balanced out and remains in a certain position with a certain size of orifice. The pressure p_1 in the front chamber of the pilot-operated valve or in the upper port of the main valve is a constant $(p_1 < p)$. The main valve core is balanced out under its upper and down pressures p_1, p, the spring force and the flow force. In this case the main valve is open with a certain orifice to ensure that the inlet pressure p of the relief valve is proportional to the electromagnetic force. A variety of inlet pressures can be obtained by adjusting the input current.

If the inlet pressure increases suddenly due to outside disturbance load, the balance state of the pilot-operated spool valve core is damaged and the valve core moves in the right. The orifice is enlarged, causing the pressure in the front chamber of the pilot valve or in the upper port of the main valve to fall down. In this case, the main valve core is no longer at a balance state. The main valve core moves up to enlarge its orifice and reduce the increased inlet pressure. If the inlet pressure returns to its setting value, the valve cores of the pilot-operated spool valve and the main valve are again under balance states. Then the valve works in a new stable position.

2.4.5.2　Proportional Pressure-reducing Valve

According to DIN – ISO1219, HYDAC proportional pressure reducing valves, type PDM, are control valves, used in oil hydraulic systems, which in the main provide a constant outlet pressure where there is a variable inlet pressure. The outlet pressure to be controlled is determined by the current signal which is supplied by appropriate control electronics and which affects an actuating solenoid.

2.4.5.3　Proportional Flow Valve

A flow control electro-hydraulic proportional servo valve is made by simply replacing the manual control parts of flow control valves by a proportional electric magnet.

New Words

1. spill [spil] v. 溢出
2. plumbing[plʌmiŋ] n. 铅制管
3. tension ['tenʃ(ə)n] n. 张力,拉力
4. unobstructed[ˌʌnəb'strʌktid] adj. 没有障碍的,畅通无阻的
5. resistance[ri'zist(ə)ns] n. 阻力
6. poppet[pɒpit] n. 提升阀
7. orifice['ɔːrifis] n. 孔口
8. gasket['gæskit] n. 垫圈
9. threaded [θredid] adj. 有线状图案装饰的
10. turbulence ['tɜːbjʊl(ə)ns] n. 湍流
11. spool[spuːl] n. 线轴,缠线框
12. fluidics ['fluːidiks] n. 应用流体学
13. rubber [rʌbə] n. 橡胶,合成橡胶

14. rigidity[ri'dʒidəti] *n.* 硬度

15. servo['sɜːvəʊ] *n.* 随动伺服系统

Phrases and Expressions

1. relief valves 安全阀，释压阀

2. pressure reducing valve 减压阀

3. sequence valve 顺序阀

4. unloading valve 卸荷阀

5. direct-acting valve 直动阀

6. direct-operated pilot-type valves 直接操作引导阀

7. adjusting screw 调整螺钉

8. direct-operated hydraulic relief valve 直接操作液压安全阀

9. cone valve core 锥形阀芯

10. check-throttling valve 检测阀门

11. reversing valve 反向阀

12. hydraulic reversing valve 电液换向阀

13. electromagnetic force 电磁力

Notes to the Text

1. When the pressure reaches a certain point, the relief valve spills the oil back to the reservoir, and the pressure is relieved. There are many modifications of these valves, depending on the system in which they are used.

当压力达到一定程度时,安全阀就会将液压油泄漏回油箱,压力就会减轻。安全阀种类很多,根据所使用的系统来选择具体的阀。

2. The relief valve may be of the direct-acting type, the direct-operated pilot-type, or the remote-actuated pilot-type of valve.

安全阀可能是直接作用式的,直接操作的导通式,或者是遥控式的阀门。

3. A sequence valve is used to direct the flow of fluid to more than one part of a circuit in sequence. The sequence valve is almost the same as a relief valve in construction.

顺序阀用于引导流体流向一个管路循环的多个部分。顺序阀在结构上与安全阀几乎是一样的。

4. After pressurized oil flows through the inlet port to overcome the force of spring and open valve core, it flows out from the outlet port via the radial orifice and the axial orifice on the valve core. As the flow direction reverses, the valve core returns against to its valve seat under pressurized oil and the spring, shutting off the flow path.

加压的油通过进口端口克服弹簧的弹力,打开阀芯后,加压油流过阀芯的径向孔和轴向孔流出出口。当流体方向相反时,阀芯在液压油和弹簧的压力下返回阀座,关闭流体通道。

5. A solenoid valve is an electromechanical valve for use with liquid or gas. The valve is controlled by an electric current through a solenoid: in the case of a two-port valve the flow is switched on or off; in the case of a three-port valve, the outflow is switched between the two outlet ports.

电磁阀是一种用于液体或气体的机电阀。这种阀门由通电螺线圈来控制,从而实现双端口阀门流体的开启或关闭;在一个三端口阀门的情况下,流出流体在两个出口端口之间进行切换。

6. A proportional relief servo valve by direct detection has a spool pilot-operated valve.

压力直接检测反馈的电液比例溢流阀为滑阀结构。

Chapter 3 Potable Water Vehicle

Section 3.1 General Description

3.1.1 General

Potable water vehicle(see Figure 3.1) is mainly for providing clean water for aircraft ordinary use, also it provides water for washing ground equipments in airport.

3.1.2 Specifications

Water tank volume:	2,800 L
Water flow:	≥120 L/min
Speed:	40 km/h (unload)
	25 km/h (load)
Gross vehicle weight:	3,100 kg (unload)
Gross vehicle weight:	5,900 kg (load)
Dimensions:	69,000 mm×1,880 mm×2,160 mm
Chassis model:	JMC JX1050TGB23

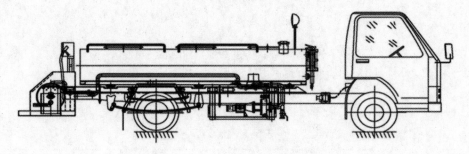

Figure 3.1 Potable water vehicle

3.1.3   Main Components

Engine	GBⅢ engine
Transmission	Manual transmission
Front axle	Steering axle
Rear axle	Drive axle
Brake system	Brake drum &. multi-circuit braking system
Steering system	Hydraulic steering system
Suspension system	Suspension spring with shock absorber
Electrical systems	System voltage 12 V
	Generator 80 W
	Starter 2.0 kW

Section 3.2   Structure Principles

Potable water vehicle consists of a commercial chassis, water tank assembly, power taking off transmission assembly, a clean water supply system, hydraulic transmission system, rear end frame, heater system and electrical system.

3.2.1   Commercial Chassis

TK-QS30 is equipped with JMC JX1050TGB23 type commercial automobile chassis.

3.2.2   Water Tank Assembly

Water tank is made of stainless steel — 304 ($1Cr_{18}Ni_9$), the capacity is about 3 m^3. The clean water tank: which has a round shape cover at the top is used to provide clean water to the aircraft; it consists of three connected tanks which are separated by a wave-shape plate. At the top of the tank, it has a round cover with three holes, which can clean and check the waste water tank. For customer in the cold regions, the tank is a two layers designed structure which is filled with insulating material so as to provide warm-keeping feature, the exposed water pipe is also wrapped by insulating material for such purpose. The tank is connected to the chassis through six rubber dampers which reduce the vibration well during vehicle movement.

Anti-slip stickers are attached at the upper surface of the tank which prevent slipping accident of the operators during operation or maintenance.

3.2.3   Power Take off Transmission Assembly

Using the manual shaft in the cab to operate the power take off at the chassis to provide power to Hydraulic Pump.

3.2.4　Clean Water Supply System

Clean water supply system has a corrosion-resistant centrifugal pump. It is used to transport clean water through reel water pipe driven by a hydraulic motor.

3.2.5　Hydraulic Transmission System

Power take off direct drives hydraulic pump, outputs high-pressure hydraulic oil to solenoid control valve to drive water pump motor, vacuum pump motor, platform lift cylinder and waste hydraulic cylinder. There should use different hydraulic oil under different environment temperature, see Table 3.1.

Table 3.1　Hydraulic oil model

Oil	Model
Hydraulic oil	L－HM46 environment temperature－5℃ or above
	L－HM32 environment temperature－20℃ or above
	L－HV32 cold regions winter environment temperature－40℃ or above

Note: This product equips with sinopec hydraulic oil.

3.2.6　Rear End Frame

Rear end frame consists of supporting framework, stepladders box and water tank.

3.2.7　Platform Lifting System

Platform lifting system consists of lifting frame, lifting column and connecting bolt.
Note: It should smear with right amount grease to sprocket, slider and track according to the lubricating condition of lifting column.

3.2.8　Heater System

There are heaters at rear water pipe reel box, flesh water pump chamber and vacuum pump chamber respectively. The heaters are individual controled and used to heat during cold days to ensure that pump chamber temperature can be maintained at 5℃ or above so as to prevent freezing of water pipe, water pump and vacuum pump, or destroying facility which may affect waterwheel operation. The water valves of the heater system locate at the rear part of cab. All water supply pipelines and drain valves are wrapped with insulation materials so as to operate under low temperate condition.
Notice: Please stop operation under－40℃.

3.2.9　Electrical System

Electrical system consists of control box, 12 V battery, alarm lamp, illumination, etc.

The control box control power switch, water supply switch, waste water suction switch, emergency stop switch, etc. and vacuum gauge of waste water box. Please press the emergency stop switch under emergency situation to turn off the machine.

There equip with lamps at both upfront and rear of the lifting frame of water chamber for night time operation.

Electrical system have warning feature, there are a low water level warning switch built in at the water tank. When the water level inside the water tank are low, buzzer will alert and the corresponding alarm signal on the control box will turn on simultaneously, operator should fill in water in time; when the hydraulic system filter is block, the buzzer will alert and the corresponding alarm signal on the control box will turn on simultaneously, operator should clean or replace filter in time.

Section 3.3   Application and Warranty

3.3.1   Check with the Following Instructions Before Used

(1) Check the fastening status of the connection of each component, especially the transmission system, steering system, braking system, suspension systems, wheels etc. If find any loose, customers should tighten these to the specified torque.

(2) Check the engine, transmission, rear axle and the oil level of steering gear.

(3) Check the condition of the grease lubrication points.

(4) Check the installation of the engine accessories and fan belt tightness.

(5) Check the battery fluid level, it must be higher than the protection plate about 10 – 15 mm.

(6) Check the working condition of brake system (shown in Figure 3.2) and any air leak observed at the pipe joints.

Figure 3.2   Brake system

1—vacuum booster (or servo brake system);   2—brake (vacuum brake)

(7) Check the working condition of electrical equipments such as lights, speaker and

instrument.

(8) Check the working condition of steering system.

(9) Measure the pressure of tire to check whether it satisfies to standard requirement.

(10) Check whether the vehicle tools and accessories are completely set.

(11) Check the hydraulic system:

1)Check the fluid level inside oil tank and check any leak observed from hydraulic pipes. Check the working condition of hydraulic pumps, hydraulic cylinders, hydraulic motors, water pumps and vacuum pumps.

2)Check the quality of hydraulic oil by opening the oil block, to observe whether any quality change of oil which may due to deterioration of oil or emulsification due to water.

3) Check whether the hydraulic oil is dirty or not, any change observed from the negative pointer table, if the pointer nears the red alarm region, please replace the hydraulic oil and filter immediately.

4)Check whether there is any air within the system. Hold the high-pressure pipeline by hand to test whether there is intermittent shaking. If there is, please process bleeding air procedure.

5)If customer finds that internal leak of oil pump is large, please take the equipment back for maintenance.

6)Check whether there is any block in the oil pipe.

(12)Check the transmission system:

1)Check whether there is any loose at both end and disks and bolt of tranfsmission shaft.

2)Check whether there is sufficient grease for the spines and universal shafts.

3)Check the extent of the port of the spine, whether sticky or not.

(13) Oil filter maintenance:

1)Filter should be replaced after using for every 2,000 hours or every 6 months. Filter may be clean by our factory with limited extent.

2)Filter should be fully checked after remove. If find that it is heavily polluted, then the possible reasons should be found out. In this case, it must shorten the checking process interval and, if suspicious, it is recommended of flushing the system.

3)Spare filters should be equipped in the machine and kept with original package to prevent pollution by dust.

(14)Air filter maintenance:

1)In general, air filter should be maintained with oil filter at the same time.

2)If there is dusty or under a humid environment, it is recommend to shorten the maintenance interval or change the position of air filter.

3)The size of filter holes of air filter should be same as oil filter.

(15)Others:

1)Replace wear parts(seals, hoses).

2)Lubricate the lubrication points which require to be lubricated manually with grease (universal shaft, plug connector).

(16)Note:

1)Note that deposition of water inside the oil tank will also cause to increase the oil level which may mislead that the oil level is too high.

2)Leaked oil is always dirty, and should not fall back to the oil tank.

3)Even if the hydraulic system had been working for a period, however, corrosion protection measure should be taken if long-term disable.

4)When restart, it should remove all preservatives carefully. The initial inspection and maintenance intervals should be shortened to half as aforementioned.

(17)Reasons of the flow become dirty:The destroy of hydraulic system is mainly due to the impurity particles in the oil. Such particles may cause burst fault. However, in general, it only will cause wearing gradually and hence decrease system's efficiency and effectiveness which eventually destroy the machine and force it to go for maintenance. If the fluid checking shows that the impurity particles in the oil are too high, the possible reasons may be as showed below:

1)Cleaning process is not sufficient or neglected before operation or after repair.

2)The transport environment is too dirty or storing site is inappropriate.

3)Have not replaced oil.

4)Maintenance equipment is too dirty.

5)Have not installed filter or filter is too dirty, block indicator.

6)Have not installed indicator or do not aware that the filter blocks the indicator.

7)The air filters of Hydraulic fluid tank are too thick.

8) Hydraulic fluid tank is not airtight (pipe entry port, clean hole, and so is not airtight).

9)From pumps, motors, valves' metal or seals wear down particles. Rust from oil tank.

10)Pour leaked oil back to oil tank.

(18)Reasons of water present in hydraulic fluid: Water in hydraulic fluid will increase wear, pollution and corrosion of system and change the characteristic of the oil which results in system failure. Below are the possible reasons of water present in the hydraulic fluid and cause system cannot work properly:

1)Cutting fluid mixed with water to enter.

2)Rainwater or cleaning water penetrates in.

3) Stream condenses due to fluctuations of temperature or fluctuations of oil tank volume.

4)Have not discharged deposition water inside the hydraulic fluid tank periodically.

(19)Hydraulic oil containing air bubbles will produce unpleasant noise (cavitations), it will cause dysfunction and damage to the system. Below are the reasons cause to air bubbles formation and leaving：

1）Fluid wrong or dirty.

2）Negative pressure zone is not airtight (for example, suction piping, pumps, throttle, etc).

3）The formation of bubbles due to pressure decrease (the pipeline sharp bends, twist on the hose).

4）The system is not fully deflated (at first time operation or after repair).

5）The strainers which are installed before and after the suction pipe are too thin or filter too small.

（20）Reasons of bubbles forming at the surface of hydraulic fluid：Bubbles forming at surface will inhale by pump or overflow from the tank. Below are reasons of large amount of bubbles forming：

1）Impurities in the oil are too much (dust, water, aging resultant, preservative and so).

2）Looping speed is too high.

3）There is no buffer clapboard installed or it is not effective.

（21）Reasons of serious external leakage：Leakage will cause a lost which the amounts of supplementary oil refill will be larger than the amounts of completely refill. Reasons of Leakage：

1）Leaking at pipe joints part of components.

2）System checking and maintenance are not enough.

3）Sealing of hydraulic tank and the shaft are damaged.

4）Pipe joints which access is difficult (therefore difficult to detect and eliminate leaks).

5）Hoses damage.

6）No sealing to protect the position where dust or metal particles may enter.

（22）In general, the temperature of oil inside hydraulic oil tank should not exceed 60℃ under industrial system. In vehicle system, temperature should not exceed 82℃ and maximum at 95℃. Oil temperature too high will shorten the life of hydraulic fluid, seals and hoses. This will accelerate residues formed, speed up wear and tear. Reasons of fluid temperature are too high：

1）Pump flow is too large (over cutting).

2）Hydraulic fluid tank is too small or poorly designed.

3）Cooler is not installed, or cooling capacity is too small, flow adjustment incorrectly, dirty, or the cooler bypass valve is opened.

4）Pipe diameter is too small.

5）The setting of pressure relief valve is incorrect, the valve is dirty or fault. The set value of pressure regulator (pressure down) is exceeding the safety valve settings.

6）Temperature increase due to external factors (sunlight, furnace, high temperature and air conditioning is faulty).

7)Air inside oil (compression heat).

8)Oil fluid level is too low.

9)Wearing on oil pump, valve, motor's metal or seal units.

10)If the fluid viscosity is incorrect, it would lead to excessive heat load. At this point, overheating is not only restricting on fuel tank, and will occur in various parts.

(23)Vacuum booster as a brake booster, it uses the vacuum created by the engine and transmits these to the booster to increase the force acting on the brake pedal by the operator. Use the vacuum at the pipes of engine to help the operator to control the brake pedal. Vacuum booster obtains vacuum from the engine, and stores these in a vacuum bottle. Please check the vacuum booster if finding it does not work. Except booster fault, it should not dismantle vacuum booster so as to avoid damage. If the vacuum booster is destroyed, it should replace the whole.

For checking the brake system which adapts vacuum booster, if finding that the pedal is too heavy, then a vacuum test for the vacuum booster should be performed. See below:

Method 1: Turn on the engine and run it for about 1 - 2 min and then turn off, press the brake pedal several times in normal way. If the pedal margin at the first time is longer and thereafter is shorter gradually, it means the brake booster is in good working condition. If not, it may have leakage occuring at vacuum booster or vacuum pipes and should be repaired.

Method 2: Before turning on the engine, press the brake pedal several times in normal way to ensure that the pedal margin remains unchanged. Next, keep the brake pedal in a pressing condition and turn on the engine, if the level of pedal is slightly decreased then it means the vacuum booster works well.

Method 3: During the engine running, press the brake pedal and then turn off the engine and keep pressing on the brake pedal about 30 s. If the level of pedal remains unchanged, it means vacuum is in good condition; if the pedal rebounds, it means the vacuum booster is fault.

3.3.2   Driving and Handling

3.3.2.1   Vehicle Handling

According to the instruction of user manual, illustrations in detail are as below:

(1)Park the vehicles at the position for adding water of aircraft, and then connect the water pipe to the water adding port of aircraft.

(2)Select the button which could turn on water supply.

(3)Press the water supply button to provide water.

(4)Switch on the light equip on vehicle during night time operation.

(5)If the filter is blocked and hence cannot add water to aircraft, operator should empty the water tank first, and then remove the filter and clean it so as to remove impurities. After that install the filter back and it can be used again.

3.3.2.2 Actual Operating Procedures for Sewage and Water Provision

When approaching to the aircraft close to 10 m distance, slow down the speed to 3 km/h. Park vehicle in suitable place, set the gearbox to neutral gear and apply parking hand brake. When oil pump is in working condition, joint the water pipe to water connector of aircraft, turn on the water supply switch to start water adding process. When process completed, turn off the water supply switch, remove the water pipe and drive away from the aircraft.

Section 3.4 Troubleshooting and Maintenance

Troubleshooting and maintenance instructions of hydraulic system and water supply system are shown in Table 3.2. Troubleshooting for chassis please refer to vehicle.

Table 3.2 Troubleshooting and maintenance instructions of hydraulic system and water supply system

Items	Period	Instruction
Oil pump noise	Every time	Hearing checking, if abnormal sound is heard, check whether oil filter is blocked or whether the suction-type tube is blocked or whether oil pump is destroied
Oil filter	6 months	Clean the oil filter and pressure oil filter, replace filters
Pressure gauge	1 month	Check whether the system's pressure is proper or not
Hydraulic oil tank oil level	1 month	Refill hydraulic oil when the oil level is low
Oil quality	6 months	Change to other oil if it is not fulfilled the standard requirement; in general, the oil should be replaced every 6 months
Whole hydraulic system	2 years	Remove components to cleaning, pipes cleaning
Water pump	Every time	Hearing test, if abnormal sound is heard or leaked, check whether hydraulic motor, water pump, motor's seal condition and blades of water pump are destroied or not

Phrases and Expressions

1. water tank 水箱
2. power take off transmission assembly 取力器传动总成
3. hydraulic transmission system 液力传动系统
4. solenoid control valve 电磁控制阀

5. vacuum pump 真空泵

Notes to the Text

1. It should smear with right amount grease to sprocket, slider and track according to the lubricating condition of Lifting column.

应根据提升降柱的润滑状况在链轮、滑块和导轨上涂抹适量的润滑脂。

2. Check the quality of hydraulic oil by opening the oil block. Observe whether any quality change of oil which may due to deterioration of oil or emulsification due to water.

打开油塞检查液压油的品质,观察因变质或水乳化引起的任何品质变化。

Chapter 4   Lavatory Service Vehicle

Section 4.1   General Description

4.1.1   General

TK – WS30 is a newly large capacity suction type airplane lavatory service truck (see Figure 4.1) designed for airport environment. It is used to collect waste water from aircraft and provide clean water for flushing and refilling.

4.1.2   Specifications

Water tank volume	1,000 L
Waste water tank volume	1,800 L
Water flow	≥120 L/min
Water pressure	0.25 – 0.35 MPa
Waste water pressure	0.25 bar
Speed	40 km/h (unload)
	25 km/h (load)
Gross vehicle weight (unload)	3,500 kg
Gross vehicle weight (load)	5,900 kg
Dimensions	6,600 mm×1,880 mm×2,020 mm
Chassis model	JMC JX1050TGB23
	(with hydraulic system)

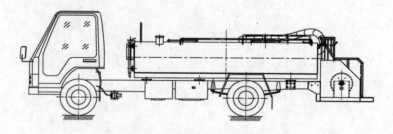

Figure 4.1   Lavatory service truck

4.1.3   Overview

TK – WS30 type airplane lavatory service truck consists of a commercial chassis, water tank assembly, power taking off transmission assembly, a clean water supply system, waste water drain system, waste system, hydraulic transmission system, rear end frame, platform lifting system, heater system and electrical system.

1. Commercial chassis

TK – WS30 is equipped with JMC JX1050TGB23 type commercial automobile chassis.

2. Water tank assembly

Water tank is made of stainless steel — 304 ($1Cr_{18}Ni_9$). It consists of three separate compartments: (a)circulating water tanks: to provide water recycling for the vacuum pump; (b)clean water tank: it has a round shape cover at the top which is used to provide clean water to the aircraft; (c)waste water tank: it is used as a storage tank to collect the waste water from the aircraft, and it consists of two connected tanks which are separated by a wave-shape plate. At the top of the tank, it has a round cover with two holes, a suction port, a waste water inlet port and a breathing hole; it has a wasting port at the bottom part. The waste water tank is not equipped with flush pipe. It connects the clean water port to the external port at the rear of the waste water tank, cleaning the waste water tank by opening the clean water pump. For customer in the cold regions, the tank is a two layers designed structure which is filled with insulating material so as to provide warm-keeping feature. The exposed water pipe is also wrapped by insulating material for such purpose. The rear part of waste water tank has a sight glass used to monitor the waste water level inside the tank. The tank is connected to the chassis through six rubber dampers which reduce the vibration well during vehicle movement.

Anti-slip stickers are attached at the upper surface of the tank which prevent slipping accident of the operators during operation or maintenance.

3. Power take off transmission assembly

Using the manual shaft in the cab to operate the power take off at the chassis to provide power to hydraulic pump.

4. Clean water supply system

It is a corrosion-resistant centrifugal pump. It is used to transport clean water through reel water pipe driven by a hydraulic motor.

5. Waste water drain system

It used hydraulic motor drive vacuum pump to create vacuum and provide certain degree of negative pressure inside waste water tank. By connecting to the waste water port at the aircraft with the waste water pipe, and turning on both waste water valve of the aircraft and vacuum pump switch, then pump working of the airplane waste water tank can be in progress in a fast and effective way. On the other hand, it ensures that leakage of waste water will not occur, and prevents pollution problem to airport and operators.

Note: When vacuum pump is destroied or suction function is not used, switching off the vacuum pump can achieve auto sewage.

6. Hydraulic transmission system

Power take off direct drives hydraulic pump, outputs high-pressure hydraulic oil to solenoid control valve to drive water pump motor, vacuum pump motor, platform lift cylinder.

There should use different hydraulic oil under different environment temperature, as shown in Table 4.1.

Table 4.1   Hydraulic oil model

Oil	Model
Hydraulic oil	L - HM46 environment temperature −5℃ or above
	L - HM32 environment temperature −20℃ or above
	L - HV32 cold regions winter environment temperature −30℃ or above

Note: This product equips with sinopec hydraulic oil.

7. Rear end frame

Rear end frame consists of supporting framework, stepladders box and water tank.

8. Platform lifting system

It consists of lifting frame, lifting column and connecting bolt.

Note: It should smear with right amount grease to sprocket, slider and track according to the lubricating condition of lifting column.

9. Heater system

There are heaters at rear water pipe reel box, flush water pump chamber and vacuum pump chamber respectively. The heaters are individual controled and used to heat during cold days to ensure that pump chamber temperature can be maintain at 5℃ or above so as to prevent freezing of water pipe, water pump or vacuum pump, or destroying facility which may affect waterwheel operation. The water valves of the heater system locate at the rear part of cab. All water supply pipelines and drain valves are wrapped with insulation materials so as to operate under low temperate condition.

Note: Please stop operation under −40℃.

10. Electrical system

It consists of control box, 12V battery, alarm lamp, illumination, etc.

The control box controls power switch, water supply switch, waste water suction switch, emergency stop switch, etc. and vacuum gauge of waste water box. Please press the emergency stop switch under emergency situation to turn off the machine.

There is equipped with lamps at both upfront and rear of the lifting frame of water chamber for night time operation. Electrical system has warning feature, where there is a low water level warning switch built in at the water tank. When the water level inside the water tank are low, buzzer will alert and the corresponding alarm signal on the control box will turn on simultaneously, operator should fill in water in time; when the hydraulic system filter is blocked, the buzzer will alert and the corresponding alarm signal on the control box will turn on simultaneously, operator should clean or replace filter in time.

11. Warm-keeping system

The warm-keeping of cleaning water pump room and vacuum pump room relies on the warm water which is heated by engine cycle in radiator, meanwhile combined with a fan to get the warm-keeping purpose. The warm-keeping of hose reel room depends on the heat of tail gas. Meanwhile, the above mentioned three rooms and all water pipes and hoses are covered by sponge to keep warm.

Phrases and Expressions

1. centrifugal pump 离心泵

Notes to the Text

1. Using the manual shaft in the cab to operate the power take off at the chassis to provide power to hydraulic pump.

操作驾驶室的手柄把输出给底盘的动力提供给液压泵。

2. When vacuum pump is destroied or suction function is not using, switch off the vacuum pump can achieve auto sewage.

当真空泵损坏或吸入功能不使用时,关闭真空泵即可实现自动排污。

3. Power take off direct drives hydraulic pump, outputs high-pressure hydraulic oil to solenoid control valve to drive water pump motor, vacuum pump motor, platform lift cylinder and waste hydraulic cylinder.

发动机动力输出直接驱动液压泵,把高压液压油输出到电磁阀以驱动水泵马达、真空马达、平台升降油缸和排污油缸。

4. There are a heater at rear water pipe reel box, flush water pump chamber and vacuum pump chamber respectively.

后水管卷筒箱、冲水泵腔和真空泵室分别装有加热器。

Section 4.2   Application and Warranty

4.2.1   Check with the Following Instructions before Used:

(1) Check the fastening status of the connection of each component, especially the transmission system, steering system, braking system, suspension systems, wheels, etc. If find any loose, customers should tighten these to the specified torque.

(2) Check the engine, transmission, rear axle and the oil level of steering gear.

(3) Check the condition of the grease lubrication points.

(4) Check the installation of the engine accessories and fan belt tightness.

(5) Check the battery fluid level. It must be higher than the protection plate about 10 – 15 mm.

(6) Check the working condition of brake system (see Figure 4.2) and any air leak observed at the pipe joints.

Figure 4.2　Brake system

1—vacuum booster (or servo brake system)；2—brake (vacuum brake)

(7) Check the working condition of electrical equipments such as lights, speaker and instrument.

(8) Check the working condition of steering system.

(9) Measure the pressure of tire to check whether it satisfies to standard requirement.

(10) Check whether the vehicle tools and accessories are completely set.

(11) Check the hydraulic system：

1) Check the fluid level inside oil tank and check any leak observed from hydraulic pipes. Check the working condition of hydraulic pumps, hydraulic cylinders, hydraulic motors, water pumps and vacuum pumps.

2) Check the quality of hydraulic oil by opening the oil block, to observe whether any quality change of oil which may due to deterioration of oil or emulsification due to water.

3) Check whether the hydraulic oil is dirty or not, and whether there is any change observed from the negative pointer table. If the pointer nears the red alarm region, please replace the hydraulic oil and filter immediately.

4) Check whether there is any air within the system. Hold the high-pressure pipeline by hand to test whether there is intermittent shaking. If there is, please process bleeding air procedure. If customer find that internal leak of oil pump is large, please take the equipment back for maintenance.

5) Check whether there is any block in the oil pipe.

(12) Check the transmission system：

1) Check whether there is any loose at both end and disks and bolt of transmission shaft.

2) Check whether there is sufficient grease for the spines and universal shafts.

3) Check the extent of the port of the spine, whether sticky or not.

(13) oil filter maintenance:

1) Filter should be replaced after using for every 2,000 hours or every 6 months. Filter may be clean by our factory with limited extent.

2) Filter should be fully checked after remove. If find that it is heavily polluted, then should find out the possible reasons. In this case, it must shorten the checking process interval and, if suspicious, it is recommended of flushing the system.

3) Spare filters should be equip in the machine and kept with original package to prevent pollution by dust.

(14) Air Filter maintenance:

1) In general, air filter should be maintenance with oil filter at the same time.

2) If there is dusty or under a humid environment, it is recommend to shorten the maintenance interval or change the position of air filter.

3) The size of filter holes of air filter should be same as oil filter.

(15) Others:

1) Replace wear parts(seals, hoses).

2) Lubricate the lubrication points which require to be lubricated manually with grease. (universal shaft, plug connector).

(16) Note:

1) Note that deposition of water inside the oil tank will also cause to increase the oil level which may mislead that the oil level is too high.

2) Leaked oil is always dirty, and should not fall back to the oil tank.

3) Even if the hydraulic system had been working for a period, however, corrosion protection measure should be taken if long-term disable.

4) When restart, it should remove all preservatives carefully. The initial inspection and maintenance intervals should be shortened to half as aforementioned.

(17) Reasons of the flow become dirty : The destroy of hydraulic system is mainly due to the impurity particles in the oil. Such particles may cause burst fault. However, in general, it only will cause wearing gradually and hence decrease system's efficiency and effectiveness which eventually destroy the machine and force it to go for maintenance. If the fluid checking shows that the impurity particles in the oil are too high, the possible reasons may be as showed below:

1) Cleaning process is not sufficient or neglected before operation or after repair.

2) The transport environment is too dirty or storing site is inappropriate.

3) Have not replaced oil.

4) Maintenance equipment is too dirty.

5) Have not installed filter or filter is too dirty, block indicator.

6) Have not installed indicator or do not aware that the filter blocks the indicator.

7) The air filters of hydraulic fluid tank are too thick.

8) Hydraulic fluid tank is not airtight (pipe entry port, clean hole, and so is not airtight).

9) From pumps, motors, valves' metal or seals wear down particles. Rust from oil tank.

10) Pour leaked oil back to oil tank.

(18) Reasons of water present in hydraulic fluid: Water in hydraulic fluid will increase wear, pollution and corrosion of system and change the characteristic of the oil which results in system failure. Below are the possible reasons of water presenting in the hydraulic fluid which cause system not work properly:

1) Cutting fluid mixed with water to enter.

2) Rainwater or cleaning water penetrates in.

3) Stream condenses due to fluctuations of temperature or fluctuations of oil tank volume.

4) Have not discharged deposition water inside the hydraulic fluid tank periodically.

(19) Hydraulic oil containing air bubbles will produce unpleasant noise (cavitations), which will cause dysfunction and damage to the system. Below are the reasons cause to air bubbles' formation and leaving:

1) Fluid wrong or dirty.

2) Negative pressure zone is not airtight (for example, suction piping, pumps, throttle, etc.).

3) The formation of bubbles due to pressure decrease (the pipeline sharp bends, twist on the hose).

4) The system is not fully deflated (at first time operation or after repair).

5) The strainers which are installed before and after the suction pipe are too thin or too small.

(20) Reasons of bubbles forming at the surface of hydraulic fluid: Bubbles forming at surface will inhale by pump or overflow from the tank. Below are reasons of large amount of bubbles forming:

1) Impurities in the oil are too much (dust, water, aging resultant, preservative and so).

2) Looping speed is too high.

3) There is no buffer clapboard installed or it is not effective.

(21) Reasons of serious external leakage: Leakage will cause a lost which the amounts of supplementary oil refill will be larger than the amounts of completely refill. Reasons of Leakage:

1) Leaking at pipe joints part of components.

2) System checking and maintenance are not enough.

3) Sealing of hydraulic tank and the shaft are damaged.

4) Pipe joints which access is difficult (therefore difficult to detect and eliminate leaks).

5) Hoses damage.

6) No sealing to protect the position where dust or metal particles may enter.

(22) In general, the temperature of oil inside hydraulic oil tank should not exceed 60℃ under industrial system. In vehicle system, temperature should not exceed 82℃ and maximum at 95℃. Oil temperature too high will shorten the life of hydraulic fluid, seals and hoses. This will accelerate residues formed, speed up wear and tear. Reasons of fluid temperature are too high:

1) Pump flow is too large (over cutting).

2) Hydraulic fluid tank is too small or poorly designed.

3) Cooler is not installed, or cooling capacity is too small, flow adjustment incorrectly, dirty, or the cooler bypass valve is opened.

4) Pipe diameter is too small.

5) The setting of pressure relief valve is incorrect, the valve is dirty or fault. The set value of pressure regulator (pressure down) is exceeding the safety valve settings.

6) Temperature increase due to external factors (sunlight, furnace, high temperature and air conditioning are faulty).

7) Air inside oil (compression heat).

8) Oil fluid level is too low.

9) Wearing on oil pump, valve, motor's metal or seal units.

10) If the fluid viscosity is incorrect, it would lead to excessive heat load. At this point, overheating is not only restricting on fuel tank, and will occur in various parts.

(23) Vacuum booster as a brake booster, it uses the vacuum created by the engine and transmits these to the booster to increase the force acting on the brake pedal by the operator. Use the vacuum at the pipes of engine to help the operator to control the brake pedal. Vacuum booster obtains vacuum from the engine, and stores these in a vacuum bottle. Please check the vacuum booster if it does not work. Except booster fault, vacuum booster should not be dismantled so as to avoid damage. If the vacuum booster is destroyed, it should be wholly replaced.

For checking the brake system which adapts vacuum booster, if find that the pedal is too heavy, then a vacuum test for the vacuum booster should be performed. As listed below:

Method 1: Turn on the engine and run it for about 1 - 2 min and then turn off. Press the brake pedal several times in normal way. If the pedal margin at the first time is longer and thereafter is shorter gradually, it means the brake booster is in good working condition. If not, it may have leakage occuring at vacuum booster or vacuum pipes and should be repaired.

Method 2: Before turning on the engine, press the brake pedal several times in normal way to ensure that the pedal margin remains unchanged. Next, keep the brake pedal in a pressing condition and turn on the engine. If the level of pedal is slightly decreased then it means the vacuum booster works well.

Method 3: During the engine running, press the brake pedal and then turn off the engine and keep pressing on the brake pedal about 30 s. If the level of pedal remains unchanged, it means vacuum bosster is in good condition; if the pedal rebounds, it means the vacuum booster is fault.

4.2.2   Driving and Handling

4.2.2.1   Vehicle Handling

Illustrations in detail as below:

(1) Park the vehicles at the position for adding water of aircraft, and then connect the water pipe to the water adding port of aircraft.

(2) Select the button which turns on water supply.

(3) Press the water supply button to provide water.

(4) Switch on the light equip on vehicle during night time operation.

(5) If the filter is blocked and hence cannot add water to aircraft, operator should empty the water tank first, and then remove filter and clean it so as to remove impurities, after that install the filter back and it can be used again.

4.2.2.2   Actual Operating Procedures for Sewage and Water Provision

When approaching to the aircraft close to 10m distance, slow down the speed to 3 km/h. Park vehicle in suitable place, set the gearbox to neutral gear and apply parking hand brake. When oil pump is in working condition, joint the water pipe to water connector of aircraft, turn on the water supply switch to start water adding process. When process completed, turn off the water supply switch, remove the water pipe and drive away from the aircraft.

4.2.3   Troubleshooting and Maintenance

Troubleshooting and maintenance instructions of hydraulic system and water supply system are shown in Table 4.2. Diagrams of electrical and hydraulic are shown in Figure 4.3 and 4.4. Troubleshooting for chassis please refer to vehicle.

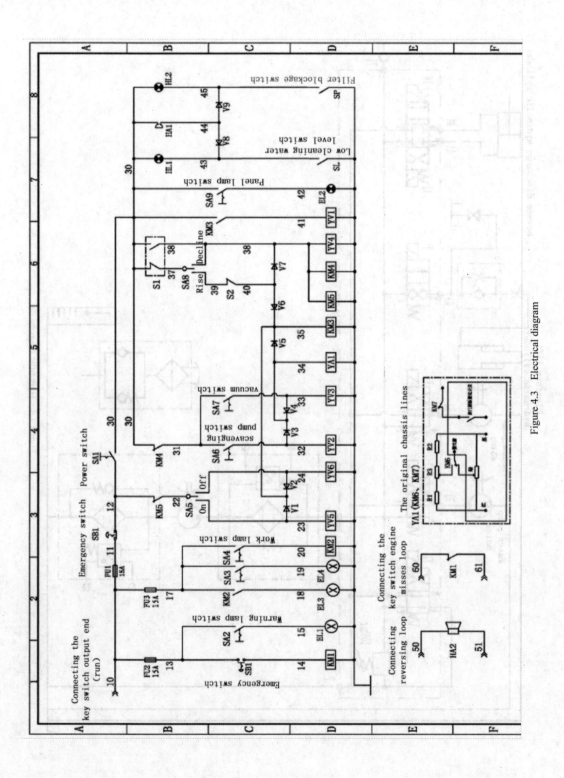

Figure 4.3 Electrical diagram

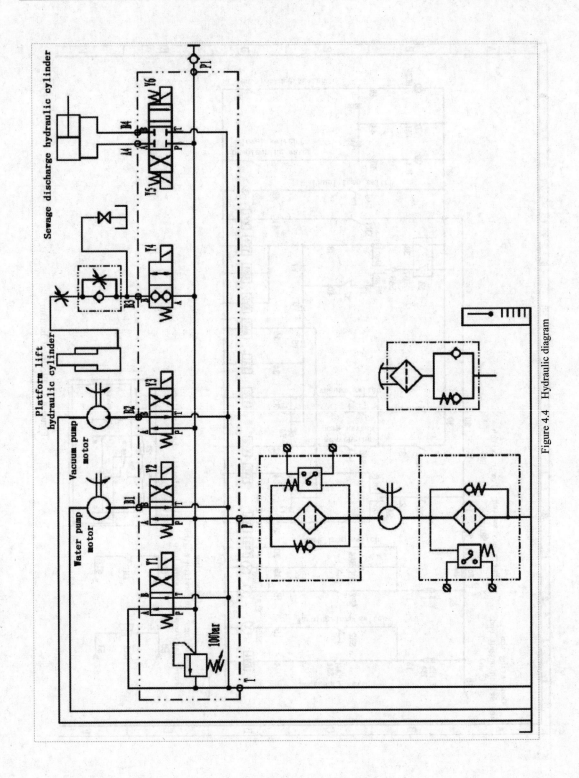

Figure 4.4 Hydraulic diagram

Table 4. 2   Troubleshooting and maintenance instructions of hydraulic system and water supply system

Items	Period	Instruction
Oil pump noise	Every time	Hearing checking, if abnormal sound is heard, check whether oil filter is blocked or whether the suction-type tube is blocked or whether oil pump is destroied
Oil filter	6 months	Clean the oil filter and pressure oil filter, replace filters
Pressure gauge	1 month	Check whether the system's pressure is proper or not
Hydraulic oil tank oil level	1 month	Refill hydraulic oil when the oil level is low
Oil quality	6 months	Change to other oil if it is not fulfilled the standard requirement; in general, oil should be replaced every 6 months
Whole hydraulic system	2 years	Remove components to cleaning, pipes cleaning
Water pump	Every time	Hearing test, if abnormal sound is heard or leaked, check whether hydraulic motor, water pump, motor's seal condition and blades of water pump are destroied or not

New Words

1. airtight ['eətait] *adj.* 密封的
2. cavitation [ˌkævi'teiʃən] *n.* 空化,气穴现象
3. strainer ['streinə(r)] *n.* 过滤器
4. sewage ['suːidʒ] *n.* 污水

Phrases and Expressions

1. vacuum booster (or servo brake system) 真空助力器(或伺服制动系统)

Chapter 5   Catering Truck

Section 5. 1   General Description

5. 1. 1   General

The catering truck is special mobile equipment used for civil aviation to deliver aircraft catering for aircraft passengers. The truck's front-end platform which is used to docking aircraft is capable to stably move forward, backwards or sideways. It is equipped with full sets of protection devices so as to ensure the safety of docking to aircraft door. Also, its support to the ground span is large, the simple and compact mechanic structure and its frame's safety factor are greater than 3. It is user friendly and reliable in performance.

5. 1. 2   Specifications

Chassis	ISUZU QL1140TNFR
Wheelbase (mm)	5,000
Platform height (from the ground) (mm)	Minimum: 2,750; Maximum: 6,000
Fixed platform (mm)	1,650 (L) × 2,300 (W)
Mobile platform (mm)	1,400 (L) × 1,000 (W)
Range of telescopic platform (mm)	650
Car carriage dimensions (mm)	7,000(L) × 2,500(W) ×2,480(H)
Range of telescopic platform (mm)	650
Car carriage dimensions (mm)	7,000(L) × 2,500(W) ×2,480(H)
Gross vehicle weight (kg)	11,980
Car carriage maximum load (kg)	4,000
Dimensions (mm)	9,500(L) × 2,530(W) × 3,760(H)
Steering type	Recirculating ball-type steering Hydraulic power steering
Minimum ground clearance(mm)	200
Fuel tank capacity(L)	200

Hydraulic fluid tank capacity(L)   140

Section 5.2   Structure Pringciples

The catering truck consists of standard chassis, auxiliary beam assembly, front-end platform assembly, catering van assembly, scissor lift frame assembly, hydraulic assembly, refrigerating unit and electrical control system (see Figure 5.1).

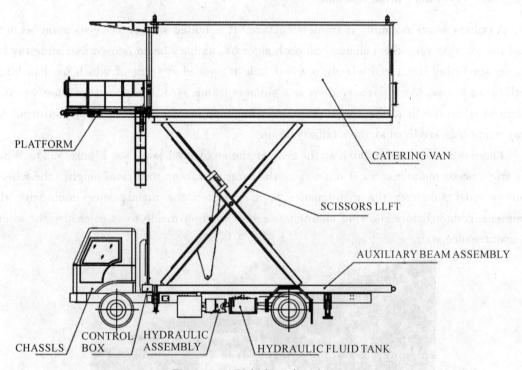

Figure 5.1   Vehicle outline drawing

5.2.1   Chassis

The catering truck is built on ISUZU chassis. The chassis specifications are as below:

Model	ISUZU QL1140TNFR
Total weight(kg)	14,000
Gross vehicle weight(kg)	4,700
Drive wheel arrangement	4×2
Wheelbase(mm)	5,000
Wheeltrack(mm)	1,960/1,855
Length(L)×Width(W)×Height(H)	
(unload) (mm)	8,785×2,465×2,780
Front overhang / rear overhang(mm)	1,395/2,390
Maximum speed(km/h)	50

Approach angle (°)/departure angle (°)	$\geqslant 22°/6°$
Maximum gradeability (%)	$\geqslant 30\%$
Engine model	4HK1 – TC
Power(kW/(r · min⁻¹))	129/2,600
Torque(N · m/(r · min⁻¹))	500/(1,500 – 2,000)
Tire	9.00 – 20 – 16PR

5.2.2 Auxiliary Beam Assembly

Auxiliary beam assembly is frame structure. It is bolted with the chassis main beam to load the catering van. Steel channel on both sides of auxiliary beam is used as a slideway for the scissor lifting frame rail wheel, tow bar hole is welded at front of which used to hinge with lifting frame. At forefront, there is a support frame used to support the front working platform when it falls to the lowest point and as a stand for the ladder of front platform. All supporting legs are fixed at the auxiliary beam.

There is a maintenance latch at the rear of the auxiliary beam (see Figure 5.2). When the truck needs maintenance, it must raise the catering van up to certain height (the wheels journey must run over the maintenance latch), fasten the maintenance latch with the maintenance bolt before allowing maintenance staff to enter inside so as to ensure the safety of maintenance staff.

Figure 5.2   Maintenance latch

Around the auxiliary beam, there are 4 telescopic hydraulic stabilizers (see Figure 5.3) used to support the whole truck when lifting and descending the catering van.

Note: The catering van will not raise or can only raise to half of its maximum height when the stabilizers are not fully downed, and this is controlled by the programs.

Figure 5.3   Telescopic hydraulic stabilizer

5.2.3   Front Working Platform Assembly

The front working platform is right above cab. It is the portal for docking plane door, and is also a food delivering path for the plane. At the vertical part of the rear of the platform, there are equipped with 4 heat treated wear-resisting wheels. It can slide up and down through the groove at the front part of the catering van. When the catering van lowers to the lowest position, the working platform will be supported by both the catering van and the front support frame at the auxiliary beam. When the catering van floor rises up to the same level of platform's surface, the platform will be driven by the limit block inside the slideway to lift synchronous with the catering van.

The front platform consists of fixed platform, mobile platform and telescopic platform. The fixed platform is above the cab and near to the catering van which is a non power-driven platform. The evenly loading capacity of front platform is 1,000 kg maximum.

Mobile platform is directly connected to fixed platform which can move sideways. The front telescopic platform is connected to the front end of mobile platform which can only move forward and backward. The main function of mobile platform is adjusting the left and right position when docking to plane door. The main function of telescopic platform is adjusting the gap when docking so as to achieve smooth and safe docking to plane door. All platforms are paved with patterned aluminum plate. The telescopic platform is equipped with buffer rubber tube at its front-end to ensure plane safety. Mobile platform is equipped with a flip-style side door. When not working, it will turn to mobile platform inside for fixing. During working, please open the side door (vertical body), which is equipped with belt made pull-type handrail to ensure operators' safety and can yield to plane door effectively.

5.2.4   Catering Van Assembly

The catering van is a Cowger structural insulating van body. The catering van wall consists of 3 layers, the inner and outer layer are made of Glass Fiber Reinforced (GFR) plastic plate and the interlayer is made of special foam processed insulating layer with light weight and good insulation features. The catering van floor is made of special aluminum alloy. Roll-up door is installed both in front and rear back of the catering van. The catering van's left and right inner walls are both equipped with 2 rows of buckle lock band and they work together with 2 fixing bars and 2 adjustable soft-straps to fix the cart. Inside the catering van, there is equipped with a control box at the front part. The control box is used to control the car carriage lifting and descending, front working platform movement and other operations. 2 soft light lamps are installed inside the catering van for night operations. 4 clearance lamps are installed on the outer side of the van body. 2 working lights are installed for night driving and operations at the front of the catering van. 3 warning beacons are installed on the top of the catering van for alerting purpose. In addition, 2 working lights are installed in the middle of the auxiliary beam to provide lighting for driving backwards at night.

5.2.5 Scissor Lifting Frame Assembly

Scissor lifting frame is the main structural element of catering van lifting, it is formed of two hinged rigid structural components to form scissors type structure. Its front external articulated to catering van bracket while its front internal is articulated to tow bar hole. The rear has two pairs of wheels rolling in the slide ways at both sides of catering van and auxiliary beam respectively. The telescopic of lifting frame is driven by hydraulic cylinder so as to achieve catering van lifting. The whole rigid frame structure is welded by high strength material (15MnTi) and is made of rectangular seamless tube session whose compressive and tensile strength is 1.6 times more compared to common materials. It is high safety factor. The wheels are made of quench-hardened steel No. 45 which is high hardness. The wheel bearing with oil added has high wear ability which ensures smooth and reliable rotation.

5.2.6 Hydraulic Assembly

The hydraulic system is the driving force delivery system for the whole truck additional actions. It consists of hydraulic fluid tank (122 L), gear pump, manual pump, electromagnetic relief valve, one way valve, hydraulic lock, manual emergency fuel drain valve, lift cylinder, stabilizers cylinder, mobile platform cylinder, telescopic cylinder and electric emergency pump hydraulic components, etc. Details please refer to hydraulic schematics of appendix.

Normally, the electrical system controls solenoid valves to change the flow direction of hydraulic fluid so as to control the cylinders. And if the solenoid valve fails, the cylinders can be controlled manually. If the hydraulic gear pump fails, please switch to emergency electric pump or emergency manual pump to withdraw the stabilizers. And the emergency descending of catering van can be manually achieved by releasing the hydraulic fluid as the requirements.

The PTO (Power Take Off) is pneumatically controlled by a solenoid valve in the cab. It is convenient and reliable. The PTO drives the hydraulic fluid gear pump to pump hydraulic fluid for the whole hydraulic system. The pressure of hydraulic fluid is built by electromagnetic relief valve. The electromagnetic relief valve is stationary-opened. When the current reaches the system settings and one of the directional electromagnetic valves works, the electromagnetic relief valve will start to build pressure for the hydraulic system. The pressure is defaulted set and locked to 13 MPa and the maximum pressure should not exceed 15 MPa.

Throttles are installed on every hydraulic fluid circuit for every action of the hydraulic systems to control the action speed. The action speed is finely tuned, and it does not adjust the throttles only if the action speed is getting slow or vibrant due to the change of viscosity of hydraulic fluid or influence of external temperature. If unusual slowness or vibration was found, please adjust throttle to round off it. Twisting the throttles clockwise will decrease flow and vice versa.

All hydraulic steel pipes are cold drawn steel tube and have been acid pickled. All hoses are double-hose wire high-pressure hoses with high precision ferrule fitting.

Note: For the first time using, please check the tightness of fittings to prevent loosing connection which may be due to transportation or other reasons that would cause hydraulic fluid leakage.

5.2.7　Electrical Control System

Electrical control system consists of operating box, operating panel, control switch, relay, relay box, position detect switch, warning lamp, lighting and outline marker lamp, details please refer to electrical schematics in appendix.

During operation, electrical control system is used to drive the alternation of solenoid valve to control each cylinders action so as to control actions of catering van, stabilizers and platform. The system has two sets of control, one set control box is located at the cab rear part which is used to control every operation except driving. The other set is in the catering van which is used to start engine and emergency stop, control catering van lifting and stabilizers retraction.

5.2.8　Protection and Warning System

(1)Platform is equipped with sensor(see Figure 5.4) and rubber tube which achieve protection by against lifting of catering van when the truck is less than 200 mm from the plane.

(2)Platform is equipped with sensors(see Figure 5.4) and rubber tube which achieve protection by terminating the extension of platform when the platform is less than 20 mm from the plane.

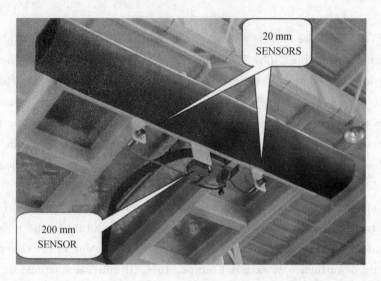

Figure 5. 4   Range sensors

(3)Stabilizers position sensors are installed on the left frontal stabilizers (see Figure 5. 6, 5. 7) which achieve protection by restricted catering van lifting to half of its maximum height when the stabilizers are not fully extended. Catering van sensor is installed in the middle of auxiliary beam (see Figure 5. 5). If the stabilizers are not fully retracted, the sensor will stop the catering van from raising up.

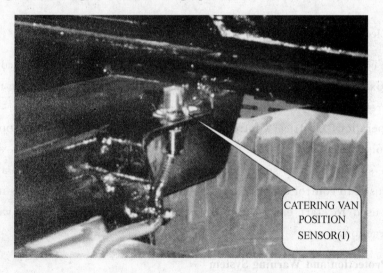

Figure 5. 5   Catering van sensor

(4)Catering van position sensors are installed on the frontal stabilizers to restrict the withdrawal of stabilizers until the catering van is fully descended.

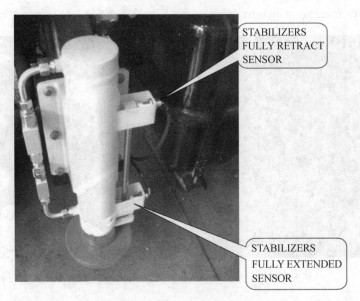

Figure 5.6　Stabilizers position sensors

Catering van fully descended sensor (see Figure 5.7): if the catering van is not fully descended, the sensor will protect the stabilizers from being retracted.

Figure 5.7　Catering van fully descended sensor

(5) Hydraulic operation only functions when the parking brake is engaged.

(6) The engine can only be started from the catering van when the parking brake is engaged.

(7) Emergency stop is equipped for emergency engine stall.

(8) Adopting catering van position sensor for lifting protection by stopping the lifting of

catering van to its maximum height (6 m) (see Figure 5.8).

Figure 5.8   Adopting catering van position sensor

(9) When the catering van descends, the catering van will contact the position sensor to stop the lifting cylinders and stop the movement.

(10) Maintenance latch is installed to keep the maintenance safety. When the truck needs maintenance, it must raise the catering van up to certain height, fasten the maintenance latch with the maintenance bolt before allowing maintenance staff to enter inside so as to ensure the safety of maintenance staff.

New Words

1. throttle [ˈθrɒtl] *n.* 节流阀

Phrases and Expressions

1. scissor lift frame assembly 剪式升降支架总成
2. maximum gradeability 最大爬坡度
3. telescopic hydraulic stabilizer 伸缩式液压支撑脚
4. insulating van body 保温厢体
5. one way valve 单向阀
6. pneumatically control 气动控制

Notes to the Text

1. All hydraulic steel pipes are cold drawn steel tube and have been acid pickling. All hoses are double-hose wire high-pressure hoses with high precision ferrule fitting.

所有液压钢管均为冷拉钢管并经酸洗处理。所有软管均为双软管钢丝高压软管，并配有高精度金属环状接头。

Section 5.3 Maintenance

5.3.1 Checking before Use

The truck had been fully checked before delivery. However, the individual parts' loose and damage may occur due to shipping etc. Thus, users are recommended to perform the following check before use.

(1) Check the fastening status of the connection of each component, especially the transmission system, steering system, braking system, suspension systems and forks of the wheels.

(2) Check the water storage of radiator and whether any leak observed at its joints.

(3) Check the liquid level of engine, transmission, rear axle, steering gear, clutch and hydraulic cylinder and the oiling condition of lubrication points.

(4) Check whether there is any leak observed at hydraulic system.

(5) Check liquid level and proportion of batteries.

(6) Check the accessories of engine and fan belt tightness.

(7) Check the working condition of electrical equipments, lights, speaker and instrument.

(8) Check the working condition of brake system and whether there is any air leak observed at the pipe joints.

(9) Measure the pressure of tire to check whether it satisfies to standard requirement.

(10) Check the working condition of steering system.

(11) Check whether the vehicle tools and accessories are completely set.

5.3.2 Driving and Handling

The working process of Catering Truck:

(1) Reverse driving the truck to the delivery platform of food storage, open the roll-up door at the catering van back, deliver the food cart into the catering van with a bridge board, fix them inside the car carriage with bars and adjustable soft-straps, remove the bridge board and then shut down roll-up door.

(2) Drive the truck to the plane sill and park it in a suitable place, apply parking hand brake, tread the clutch, set the gearbox to neutral gear and engage power take off, then release the clutch to make gear pump in working condition.

(3) Down the stabilizers fully and let operators enter into the truck. Turn on the switch of lifting catering van and raise it to the same level of the plane door. Move the platform and dock it to the plane door, then adjust platform fence to make sure that it will not hinder the plane door to open. When the plane door open, put away the bridge board to let operators to deliver the food cart into or out the plane.

(4)When task completed, withdraw the bridge board, retract platform, and fix the food cart well inside catering van, then down the catering van and retract stabilizers, disengage power take off, select reverse gear and drive away the plane.

(5)Reduce the truck's load to half of its maximum when the wind speed exceeds 80 km/h and stop to use the catering truck when the wind speed is more than 130 km/h.

Points to note:

(1)Do not drive the truck if the catering van is not in the lowest level.

(2)Do not drive the truck if the stabilizers haven't been fully withdrew.

(3)Disengage the power take off before driving after finishing the task.

(4)Do not store the catering truck for a long time. Use or lift the platform at least once a month to maintain it in a good condition.

(5)Lower down the catering cabin and extend the stabilizers to relieve the wheel load when garage catering truck.

5.3.3　Maintenance

(1)Maintenance of chassis please refer to manual.

(2)Make sure that hinge and sliding part are under lubrication. Lubrication is suggested as once every 3 months.

(3)Check the hydraulic fluid level frequently and refill it in time.

(4)Keep the catering truck clean.

(5)Maintenance of hydraulic system should be performed as listed in Table 5.1.

Table 5.1　Maintenance of hydraulic system

Items	Period	Instruction
Hydraulic fluid pump noise	Every day	Hearing checking, if abnormal sound is heard, check whether hydraulic fluid filter is blocked or the hydraulic fluid suction tube is blocked or the oil pump is damaged
Hydraulic fluid filter	6 months	Cleaning
Pressure gauge	1 month	Check whether the system's pressure is normal or not
Hydraulic fluid leakage	Every day	Check whether leak observed. If it is, fasten it or replace gasket
Hydraulic fluid tank level	1 month	Refill hydraulic fluid when the level is low
Hydraulic fluid quality	6 months	Replace the hydraulic fluid if it is not up to the standards. It is suggested to change the hydraulic fluid every 6 months
Whole hydraulic system	2 years	Check and clean components, pipes and replace the damaged ones

5.3.4　Renewal of Hydraulic Fluid

The renewal of hydraulic fluid should be performed after the first 500 hours, otherwise it will cause hydraulic system contamination and this is vital to the equipment safety as well as personnel safety.

According to statistics, 90% of the hydraulic faults are caused by hydraulic system contamination. Non-renewal of hydraulic fluid would result as the following:

1. Hydraulic fluid gear pump

(1) The temperature of hydraulic fluid gear pump would heat up abnormally quickly during the operation.

(2) The hydraulic fluid gear pump would give out abnormal noise.

(3) The hydraulic fluid gear pump outputs insufficiently.

2. Cylinders

(1) Hydraulic fluid leakage.

(2) Moves slowly and this will affect the operation.

(3) The piston moves roughly or slips.

3. Valves

(1) Hydraulic fluid leakage.

(2) Heats up quickly during operation and gives out noise.

(3) Plunger sliding valve gets stuck and moves roughly.

4. Hydraulic fluid motor

(1) Hydraulic fluid motor works inefficiently.

(2) Hydraulic fluid motor heats up abnormally quickly.

(3) Hydraulic fluid leaks at hydraulic fluid motor output axis.

Note: Noises and overheating may be caused by other problems (fluid cavitations) or may be caused by a series of problems (hydraulic fluid contamination, air in the hydraulic system, filter clogging, etc.): Please make synthetic judgment when performing maintenance.

Renewal must be performed when the hydraulic fluid reaches its life time. The hydraulic fluid tank contains 3/4 of all the hydraulic fluid that a truck needs and the other 1/4 are stored in other hydraulic components like pipes, motors, valves and cylinders.

Caution: Mixture of 2 different hydraulic fluids would deteriorate the hydraulic fluid.

Park the truck on flat ground and chock at least 2 of its wheels when renewing the hydraulic fluid.

Clean the whole hydraulic system when changing to a different hydraulic fluid.

Keep the working environment 100% clean when performing hydraulic fluid renewal.

Hydraulic fluid are suggested as below (see Table 5.2).

Table 5. 2　Hydraulic fluid types

Item	Types	Related Assembly or Components
Hydraulic fluid	L－HM46 for－5℃ or above L－HM32 for－20℃ or above L－HV32 for－30℃ or above (Cold regions)	Hydraulic system
Lubricating oil	General vehicle lithium grease	Bearing, axle sleeve and relate positions

Notes: This product has been refilled with hydraulic fluid (brand: Great Wall) before delivery, Users can replace with other brands of anti-wear hydraulic oil if necessary.

5.3.5　Hydraulic Fluid Renewal Procedure

1. Preparation

Please make the following preparations when renewing the hydraulic fluid:

(1) Enough hydraulic fluid for the catering truck (more than 1. 5 times than the hydraulic fluid capacity).

(2) Various hydraulic filters.

(3) Cleaning-use hydraulic fluid.

(4) Plugs used to seal the hydraulic pipes.

(5) Air pump and empty barrels.

2. Clean the hydraulic fluid tank

(1) Empty the hydraulic fluid tank and remove the filters. Disconnect hydraulic fluid pipe between the hydraulic fluid pump and hydraulic fluid tank on the hydraulic fluid pump connector to empty the hydraulic fluid in the pipe.

(2) Clean the hydraulic fluid tank and pipes with cleaning-use hydraulic fluid. Wipe the hydraulic fluid tank and then clean the hydraulic fluid with cleaning-use hydraulic fluid again. Dry the outer side of hydraulic fluid tank with compressed air. Check the hydraulic fluid tank and clean the dirt if there is any and flush it with new hydraulic fluid again. (Sticky dough is suggested to clean the hydraulic fluid tank if there is too many dirt in it).

(3) Renew the filters and cover it.

3. Clean the hydraulic fluid pump

Pull down the hydraulic fluid gear pump and clean it with cleaning-use hydraulic fluid then dry it with compressed air and flush it with new hydraulic fluid and change the outlet O-rings.

4. Renew the filters

5. Connect all the hydraulic fluid pipes

6. Refill the hydraulic fluid tank with new hydraulic fluid via the inlet filter

5.3.6　Notes when Renew the Hydraulic Fluid

(1) Old hydraulic fluid and cleaning-use fluid must be fully emptied, keep the hydraulic

drain open for 3 − 5 min when the hydraulic fluid stopped to drain out to fully empty the hydraulic fluid.

(2) Keep the hydraulic fluid flow to one direction when flushing.

(3) Taping the pipes would vibrant the oxydum off and it helps to flush it out when flushing.

(4) All the disassembled hydraulic pipes should get its gaskets changed. Gasket cement and PTFE tape must be applied to avoid leakage when connecting pipes.

(5) Hydraulic fluid quality is very important, please choose qualified hydraulic fluid only and refill the hydraulic fluid 4 hours later after the purification.

(6) After renewing the hydraulic fluid please keep the truck running idle for a while and then start from low pressures movement to normal movement. This running-in procedure should last for 1 − 1.5 h.

(7) While running-in the hydraulic fluid will get pumped to the oil pipes and cylinders and hydraulic fluid level in the hydraulic fluid tank will get lower, please do not add hydraulic fluid.

(8) Hydraulic fluid life time:

The life time of static hydraulic fluid is about 2 years. The hydraulic gets mixed with water due to the condensation. And with time going by, the water will get acidified and it will corrode the metal surface and cause prills. Meantime with the mechanic vibration and hydraulic fluid flow pressure, the drops of prills will get mixed into the hydraulic fluid. Even some of the big prills get filtered the small ones which are less than 10 remain and aggravate the wear and tear. It is recommended to garage the truck indoor and change the hydraulic filter regularly to avoid condensation and prills, which is particularly important for cold area users.

(9) Hydraulic fluid filter recommendation :

10 μm filter fineness is recommended, and it is suggested renewing the filters every 500 h.

Section 5.4   Troubleshooting

Troubleshooting for hydraulic system is shown in Table 5.3.

Table 5.3   Troubleshooting for hydraulic system

Fault	Reasons	Solution
Hydraulic cylinders do not work	Power take off does not engage; hydraulic fluid pump damaged; solenoid valve does not work	Engage power take off; check and replace hydraulic fluid pump; check wirings; check and clean solenoid pin

Continued

Fault	Reasons	Solution
No pressure in the hydraulic system	Hydraulic fluid pump damaged; solenoid valve does not work	Check wirings, solenoid; check and clean solenoid pin
Hydraulic locks cannot fully lock the cylinders	Jamming of check valve of hydraulic lock	Remove hydraulic lock and clean it or grind valve seat
Lifting component vibrate when descending	Cylinder descending speed is too fast; air in hydraulic cylinder	Adjust throttle of lifting cylinder to slow down the speed; reciprocating slowly to exhaust air
Lifting hydraulic cylinder is insufficient in pressure	Lack of hydraulic fluid inside hydraulic cylinder; filter blocked; relief valve is not adjust properly; pipe leaking; oil pump leaking; hydraulic cylinder leaking	Check hydraulic cylinder, refill oil if it is insufficient; clean filter; adjust relief valve to standard value; fasten pipes; check or replace hydraulic fluid pump; check hydraulic cylinderor replaced piston seal

If there is a fault in catering truck before driving it away from the plane, stabilizers won't fully withdrew, please follow the following instructions:

When hydraulic gear pump does not work (engine fault, hydraulic fluid gear pump fault), and the emergency electric pump (see Figure 5. 10) works normal, start the emergency electric pump to low the catering van and withdraw the stabilizers and then drive or tow away the catering truck from plane.

When hydraulic fluid gear pump and electric pump fails, switch off the bypass stop valve and use the manual pump (see Figure 5. 11) to low the catering van and then open the emergency stabilizers hydraulic fluid drain valve (see Figure 5. 9) to release the hydraulic fluid to withdraw the stabilizers and then drive or tow away the catering truck from plane.

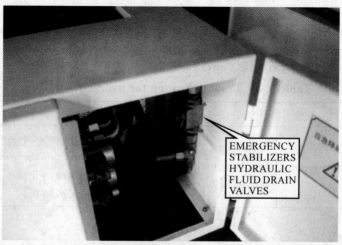

Figure 5. 9   Emergency stabilizers hydraulic fluid drain valve

Figure 5.10   Emergency electric pump

Figure 5.11   Manual pump

If it is just electric fault, please open the lifting cylinder's hydraulic fluid releasing valve to low catering van slowly and then press the corresponding solenoid spool to achieve corresponding movements.

After emergency operation, please make sure that all hydraulic fluid drain valves are turned off. Turn on the bypass stop valve and restore the hydraulic system to normal working condition.

New Words

1. prill[pril] *vt.* 使变颗粒状

Phrases and Expressions

1. hearing checking 听音检查
2. air pump 气泵,抽气机

Notes to the Text

1. Make sure that hinge and sliding part are under lubrication. Lubrication are suggested as once every 3 months.

确保铰链和滑动部件处于润滑状态。建议每 3 个月润滑一次。

2. The hydraulic fluid gear pump would give out abnormal noise.

液压齿轮泵会发出异响。

3. The piston moves roughly or slips.

活塞运动粗暴或滑脱。

4. Hydraulic fluid leaks at hydraulic fluid motor output axis.

液压油在液压马达输出轴处泄漏。

Chapter 6　Pallet Loader

Section 6.1　General Information

6.1.1　General

The COMMANDER 30i loaders (see Figure 6.1) are single-operator, self-propelled vehicles capable of lifting and conveying cargo weighing up to 15,000 kg (33,000 lb). They can handle containers or pallets and service a variety of aircraft.

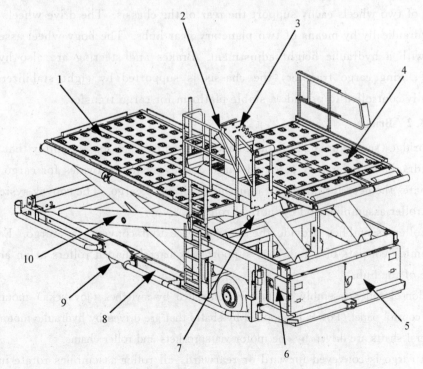

Figure 6.1　Typical COMMANDER 30i container/pallet loader

1—platform;　2—driver's panel;　3—operator's panel;　4—bridge;　5—power unit;
6—main panel (gauges);　7—forward scissors;　8—chassis;　9—body wheel;　10—rear scissor

Design concept utilizes the latest in technology and incorporates modular power units, improved conveying system, electrical systems, and hydraulic components. Power units can be a variety of diesel engines or electric motor. The electrical system is 24 V DC PLC controlled, and the hydraulic system is closed-center and load-sensing. Two hydraulic motors power the planetary drive wheels to propel the loader.

The COMMANDER 30i loaders are available in different configurations. The various configurations and features available are described in this section.

6.1.2 Capabilities

The minimum height of 0.50 m (19.5 in) of the platform facilitates transfer of cargo from surface vehicles. The turning radius of 11.7 m (38.5 ft) and inching capability of the propulsion system provide safe and precise control for positioning the loader. The maximum height to which cargo can be lifted is 5.59 m (220 in).

6.1.3 Major Components

6.1.3.1 Chassis

The chassis is a rigid steel framework on which all other components are mounted. Two steerable drive wheels support the chassis at the front and two bogy wheel assemblies, consisting of two wheels each, support the rear of the chassis. The drive wheels propel the chassis hydraulically by means of two planetary gear hubs. The bogy wheel assemblies are supplied with a hydraulic height adjustment. Brakes and steering are also hydraulically powered. During cargo transfer, the chassis is supported by eight stabilizers that are hydraulically controlled to provide a stable platform for cargo transfer.

6.1.3.2 Bridge

The bridge (see Figure 6.2) is raised and lowered by a scissors assembly that is powered by two hydraulic cylinders. A patented cargo convey system provides for cargo movement and eliminates the need for manually adjusting position of cargo. The convey system consists of cluster roller assemblies and cylindrical rollers.

The roller assemblies provide the motive force that conveys the cargo. Each cluster roller assembly consists of a hub that supports six barrel-shaped rollers at an angle to the center line of the hub.

Direction of roller assembly rotation is controlled by switches (joysticks) mounted on the operator's control panel. Power is supplied by shafts that are driven by hydraulic motors. In some cases, several shafts are driven by one motor via sprockets and roller chains.

When cargo is conveyed forward or rearward, all roller assemblies rotate in the same direction. For movement to either side, some roller assemblies are driven in one direction, and others are driven in the opposing direction. The various combinations of rotation allow the operator to control cargo position without being required to manually shift it.

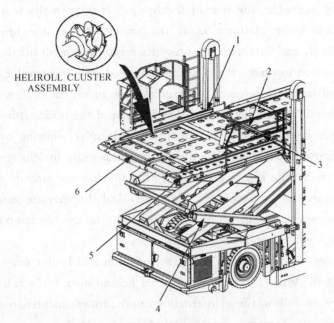

Figure 6.2　Bridge components (typical)

1—side guide (2);　2—load stop;　3—handrail;　4—scissors assembly;　5—wing;　6—cylindrical roller

Two guides on the bridge are hydraulically adjustable from side to side to assist in aligning cargo for transfer onto the aircraft. The front of the bridge is equipped with three folding wings (see Figure 6.3) so that the loader can be used to transfer cargo to or from aircraft with varying door widths. Three folding wings are raised and lowered hydraulically.

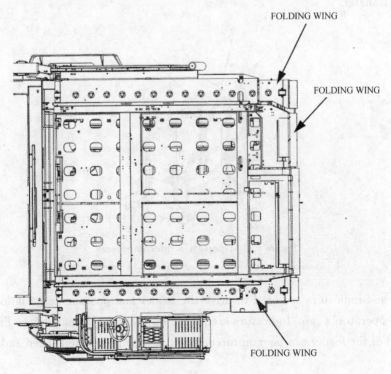

Figure 6.3　Bridge wings

A load stop is located at the rear of the bridge. It is normally in the extended (up) position, except when the platform is at the same level as the bridge. The stop is mechanically operated, and automatically prevents cargo movement off the bridge unless the platform is in a position to accept the cargo. A powered cylindrical roller at the front of the bridge supports and transfers cargo as it is conveyed on or off the bridge. A fixed or hinged and telescoped handrail is installed on the opposite side of the bridge from the cab.

The bridge can be equipped with an optional aircraft following assembly (tracking sensor) that automatically adjusts bridge height to compensate for change in aircraft height as cargo is transferred. The sensor roller assembly touches the aircraft at one point only. The automatic feature can be bypassed, if desired, and the operator can manually change bridge height as necessary. Bridge tilt provides flexibility for uneven ramp conditions. Bridge can be tilted to align with aircraft doorway.

Warning: Bridge must be in the full down position and loader must be parked before boarding or exiting the loader. Use ladder located behind operator's cab to board and exit the loader. Failure to follow these instructions and proper maintenance procedure could result in serious property damage, personal injury and/or death.

6.1.3.3 Operator's Cab

The operator's cab (see Figure 6.4) contains all controls required to drive the loader and transfer cargo. The stand-up design offers maximum visibility as well as safe, convenient, and comfortable access to loader and aircraft controls. The operator's cab is hydraulically adjustable fore and aft to allow the operator to gain access to aircraft controls during cargo transfer.

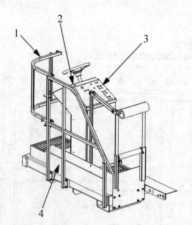

Figure 6.4　Operator's cab components
1—handrail;　2—driver's;　3—operator's;　4—operator's cab

Controls and indicators are used to drive the loader and position cargo is located on two panels on the operator's cab. Indicators are placed on the driver's panel (see Figure 6.5) so that operation of the loader can be monitored. Controls for propulsion speed and direction are also included. Located on the operator's panel (see Figure 6. 6) are the switches used to position and transfer the cargo to raise and lower the bridge and platform, and to operate the side and rear stops.

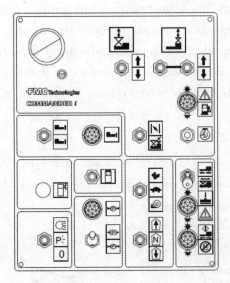

Figure 6. 5   Driver's panel

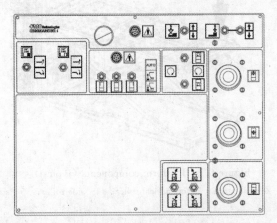

Figure 6. 6   Operator's panel

Emergency pump switch is located on the driver's console in the cab. It is used in case of engine failure to supply hydraulic oil and control power so that the platform, bridge can be lowered and stabilizers can be raised. This pump must not be operated for more than 60 s at a time. At least ten minutes must be allowed for cooling time among use periods.

Also included is an accelerator pedal that proportionally controls the speed of the loader. The proportional control feature allows precise positioning of the loader and provides an inching capability as the aircraft is approached. A pedal actuates the hydraulic brakes. Handrails are an integral part of the platform for operator safety during operation of the loader.

Warning: Bridge must be in the full down position and loader must be parked before boarding or exiting the loader. Use ladder located behind operator's cab to board and exit the loader. Failure to follow these instructions and proper maintenance procedure could result in serious property damage, personal injury and/or death.

Caution: Do not operate electrical pump continuously for more than 1 min. Operation for more than 1 min will overheat motor and may cause damage. If emergency procedures cannot be completed within 1 min, wait at least 10 min to allow motor to cool, then continue.

6.1.3.4 Platform

The platform (see Figure 6.7) is also raised and lowered by a scissors assembly; however, it is powered by two hydraulic cylinders that position the platform.

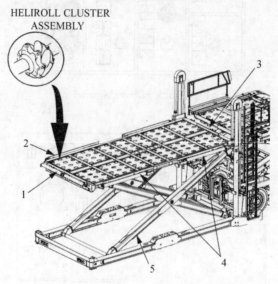

Figure 6.7　Platform components (typical)

1—rear roller; 2—side stop; 3—cylindrical roller; 4—side roller; 5—scissor assembly

Depending on the configuration of the platform, a different combination of ball mats or HeliRoll cluster assembly is used. Hydraulically operated stops prevent unintentional off loading of cargo. The stops can be automatically or manually operated. Proximity switches prevent manual operations when the platform is not in the proper position for loading or unloading. Proximity switches on the bridge must also sense correct position of the platform before cargo can be transferred to or from the platform.

All configurations allow the operator to side shift and rotate containers or pallets on the platform.

Warning: Platform is not a personnel lift or transport! No personnel are allowed on platform while loader is in motion! Failure to follow these instructions and proper maintenance procedure could result in serious property damage, personal injury and/or death.

6.1.3.5　Power Unit

The power unit (see Figure 6.8) is located at the front of the loader. It is a modular unit that is hinged on the right side of the loader. A single bolt on the left side can be removed to permit the module to swing out for complete access to components when maintenance is required. A power panel on the right side of the module contains controls and indicators used to start and operate the power unit at ground level.

A diesel engine and an optional electrical power unit are available as the primary source of power for the loader.

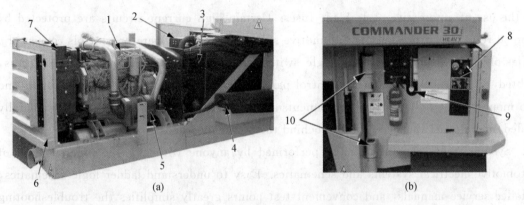

Figure 6.8　Power unit (typical)

1—engine;　2—main panel;　3—hydraulic reservoir;　4—batteries;　5—air cleaner;　6—tee bolt holder;
7—radiator;　8—gauges and start panel;　9—hydraulic fluid level/temperature;　10—hinges

6.1.3.6　Hydraulic System

A closed-center, load-sensing hydraulic system is used on the COMMANDER 30i loaders. It provides hydraulic power for the cargo transfer, raising and lowering the bridge and platform, proportional propulsion, steering and braking, and operation of the various guides that are used to insure safe handling of cargo. A dynamic braking feature is also incorporated into the hydraulic system to provide smooth deceleration when the operator releases the accelerator.

The axial piston pumps are directly driven by the power unit and deliver 242 liters per minute at 241 bar (64 g/min at 3,500 psi).

Solenoid valves control fluid flow at correct pressure to operate the loader's hydraulic components. Checking valves prevents load-bearing hydraulic cylinders from retracting if hydraulic pressure is not properly maintained in the system. An electrically driven emergency pump is included to allow the operator to perform emergency procedures if the power unit or main hydraulic pump should fail.

6.1.3.7 Electrical System

The COMMANDER 30i loaders use a 24 V automotive style electrical system to power the engine starter and ignition controls, hydraulic valves, lights, signals, and other accessories. Diesel powered loaders utilize two heavy duty 12 V automotive type batteries, connected in series, to provide 24 V DC. An engine driven alternator maintains battery charge and system load requirements.

The electrical control system utilizes a combination of common relays and PLC controllers. Wherever possible, all control logic functions are performed by the PLC. Relays are utilized for switching of higher current circuits. Individual circuit protection is achieved by the use of automotive style blade fuses. Certain high current circuits are protected by manual reset circuit breakers or automotive style mega-fuses. Operator controls consist of a series of environmentally sealed toggle switches or joysticks mounted on easy to access, lighted, and permanently marked control panels. With exception of the control panels and cab mounted PLC, all electrical components are housed in an easy to access, environmentally sealed, main panel enclosure located behind the power mod doors.

System troubleshooting can be performed by anyone with a basic understanding of automotive electrical systems and schematics. Easy to understand ladder logic schematics, detailed service manuals, and convenient test points greatly simplifies the troubleshooting process. Additionally, a display module mounted in the main electrical enclosure, will provide a complete system status indicating the presence of operator selected input signals and controller output signals.

6.1.4 Miscellaneous Components

6.1.4.1 Lights

Sealed beam headlights and front and rear taillights are supplied for night operations. Rear reverse lights are supplied to indicate the loader is in reverse gear and light the ground area around the loader. Front and rear turn signals and side marker lamps are available as options.

6.1.4.2 Horn

An electrical automotive type horn is included.

6.1.4.3 Audible Alarms

An Alarm sounds when the loader is propelled in reverse.

6.1.4.4 Powered Bogy Wheels

Hydraulically powered bogy wheel assemblies increase the ground clearance at the rear of the chassis while propelling.

6. 1. 4. 5　Handrail

Folding and extendable handrail are located on left side of forward bridge for added operator protection.

New Words

1. convey [kənˈvei] vt. 运送,传送,输送
2. cargo[ˈkɑ:(r)gəu] n.（车、船、飞机等运输的）货物
3. container[kənˈteinə] n. 集装箱
4. pallet[ˈpælət] n. 货盘
5. scissor [ˈsizə] n. 剪刀
6. planetary [ˈplænət(ə)ri] adj. 行星齿轮的
7. inching[ˈintʃiŋ] n. 微调
8. propulsion [prəˈpʌlʃən] n. 推进力
9. stabilizer[ˈsteibəlaizər] n. 支脚,支柱
10. handrail[ˈhandreil] n. 栏杆,扶手
11. roller[ˈrəulə(r)] n. 滚筒
12. cluster[ˈklʌstə(r)] n. 群,组
13. sprocket [ˈsprɒkit] n. 链轮,链齿
14. visibility [vizəˈbiləti] n. 可见度,能见度
15. accessory [əkˈsɛsəri] n. 附件

Phrases and Expressions

1. pallet loader 平台车
2. PLC (Programmable Logic Controller) 可编程序逻辑控制器
3. turning radius 转弯（向）半径
4. gear hub 齿轮毂
5. roller chain 滚子链
6. accelerator pedal （汽车等的）油门踏板
7. proximity switch 接近开关
8. blade fuse （汽车等）插片保险丝

Notes to the Text

1. Design concept utilizes the latest in technology and incorporates modular power units, improved conveying system, electrical systems, and hydraulic components. Power units can be a variety of diesel engines or electric motors. The electrical system is 24 V DC PLC controlled, and the hydraulic system is closed-center and load-sensing. Two hydraulic motors power the planetary drive wheels to propel the loader.

此平台车的设计是基于最新的技术和各厂商的紧密合作，从而提高了运送系统、电气系统及液压元件的功能。动力装置可选用多种柴油机，电气系统由 24 V 直流可编程逻辑控制器控制，采用闭心式及负载感应式的液压系统。通过双液压马达驱动行星齿轮系统驱动平台车。

2. The chassis is a rigid steel framework on which all other components are mounted. Two steerable drive wheels support the chassis at the front and two bogy wheel assemblies, consisting of two wheels each, support the rear of the chassis. The drive wheels propel the chassis hydraulically by means of two planetary gear hubs. The bogy wheel assemblies are supplied with a hydraulic height adjustment. Brakes and steering are also hydraulically powered. During cargo transfer, the chassis is supported by eight stabilizers that are hydraulically controlled to provide a stable platform for cargo transfer.

车架是坚固的钢架结构，用来安装其他的元件。在车架前部装有两个驱动转向轮，车架后部装有两组车架轮，每组有两个轮子。两个驱动轮上装有液力驱动的行星齿轮的轮毂，用于驱动车架行驶。车架轮通过液压装置调节高度，制动及转向系统也是液压驱动的。在平台运送货物时，底盘由 8 个液压控制的支脚支撑，以保持平台的平稳。

3. The bridge is raised and lowered by a scissors assembly that is powered by two hydraulic cylinders. A patented cargo convey system provides for cargo movement and eliminates the need for manually adjusting position of cargo. The convey system consists of cluster roller assemblies and cylindrical rollers.

桥台是通过剪式臂进行升降的，剪式臂由两个液压作动筒驱动。具有专利的运送系统运送货物，并且无须人工调节货物位置。运送系统包括滚轮组总成及柱型滚轴。

4. A load stop is located at the rear of the bridge. It is normally in the extended (up) position, except when the platform is at the same level as the bridge. The stop is mechanically operated, and automatically prevents cargo movement off the bridge unless the platform is in a position to accept the cargo.

桥台后部装有货物挡块，正常情况下它是伸出状态，只有桥台和平台在同一水平面时，挡块缩回。挡块由机械控制并能自动阻止货物移出过桥，除非平台处于接收货物的位置。

5. Bridge tilt provides flexibility for uneven ramp conditions. Bridge can be tilted to align with aircraft doorway.

桥台的倾斜可适应不平坦的地面条件。桥台通过倾斜来保持与飞机舱门的平行。

6. Also included is an accelerator pedal that proportionally controls the speed of the loader. The proportional control feature allows precise positioning of the loader and provides an inching capability as the aircraft is approached. A pedal actuates the hydraulic brakes. Handrails are an integral part of the platform for operator safety during operation of the loader.

同时还有一个加速踏板来控制平台的速度。比例控制特性可以精确控制平台车的位置，当接近飞机时还可使用爬行性能。另一个踏板可提供液压制动。操作平台时，护栏是保护操作者不可缺少的一部分。

7. The power unit is located at the front of the loader. It is a modular unit that is hinged on the right side of the loader. A single bolt on the left side can be removed to permit the module to swing out for complete access to components when maintenance is required. A power panel on the right side of the module contains controls and indicators used to start and operate the power unit at ground level.

动力装置安装在平台车的前部,它通过铰链装在平台车的右侧。进行维护时可将右侧的螺栓拆下,将动力装置完全旋转出来,以便检查所有元件。动力装置右侧的面板上装有操作开关指示器,以便在地面进行动力装置的启动和操作。

8. A closed-center, load-sensing hydraulic system is used on the COMMANDER 30i loaders. It provides hydraulic power for the cargo transfer, raising and lowering the bridge and platform, proportional propulsion, steering and braking, and operation of the various guides that are used to insure safe handling of cargo. A dynamic braking feature is also incorporated into the hydraulic system to provide smooth deceleration when the operator releases the accelerator.

COMMANDER 30i 平台车上使用的是闭心式的、负载感应式的液压系统,它为平台和桥台的升降、比例驱动、转向和制动提供动力。通过它可插装阀各种护栏以保证货物的安全。液压系统还有动态制动的特性,可在驾驶员松开加速踏板时提供平稳的减速。

Section 6.2   Control and Indication

6.2.1   Driver's Panel Controls

Note: Panel controls are shown in Figure 6.9. Symbols are shown in Figure 6.10, 6.11.

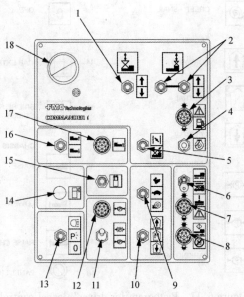

Figure 6.9   Driver's panel

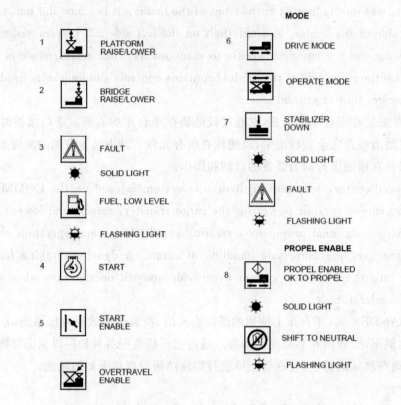

Figure 6.10 Pictogram for driver's panel controls

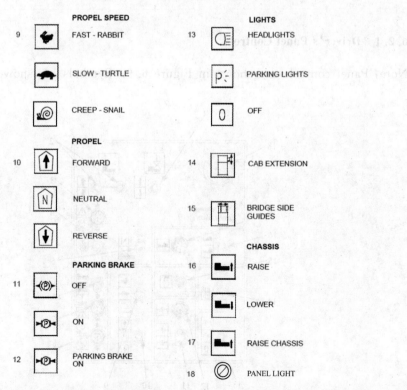

Figure 6.11 Pictogram for driver's panel controls

In Figure 6. 9:

1— Platform raise/lower switch — three-position switch spring-loaded to OFF position; raises and lowers platform.

2— Bridge raise/lower switches — two three-position switches spring-loaded to OFF position: raises bridge when switch is up, and lowers bridge when switch is down. Both switches must be activated to move bridge.

3— Low fuel indicator (flashing red) — illuminates when the control system detects low fuel (19 liters (5 gallons)). Fault indicator (steady red) — illuminates when the control system detects an alarm condition. Alarms include loss of oil pressure, loss of generator, loss of communication, and other system faults.

4— Auxiliary start switch — two-position toggle switch starts power unit when set to START and the spring-loaded start enable switch is held in NORMAL position. The master start switch on the main panel downstairs must be ON.

5— Start enable switch — OVERRIDE to start, NORMAL when engine running. Hold the start enable switch in OVERRIDE in the event of platform overtravel to bypass shutdown during cranking. Switch must remain in OVERRIDE to keep engine running.

6— Mode switch — two-position switch: extends (lowers) stabilizers and rear chassis when placed in operate position, and retracts stabilizers when placed in drive position. Amber indicator illuminates when stabilizers are fully extended. Locking switch standard. Switch must be lifted out of position prior to change selection or damage could result to switch if excessive force is used.

7— Stabilizers down indicator (amber) — illuminates steady when stabilizers are down. Fault indicator (flashing amber) — illuminates flashing when a stabilizer input disagreement fault (PRS1 and PRS3) is detected.

8— Propel enable indicator (green) — illuminates steady when loader is ready to drive. Shift to NEUTRAL (flashing green) — illuminates flashing for shift to neutral.

9— Propel speed switch — three-position control used to select ranges of speed in forward and reverse direction. The fast (rabbit) position provides maximum speed for direction selected with drive control, slow (turtle) position provides a medium speed, and creep (snail) position provides minimum speed control.

10— Propel switch — three-position control used to select a forward direction, a neutral position, and a reverse direction.

11— Parking brake switch — two-position switch. Applies parking brake when set to ON and illuminates red indicator to show that brake is applied. Releases brake when set to OFF and causes red indicator to go out. Parking brake is automatically applied when mode switch is set to operate position.

12— Parking brake indicator (red) — illuminates when parking brake is ON.

13— Lights switch — three-position switch turns head lamps and running lights on when switch is up; turns parking lights on when switch is in center position; turns lights off when switch is down.

14— Cab extension switch — three-position switch spring-loaded to OFF position; extends operator's cab when switch is momentarily up; retracts operator's cab when switch is momentarily down.

15— Bridge side guides switch — three-position switch spring-loaded to CENTER (OFF) position; shifts guides left when momentarily placed in left position, or to right when placed in right position.

16— Chassis switch — three-position momentary switch (CENTER position = OFF). Raises or lowers the rear chassis by extending or retracting bogy wheel cylinders.

17— Raise chassis indicator (flashing red) — illuminates unless chassis is raised.

18— Panel light — illuminates panel when headlights are on.

6.2.2 Operator's Control Panel

Note: Panel controls are shown in Figure 6.12. Pictograms are shown in Figure 6.13, 6.14.

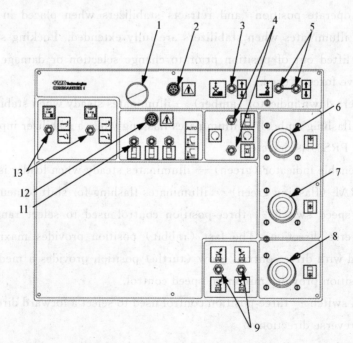

Figure 6.12 Operator's control panel

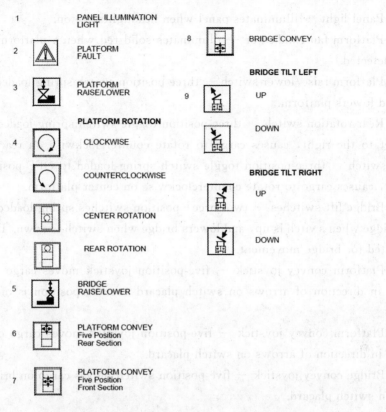

Figure 6. 13　Pictogram for operator's control panel

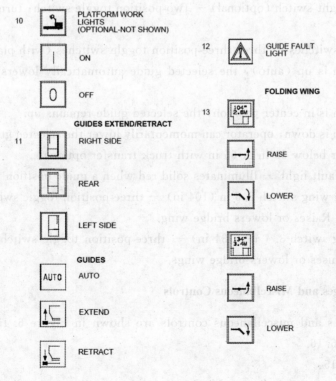

Figure 6. 14　Pictogram for operator's control pane

1— Panel light — illuminates panel when headlights are on.

2— Platform fault indicator — illuminates solid red when a platform proximity switch fault is detected.

3— Platform raise/lower switch — three-position switch spring-loaded to OFF position; raises and lowers platform.

4— Rear rotation switch — three-position toggle switch spring-loaded to OFF position. When set to the right, causes cargo to rotate counterclockwise on rear platform. Center rotation switch — three-position toggle switch spring-loaded to OFF position. When set to the right, causes cargo to rotate counterclockwise on center platform.

5— Bridge lift switches — two three-position switches spring-loaded to OFF position; raises bridge when switch is up, and lowers bridge when switch is down. Both switches must be activated for bridge movement.

6— Platform convey joystick — five-position joystick moves cargo on rear section of platform in direction of arrows on switch placard (center position is off, fwd, aft, left, right).

7— Platform convey joystick — five-position joystick moves cargo on front section of platform in direction of arrows on switch placard.

8— Bridge convey joystick — five-position switch moves cargo on bridge in direction of arrows on switch placard.

9— Bridge tilt — two-position toggle switch operation.

10— Floodlight switch (optional) — two-position toggle switch; turns floodlights on or off.

11— Guide switches — three three-position toggle switches (with platform down):

When switch is up (auto), the selected guide automatically lowers when platform is below 24 in.

When switch is in center position, the selected guide remains up.

When switch is down, operator can momentarily lower the selected guide when platform is below 24 in (or below 60 in or 65 in with truck transfer option).

12— Guide fault light — illuminates solid red when a guide position fault is detected.

13— Folding wing switch 2. 6 m (104 in) — three-position toggle switch; spring-loaded to OFF position. Raises or lowers bridge wing.

Folding wing switch 3. 4 m (134 in) — three-position toggle switch; spring-loaded to OFF position. Raises or lowers bridge wings.

6.2.3 Gauges and Miscellaneous Controls

Note: Gauges and miscellaneous controls are shown in Figure 6. 15. Pictograms are shown in Figure 6. 16.

In Figure 6. 15:

1— Shift to neutral indicator (flashing amber) — illuminates flashing for "shift to neutral". Ignition lamp (steady amber) — illuminates steady when master start switch is ON.

2— Low fuel indicator (flashing red) — illuminates on machine faults including low fuel (19 liters (5 gallons)). Fault indicator (steady red) — illuminates when oil pressure, engine coolant, hydraulic oil temperature or generator is at fault.

3— Master start switch — starts engine. ON: engine running; OFF: engine shutdown.

4— Wait to start lamp (amber) — illuminates when engine is in preheat mode.

5— Start enable switch — NORMAL — override to start; COLD to preheat engine if so equipped.

6— Hourmeter — indicates total hours of engine operation.

7— Emergency stop switch — when pushed in, shut off power unit. Switch is not to be used for routine shutdown of power unit. Proximity switches are de-energized during an emergency stop.

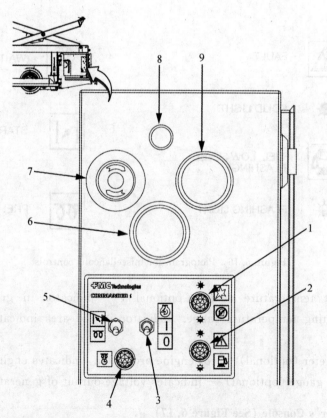

Figure 6.15   Gauges and miscellaneous controls

8— Panel light — illuminates panel when headlights are ON.

9— Fuel level gauge — indicator shows approximate level of fuel in tank. Red area at lower left shows limited amount of fuel available and indicates that loader should be refueled to insure that operations can be continued without interruption.

10— Oil pressure gauge (optional) — indicator in green area shows satisfactory engine oil pressure. Indicator in red area shows that pressure is low.

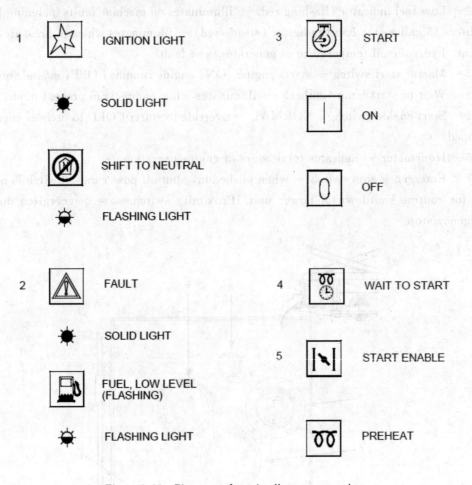

Figure 6.16 Pictogram for miscellaneous controls

11— Coolant temperature gauge (optional) — indicator in green area indicates satisfactory operating temperature range. Indicator in red area indicates excessively hot temperature.

12— Tachometer (optional) — for engine protection, indicates engine rotate speed.

13— Voltage gauge (optional) — indicates voltage output of generator.

6.2.4 Driver's Console (See Figure 6.17)

In Figure 6.17:

1— Horn switch — push button at center of steering wheel, sounds horn when pressed.

2— Emergency stop switch — push button shuts down loader and applies parking brake.

3— Emergency pump switch — located under right corner of driver's console. Provides emergency hydraulic power for all loader functions except driving and platform liftting when ignition start switch is ON.

4— Accelerator — foot pedal, regulates drive speed of loader.

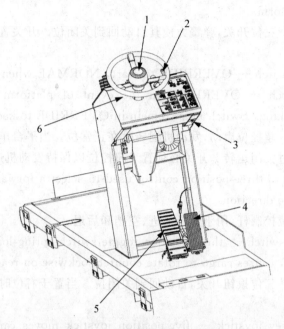

Figure 6.17   Driver's console

5— Brake — foot pedal, applies service brakes.

6— Turn signal switch (optional) — indicates direction of turn, flasher included.

New Words

1. symbol['simbəl] *n.* 记号,符号
2. override[əuvə'raid] *vt.* 使过量负载
3. pictogram['piktəgræm] *n.* 象形图
4. headlight['hedlait] *n.* 前灯
5. placard['plæka:(r)d] *n.* 招牌,名牌
6. floodlight['flʌdlait] *n.* 泛光灯,探照灯
7. preheat[pri:'hi:t] *vt.* 预热

Phrases and Expressions

1. spring-loaded 弹簧承载的,受弹簧支撑的
2. toggle switch 拨动开关
3. proximity switch 接近开关
4. steering wheel 方向盘

Notes to the Text

1. Platform raise/lower switch — three-position switch spring-loaded to OFF position; raises and lowers platform.

平台举升开关——三位开关,弹簧力使其自动回到关闭位。开关置于上位升起桥台,开关置于下位降下桥台。

2. Start enable switch — OVERRIDE to start, NORMAL when engine running. Hold the start enable switch in OVERRIDE in the event of platform overtravel to bypass shutdown during cranking. Switch must remain in OVERRIDE to keep engine running.

启动工作开关——超控位启动,发动机运转时在正常位。当平台车发生超行程时,将启动开关置于超控位可绕过关闭运转。开关必须置于超控位以保持发动机运转。

3. Propel switch — three-position control used to select a forward direction, a neutral position, and a reverse direction.

行驶开关——三位控制杆,用于选择前进、空挡和后退。

4. Rear rotation switch — three-position toggle switch spring-loaded to OFF position. When set to the right, causes cargo to rotate counterclockwise on rear platform.

后部旋转开关——三位扳钮开关,自动回到关闭位。当置于右位时,货物在后平台上逆时针旋转。

5. Platform convey joystick — five-position joystick moves cargo on rear section of platform in direction of arrows on switch placard (center position is off; fwd; aft; left; right).

平台运送控制杆——五位控制杆,使货物在平台后部按开关牌上箭头方向移动(中位关闭及前、后、左、右四个方向)。

6. Emergency stop switch — when pushed in, shuts off power unit. Switch is not to be used for routine shutdown of power unit. Proximity switches are de-energized during an emergency stop.

应急停止开关——按下开关,关闭动力装置。该开关不能用来正常关闭动力装置。应急停车时接近开关断电。

Section 6.3　Procedure

6.3.1　Starting Power Unit — Diesel Engine

Warning: Before starting any type of power unit, observe all precautions below:

Insure power module tee bolt is securely fastened.

Insure that all personnel are clear of loader.

(1) Set propel switch to ON.

(2) Insure that both emergency stop switches (on main panel and driver's console) are pulled out.

(3) Insure that parking brake switch is set to ON.

(4) Set master start switch to ON and wait for cold start light to go out (if equipped with COLD START option).

(5) Set master start switch to START until engine starts, then release switch to ON after engine starts.

(6) Observe gauges (optional) to insure that all indicators show normal operation.

(7) Allow engine to warm up for several minutes.

(8) Refer to paragraph 6.3.2 through 6.3.5 below for required operational procedures.

6.3.2　Stopping Power Unit — Diesel Engine

(1) Set parking brake switch to ON and note that indicator illuminates.

(2) Shift propel switch to ON.

(3) Let engine idle for 3 or 4 min.

(4) Set Master start switch to OFF.

6.3.3　Bridge Tilt (Optional)

Warning: When using bridge tilt, ensure adequate clearance between loader and aircraft. Failure to do so could cause damage to aircraft or loader.

With loader in position at the aircraft doorway, operate either bridge tilt switch to obtain desired interface angle.

6.3.4　Approaching Aircraft for Cargo Transfer

Warning: The COMMANDER 30i is not designed for use as a transporting vehicle. Any attempt to use it for operations other than cargo transfer may result in injury to personnel or damage to equipment.

(1) Start power unit:

1) Set propel switch to ON.

2) Insure that emergency stop switches (on main panel and driver's console) are pulled out.

3) Insure that parking brake switch is set to ON.

4) Set master start switch to ON and wait for cold start light to go out (if equipped with COLD START option).

5) Set master start switch to START while holding start enable switch on until engine starts, then release switch to ON after engine starts. It may be necessary to hold start enable on until engine develops oil pressure.

6) Observe gauges (optional) to insure that all indicators show normal operation.

7) Allow engine to warm up for several minutes.

(2) Retract the operator's cab fully.

(3) Retract the L. H. handrail.

(4) Actuate bridge tilt switches (optional) so that both sides are at their lowest height.

(5) Check area to be sure that intended drive path is free of obstructions.

(6) Set mode switch to DRIVE.

(7) Hold chassis switch up until red flashing raise chassis indicator (optional) goes out.

(8) Insure that stabilizers down indicator goes out.

(9) Set parking brake switch to OFF and note that indicator goes out.

(10) Set propel switch to the forward or reverse position.

Caution: Only use snail speed if loader is closer than 3 m (10 ft) to aircraft.

(11) Set propel speed switch to desired speed. Start with switch in low range (snail), then set to mid-range (turtle), then set to high range (rabbit) at about 5 km/h (3 mile/h).

(12) Press accelerator pedal; as soon as loader moves, release accelerator pedal and press brake pedal to check for smooth and positive brake operation.

(13) Drive loader to within 3 m (10 ft) of aircraft.

(14) Stop loader; set propel speed switch to low range (snail).

(15) Lower chassis fully; raise chassis indicator will flash.

(16) Adjust height of bridge as required so that aircraft cargo door will clear bridge when loader is positioned and door is opened.

(17) Lower bridge and platform and safety rails (where applicable) to provide appropriate clearance to aircraft fuselage.

Note: Some aircraft configurations require bridge to be fully lowered.

(18) Slowly drive loader toward aircraft; stop when bridge is approximately 0.3 m (12 in) from aircraft. Be sure that loader is squarely positioned relative to aircraft fuselage.

(19) Set parking brake switch to ON and note that indicator illuminates.

(20) Lower (fold down) left hand handrail (if required for aircraft door clearance).

(21) Open aircraft cargo door.

(22) Adjust height of bridge to aircraft.

(23) Set parking brake switch to OFF and drive loader forward until in position with aircraft door.

(24) Move propel switch to ON. Set mode switch to OPERATE (This will automatically set parking brake and extend stabilizers). Note that parking brake indicator and stabilizers down indicator illuminate.

(25) Move bridge side guides switch to left or right position to align side guides and door for cargo transfer.

(26) Position bridge height for proper alignment with aircraft.

(27) Extend operator's platform as required for cargo transfer (optional).

(28) Raise safety rails where applicable.

(29) Transfer cargo from aircraft to ground equipment (see paragraph 6.3.5) or to aircraft (see paragraph 6.3.6).

6.3.5　Transferring Cargo from Aircraft

Note: It is assumed that all procedures for approaching the aircraft (see paragraph 6.3.4) have been performed.

(1) Hold platform lift switch up until platform automatically stops at same level as bridge. Note that stop at rear of bridge retracts.

(2) When cargo is in position to be moved onto bridge, hold bridge joystick to rearward convey position until cargo is completely on bridge.

(3) Adjust position of cargo on bridge by operating bridge joystick in required directions until cargo is laterally centered on bridge.

(4) Insure that side and rear stops on platform are extended.

(5) Hold all three joysticks simultaneously to rearward position until cargo is as far back as possible on platform.

Caution: It is necessary to adjust height of bridge as aircraft position changes during cargo transfer. Failure to maintain alignment of aircraft and bridge may result in damage to equipment.

Note: It may be necessary to rotate cargo on platform. Use rotation switches as required.

(6) If another container is to be transferred, repeat step (3) through (5) but do not operate rear platform joystick; then continue to step (7) below.

(7) Hold platform lift switch down until platform is at same level as ground vehicle.

(8) Set guides switch to AUTO or DOWN to retract left, right, or rear stop on platform (whichever is closest to ground vehicle).

(9) Hold platform joystick to move cargo onto ground vehicle, then release platform joystick and guides switch (if not in AUTO position).

(10) If more cargo is to be transferred, repeat step (1) through (9). If transfer is complete, continue to step (11) below.

(11) Hold platform lift switch down until platform stops automatically, then perform procedures in paragraph 6.3.7.

6.3.6　Transferring Cargo to Aircraft

Note: It is assumed that all procedures for approaching the aircraft (paragraph 6.3.4) have been performed.

(1) Operate platform lift switch until platform is at same level as ground vehicle that contains cargo.

(2) Set guides switch to AUTO position or hold in retracting position to retract left, right, or rear stop on platform (whichever is closest to ground vehicle).

(3) Hold platform joystick to move cargo onto platform, release joystick when cargo is approximately centered laterally on platform, then release guides switch (if not in AUTO).

(4) Hold platform lift switch up until rear platform automatically stops at same level as bridge. Note that stops at rear of bridge retracting.

(5) Hold center and forward joysticks to forward position until container is on forward platform.

(6) Hold bridge joystick to forward position until cargo is on aircraft.

Caution: It is necessary to adjust height of bridge as aircraft position changes during cargo transfer. Failure to maintain alignment of aircraft and bridge may result in damage to equipment.

Note: For the second container, operate joysticks simultaneously to move container onto bridge, then repeat step (5) and (6) above.

(7) Hold platform lift switch down until platform is again at same level as ground vehicle that contains cargo.

(8) If more cargo is to be transferred, repeat step (2) through (7). If transfer is complete, go to step (9) below.

(9) Hold platform lift switch down until platform stops automatically, then perform procedures for departing from the aircraft in paragraph 6.3.7 below.

6.3.7 Departing from Aircraft

(1) Check area to be sure that intended drive path is free of obstructions.

(2) Amber parking brake indicator and stabilizers down indicator should be illuminated.

(3) Retract operator's platform so that it will clear aircraft when bridge is moved.

(4) Lower bridge so that it will clear cargo door when it is closed.

(5) Carefully close and latch cargo door.

(6) Set mode switch to DRIVE. Note that stabilizers retract and stabilizers down indicator goes out.

(7) Raise chassis. (Raise chassis indicator flashes when rear chassis is raised with switch.)

(8) Set propel switch to reverse position.

(9) Set propel speed switch to low range (snail).

(10) Set parking brake switch to OFF and note that indicator goes out.

(11) Slowly back loader away from aircraft.

(12) Lower bridge until it stops automatically.

(13) Drive loader to assigned location with propel speed switch set to mid-range (turtle) or high range (rabbit).

(14) If no further operations are required, park loader and stop power unit as instructed in paragraph 6.3.8 below.

6.3.8 Parking Loader

(1) Set propel switch to ON.

(2) Set the parking brake switch to ON and note that the parking brake indicator illuminates.

Caution：To prevent damage to the stabilizer cylinder assemblies, do not set the mode switch to "OPERATE" when the unit is parked as this will extend the stabilizers.

(3) Let engine idle for 3 to 4 min.

(4) Set master start switch to OFF.

New Words

1. precaution [pri'kɔːʃ(ə)n] *n.* 预防/防范措施(方法)
2. fasten ['faːs(ə)n] *vt.* 夹紧,结牢
3. shift[ʃift] *vt.* 换挡
4. splash[splæʃ] *vt.* 溅,泼,洒
5. obstruction [əb'strʌkʃ(ə)n] *n.* 阻碍,妨碍
6. laterally['lætərəli] *adv.* 侧面地,横向地
7. simultaneously [siməl'teiniəsli] *adv.* 同时地

Phrases and Expressions

1. tee bolt T形螺栓
2. fuel spill 燃料溢出

Notes to the Text

1. Warning：Before starting any type of power unit, observe all precautions below：

Insure power module tee bolt is securely fastened.

Insure that all personnel are clear of loader.

警告:启动任何型号的发动机前,必须遵守下列预防措施:

保证发动机的 T 形螺栓紧固。

确保全体人员与平台保持安全距离。

2. Warning：When using bridge tilt, ensure adequate clearance between loader and aircraft. Failure to do so could cause damage to aircraft or loader.

警告:当使用桥台倾斜功能时,要确保飞机与平台间有足够的空间,否则会对飞机或平台造成损伤。

3. Press accelerator pedal；as soon as loader moves, release accelerator pedal and press brake pedal to check for smooth and positive brake operation.

踩下加速踏板,平台车开始移动后,松开加速踏板并踩下制动踏板检查制动操作的平稳性。

4. Adjust position of cargo on bridge by operating bridge joystick in required directions until cargo is laterally centered on bridge.

通过桥台控制杆,在要求的方向上将货物调整到桥台横向的中心

5. Caution：It is necessary to adjust height of bridge as aircraft position changes during

cargo transfer. Failure to maintain alignment of aircraft and bridge may result in damage to equipment.

注意:在运送过程中,要随着飞机位置的变化,及时调整桥台的高度,如果不能保持飞机和桥台的水平,可能会对设备造成损伤。

6. Caution:To prevent damage to the stabilizer cylinder assemblies, do not set the mode switch to "OPERATE" when the unit is parked as this will extend the stabilizers.

注意:为防止支撑脚作动筒损坏,在停车后不能将模式开关置于操作位上,因为在此位置上,支撑脚将会自动伸出。

Chapter 7　Mobile Belt Loader

Section 7.1　General Description

7.1.1　Application

The TUG model 660 is a self-propelled vehicle designed to load and unload baggage, light freight and mail into and out of the lower hold of aircraft. The conveyor may also be used to transfer freight from trucks or between any two points at the same or differing elevations.

7.1.2　Component/System

1. Power package

The tractor is equipped with an industrial type engine (4-cylinder diesel or 6-cylinder gasoline) with an automatic transmission. The engine will be certified by the manufacturer to meet the emissions standards required by Federal Law which are applicable at the date of manufacture. The gasoline engine is designed to operate satisfactorily using 87 Octane no-lead fuel, and the diesel engine operates satisfactorily using diesel fuel number 2 and other fuels (see engine manual for list of acceptable fuels).

(1) Engine

1) Gasoline & CNG: The gasoline engine is a heavy duty, industrial type with a short stroke and exhaust valve rotators.

2) Diesel: The diesel engine is a heavy duty, industrial type with 4-stroke cycle and direct injection.

(2) Transmission

1) Ford: The C6 3-speed (diesel) or 2-speed (gasoline) automatic transmission is equipped with a torque converter and park brake assembly.

2) Durst (option): The GSE - 1 Velvet Drive single-speed transmission, its torque converter, output coupling and brake actuation mechanism have the same mounting points to the engine, vehicle and axle drive shaft as a Ford C6 transmission fitted with a parking

brake. The bell housing on the Durst GSE - 1 transmission is detachable from the main transmission case. The parking brake is a drum with shoe design which is mechanical actuation. SAE ♯4 bell housing is available for connecting to the Cummins B3. 3L engine.

2. Drive shaft

The tractor's drive shaft is an automotive type with double universal joints and a spline coupling.

3. Drive axle

The rear (drive) axle is a Dana Model 60 with full floating axles and drum brakes.

4. Steer axle

The front (steer) axle is a Dana Model 44 with disc brakes.

5. Brakes

(1) Hydraulic service brakes are provided on all four wheels. The master cylinder is mounted in front of the instrument panel. Pressure is transmitted to all four brake assemblies by depressing the brake pedal.

(2) A rear wheel cable-actuated parking brake is provided, and is engaged by the brake lever mounted to the right of the seat in the operator's compartment.

6. Chassis

The 660 Mobile Belt Loader is built on a heavy duty chassis which includes the power package and running gear. The chassis has a 110 in wheel base, formed steel channel frame supported by the front and rear axles.

7. Body

(1) The vehicle's body panels are formed from 3/16 in steel and bolted to the chassis. The front and rear fenders are installed with standard mounting hardware and are independently replaceable from the center section.

(2) The 3/16 in formed steel body panels are fabricated in 6 sections (3 sections on right and 3 sections on the left hand side of vehicle). The front and rear sections (representing the front and rear fenders) are mounted separately from the center section and, if necessary, may be replaced independently.

(3) A body panel support structure is integrated with the vehicle chassis which incorporates the pivot and attaching points for the front and rear lift frames and the lift cylinders.

(4) Protective rub strips can be attached to the full length of both sides of the vehicle to afford further protection for the body panels.

8. Conveyor

(1) The conveyor is fabricated from standard structural shapes and formed steel, and is attached to the chassis through front and rear lifting frames.

(2) The front cylinder, supporting the front lifting frame, raises the conveyor to 170 in on the Model 660, and 181 in on the Model 661.

(3) The conveyor belt, 24 in wide, is supported by 2 in diameter rollers and is hydraulic

motor driven through a roller chain reduction. The front roller is mounted in a flange block with a screw-adjustable take-up that allows 6 in of adjustment. Crowder rollers under the support rollers direct the return side of the belt into areas where required clearances around fixed components is maintained. An inverted angle iron track on the lower forward section of the conveyor maintains lateral stability through V-shaped rollers mounted to the front lift frame.

(4) Operator stations for controlling the conveyor belt are located at each end of the conveyor. When electric belt controls are supplied, remote actuation of conveyor functions is available.

9. Operator's compartment

(1) The operator's compartment contains all the required accessories to drive the vehicle, and to control the lifting and lowering of the conveyor.

(2) Engine controls are mounted in the operator's compartment instrument panel and include oil pressure gauge, water temperature gauge, ammeter (charge-discharge), hour meter, ignition switch and light switches.

10. Electrical system

(1) Electrical power is supplied by an alternator, belt-driven, mounted on the side of the engine. The system is a 12 V, direct current system with negative ground.

(2) All operating circuits are wired through the ignition switch and are protected by 20 amp circuit breakers as follows:

CB-1 20 amp ignition.

CB-2 20 amp electric conveyor controls (optional).

CB-3 20 amp lighting.

11. Hydraulic system

(1) The hydraulic schematic represents an open center system that supplies power to raise and lower the conveyor and to drive the conveyor belt.

(2) Hydraulic power is supplied by a hydraulic pump, belt-driven, mounted on the side of the engine. The pump supplies 9 g/min at 1,500 r/min to raise and lower the conveyor and to drive the conveyor belt.

(3) Hydraulic oil from the 10-gallon reservoir (located in the left center section body panel) delivered by the pump passes through the relief valve (factory set at 1,500 psi). Past the relief valve, oil is directed to the two-spool, manually-controlled valve (with built-in relief and power beyond feature) that controls the front and rear lift cylinders. The valve is located beneath the driver's seat for convenience and accessibility. Oil flow is then directed to the valve that controls the direction of the conveyor belt motor. For electrical controls, the solenoid valve is located on the left hand frame channel of the chassis. For mechanical controls, the manual valve is located on the underside of the conveyor near the forward end. From the belt valve, oil flows back to the brake booster, through the steering valve, and back to the tank through the return line filter.

(4) Both front and rear cylinders are locked in any position by holding valves. The orbital motor has incorporated a counter-balance and cushion valve in the circuit.

New Words

1. conveyor [kən'veiə(r)] *n.* 传送装置，传送机

Phrases and Expressions

1. self-propelled 自行式
2. spline coupling 花键联轴器

Notes to the Text

1. The conveyor belt, 24 in wide, is supported by 2 in diameter rollers and is hydraulic motor driven through a roller chain reduction. The front roller is mounted in a flange block with a screw-adjustable take-up that allows 6 in of adjustment. Crowder rollers under the support rollers direct the return side of the belt into areas where required clearances around fixed components is maintained. An inverted angle iron track on the lower forward section of the conveyor maintains lateral stability through V-shaped rollers mounted to the front lift frame.

输送带宽 24 英寸，由两个直径 2 英寸的滚筒支撑，并由液压马达通过滚子链减速驱动。前滚筒安装在一个带有可调节螺钉的凸缘座上，并允许向上调整 6 英寸。支承滚筒下的克劳德滚筒引导返回侧输送带，使其与固定部件保持在要求的间隙范围内。传送机下部前方的倒转角钢轨道通过安装在前升架上的 V 形滚筒保持横向稳定性。

Section 7.2   Operation and Procedures

7.2.1   Instrument Panel

The instrument panel in the operator's compartment contains the following gauges and electromechanical devices：

(1) Oil temperature.

(2) Ammeter (charge/discharge).

(3) Oil pressure.

(4) Headlight switch.

(5) Override switch.

(6) Ignition switch.

(7) Hour meter.

7.2.2   Maneuvering Controls

(1) Steering wheel.

(2) Brake pedal.

(3) Foot throttle.

(4) Turn signal.

(5) Hand throttle.

(6) Direction selector.

7.2.3   Conveyor Controls

(1) Conveyor height controls in the operator's compartment (located below driver's seat) are limited to:

1) Front cylinder extension/retraction.

2) Rear cylinder extension/retraction.

(2) Electrical belt drive controls, when provided, are located on the left side of the conveyor at the front and on both sides of the conveyor at the rear.

7.2.4   Main Electrical Supply

The main electrical supply panel is located below the instrument panel.

Control relays for the electrical conveyor belt drive solenoid control valves are located on the conveyor frame.

7.2.5   Preliminary Checks

Procedures for preliminary checks before maneuvering the vehicle to an aircraft shall include (but may not be limited to) the following items:

(1) Fuel level.

(2) Tire inflation.

(3) Engine oil level.

(4) Transmission oil level.

(5) Service brake effectiveness.

(6) Parking brake effectiveness.

(7) Vehicle drive forward.

(8) Vehicle drive reverse.

(9) Conveyor belt drive forward.

(10) Conveyor belt drive reverse.

(11) Front cylinder extension.

(12) Front cylinder retraction.

(13) Rear cylinder extension.

(14) Rear cylinder retraction.

7.2.6   Maneuvering Procedures

Procedures for maneuvering the vehicle to an aircraft are outlined below:

1. Start engine

(1) Apply parking brake..

(2) Move transmission selector to NEUTRAL. (Starter motor will not activate unless selector is in NEUTRAL).

(3) Move ignition switch to "START".

(4) Allow ignition switch to return to "ON".

(5) Switch must be returned to "OFF" before it can be moved to start again.

Note: Refer to dash decal for cold starting instructions.

2. Select transmission drive (FWD/REV)

A Morse control is connected to the transmission shift lever through a flexible push/pull cable. The Morse control incorporates the hand throttle and the transmission drive selector in two levers. The lever to the left (when seated in the driver's compartment) is the hand throttle and the lever to the right is the transmission drive selector. To select a vehicle drive direction:

(1) Move transmission drive selector lever up (toward instrument panel) to select reverse drive direction.

(2) Move transmission drive selector lever down (toward driver's seat) to select forward drive direction.

(3) Neutral is located midway between forward and reverse and is indicated by a positive stop notch into which the spring loaded lever will fall when shifting.

(4) The hand throttle lever is used to set the engine speed which directly controls the conveyor belt speed. Move the hand throttle lever down (toward the driver's seat) to increase engine/belt speed.

(5) Transmission lever will not move into forward or reverse with the hand throttle advanced.

3. Release parking brake

The parking brake is a cable-actuated type attached to the transmission tail shaft. The brake-actuating lever is located to the right of the driver's seat in the driver's compartment.

(1) To apply parking brake, lift actuating lever up.

(2) To release parking brake, push actuating lever down.

4. Depress foot throttle

The foot throttle is a conventional automobile or truck accelerator and is connected to the carburetor or ECM through a direct coupled pull cable.

(1) To increase speed, depress foot throttle.

(2) To decrease speed, release foot throttle.

5. Maneuver vehicle with steering wheel

The steering system is a hydraulic type powered from the belt-driven hydraulic pump.

(1) To turn left, rotate the steering wheel counter-clockwise.

(2) To turn right, rotate the steering wheel clockwise.

7.2.7　Conveyor Safety Prop

Warning：Always use the prop when working under the conveyor.

(1) The prop is provided to support the front of the conveyor for maintenance, checking oil, etc.

(2) To install the prop, raise the front of the conveyor to a height that permits the prop to be rotated to the "UP" position. Raise the prop to the "UP" position. Slowly let the conveyor down until prop aligns with sockets on bottom of the lift arm. Be sure the prop is fully engaged inside sockets before working under the conveyor.

(3) Power conveyor down gently onto prop or structural failure may result.

New Words

1. maneuvering [məˈnuːvəriŋ] v. 操纵,移动
2. prop [prɒp] n. 支柱,支撑杆

Phrases and Expressions

1. instrument panel 仪表板
2. dash decal 小贴纸,小标记牌
3. cold starting 冷启动
4. cable-actuated 钢索驱动

Notes to the Text

1. Control relays for the electrical conveyor belt drive solenoid control valves are located on the conveyor frame.

电动传送带驱动电磁控制阀的控制继电器位于传送架上。

Section 7.3　Specifications

7.3.1　General

1. Dimensions

Length	300 in
Width (excluding side bumpers)	78 in
Height	59 in
Wheel base	110 in

Conveyor length (less front bumper)	294 in
Conveyor width (less side rails)	34 in
Belt width	24 in
Gross vehicle weight	7,200 lb

2. Performance

Top speed	27 mile/h
Conveyor capacity (to 15° angle)	2,000 lb
Max conveyor load	2,000 lb, not to exceed 200 lb per square foot
Belt control	Manual, electric, or rotary
Conveyor belt speeds	45 to 90 ft/min
Turning radius	27 ft, measured to the outside front wheel
Aircraft capability	Will service aircraft with door sill heights from 40 in to 170 in

3. Other specifications

Battery	12 V
Ground	Negative
Fuel tank capacity	15 gal (U.S.)
Hydraulic reserve capacity	10 gal (U.S.)
Radiator capacity	83/4 qt(1 qt=1.136 L)
Service brakes	
Rear	Hydraulic, internal-expanding, drum-type, self-adjusting
Front	Hydraulic disc
Rear axle	Dana, single-speed, full differential, semi-floating

7.3.2 Transmission

(1) Ford: The C6 3-speed (diesel) or 2-speed (gasoline) automatic transmission is equipped with a torque converter and park brake assembly.

(2) Durst (option): The GSE - 1 Velvet Drive single-speed transmission, its torque converter, output coupling and brake actuation mechanism have the same mounting points to the engine, vehicle and axle drive shaft as a Ford C6 transmission fitted with a parking brake. The bell housing on the Durst GSE - 1 transmission is detachable from the main transmission case. The parking brake is a drum with shoe design which is mechanical actuation. SAE #4 bell housing is available for connecting to the Cummins B3. 3L engine.

7.3.3 Heights and Angles

The height of the conveyor at different angles is shown in Figure 7.1, Figure 7.2, Table 7.1, Figure 7.3, Figure 7.4 and Table 7.2.

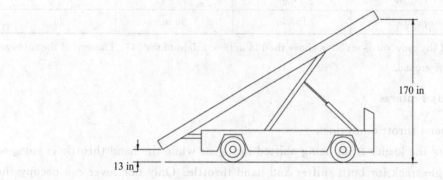

Figure 7.1   Full down rear, full up front

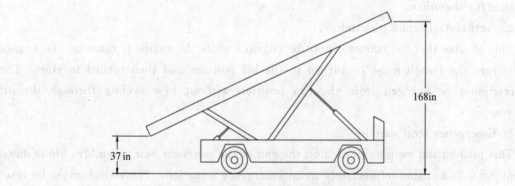

Figure 7.2   Full up rear, full up front

Table 7.1   Conveyor heights and angles

Conveyor Attitude	Front	Rear	Angle
Front up, rear down	170 in	13 in	35°
Front up, rear up	168 in	37 in	27°

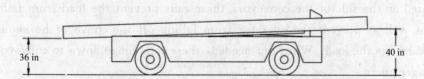

Figure 7.3   Full down rear, full down front

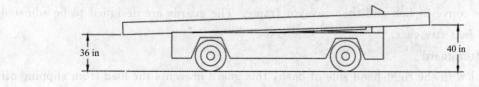

Figure 7.4   Full up rear, full down front

Table 7.2　Conveyor heights and angles

Conveyor Attitude	Front	Rear	Angle
Front down, rear down	40 in	36 in	1°
Front down, rear up	31 in	61 in	−6°
Nominal 747 operation	158 in	30 in	26°

* The front of the conveyor is set 30 in above the 128 in floor height of the 747. The rear of the conveyor is set 30 in above the ground.

7.3.4　Safety Features

1. Shifter/hand throttle interlock

This prevents the loader from being shifted into gear while the hand throttle is engaged by utilizing a single track for both shifter and hand throttle. Only one lever can occupy the track at a time. A wider neutral position allows one lever to move sideways to provide clearance for the other.

2. Anti-restart ignition switch

This assures that the starter cannot be engaged while the engine is running. To engage the starter, the switch must be turned to the off position and then turned to start. The starter cannot be engaged from the run position without first cycling through the off position.

3. Emergency stop switch

This pushbutton switch, located on the end of the conveyor belt assembly, shuts down the conveyor and engine immediately in an emergency situation. The switch must be reset before the unit can be restarted (optional).

4. Protect-start

This is a relay wired between the alternator and starter. If an AC voltage of 3 V or larger is detected in the alternator, the relay opens, preventing the starter from being engaged while the alternator is turning (optional).

5. Side/hand rails

Mounted on the side of the conveyor, these rails prevent the load from falling off the conveyor as well as prevent the operator from falling off the conveyor he should have to climb to rearrange the load. When not needed, these rails hinge down to conveyor level.

6. Finger guard

Finger guards are located on the front, rear and bottom of the conveyor. When used correctly, the adjustable guards help to prevent the operator from getting his fingers caught between the conveyor belt and the conveyor frame. The guards are designed to be adjusted open as the belt stretches.

7. Shifter guard

Integral with the right-hand side of dash, this guard prevents the load from slipping off and knocking the loader in gear, if the throttle interlock has not been positioned to lock the shift lever in neutral position.

8. Safety props

Painted yellow and located at the front and rear of the vehicle, these manually engaged safety devices prevent the conveyor from drifting down during maintenance work. They should be engaged any time the conveyor is in the raised position for maintenance work.

9. Neutral safety start switch

Located in the shifter/throttle assembly, this switch, wired between the ignition and the starter, closes when the shifter is in the neutral position, allowing the starter to engage. If the shifter is in the forward or reverse position, the switch is open and the starter will not engage.

10. Parking brake

The parking brake is located on the drive shaft and is activated by an Orchlin over-center lever. When parking brake is engaged, the loader will not move, even if shifted into gear.

11. Pilot operated hydraulic holding valves

Located on both lift cylinders and reverse travel direction of the belt, these valves hold the cylinders and the conveyor belt at the set position. Conveyor belt valve prevents free wheeling of the belt if the load is stopped with the conveyor at an incline.

12. Grit paper

Grit paper is positioned on the pontoons for better traction and safer operation. The grit paper on pontoons should be inspected monthly and replaced when found to be torn or worn off.

Phrases and Expressions

1. instrument panel 仪表板
2. power booster 助力器
3. cable-actuated 钢索驱动
4. shifter/hand throttle interlock 换挡手柄/手动油门互锁
5. pilot operated hydraulic holding valves 先导式液压保持阀

Notes to the Text

1. This prevents the loader from being shifted into gear while the hand throttle is engaged by utilizing a single track for both shifter and hand throttle. Only one lever can occupy the track at a time.

当手油门正在使用换挡手柄与手油门共用的导轨时,这(互锁装置)用来防止行李传送车挂挡。该导轨一次只能容纳一个手柄。

Section 7.4   Shipping

7.4.1   Preparation

Warning: Hazard exists when refueling and defueling the tractor — no smoking, no open flames, no electrical device operations.

1. Radiator

Drain engine coolant from radiator and engine block.

2. Fuel cell

(1) Drain fuel from fuel tank into appropriate container.

Note: Drain is located under unit immediately below tank.

(2) Disconnect fuel line at lowest readily accessible connection and drain fuel from fuel line.

(3) Reconnect fuel lines.

(4) Leave fuel cap on tank, but loosened.

3. Battery

(1) Use emergency hydraulic pump to lower conveyor, if required.

(2) Disconnect battery terminals from battery posts.

(3) Leave battery in battery box.

4. Checks

(1) On tractors so equipped, make certain in-line fuel shut-off valve is open.

(2) Make certain fuel line is connected.

(3) Make certain shut-off valve on radiator is closed.

(4) Make certain fuel cell drain valve is closed.

7.4.2   Preparation for Operation

Warning: Make certain all fuel lines are properly connected from tank to fuel pump to carburetor (gasoline) or injector pump (diesel).

1. Battery

(1) Use emergency hydraulic pump to raise conveyor, if required.

(2) Connect battery terminals to battery posts.

2. Radiator

(1) Make certain that radiator shut-off valve is closed.

(2) Add required amount of coolant/water mixture to radiator

3. Fuel cell

(1) Add fuel to fuel tank.

(2) Check all fuel lines for leaks.

Section 7.5　Storage

7.5.1　For One Month

7.5.1.1　Engine

1. Gasoline

(1) Run the engine at 1,500 r/min and treat the upper cylinder by spraying an engine preservative oil into the carburetor (if equipped) air intake for about two minutes. This oil should be SAE #10, formulated for antirust and anticorrosion protection, along with being a high detergent type that meets requirements for most severe (M. S.) service and also Ford Specification M – 2C35. Open throttle for a short burst of speed, then shut off the ignition and allow the engine to come to a stop while continuing to spray the oil into the carburetor air intake.

(2) Leave the spark plugs installed and cover all engine openings with dust-proof caps or shields.

(3) Drain oil from crank case.

(4) Drain water from radiator and engine block.

(5) Drain fuel from tank, carburetor and fuel lines.

(6) Attach a tag on steering wheel stating.

Caution: Oil and water has been removed from engine.

2. Diesel

(1) Run the engine until it reaches normal operating temperature and then shut the engine down.

Caution: Oil will be hot.

(2) Drain oil from crank case and refill with required amount.

(3) Drain water from radiator and engine block.

(4) Attach a tag on steering wheel stating.

Caution: Water has been removed from engine.

7.5.1.2　Transmission

No special attention needed.

7.5.1.3　Axle and Reduction Gear Box

No special attention needed.

7.5.1.4　Tires

The tractor should be raised and axles chocked to prevent tire contact with ground. The pressure should be reduced to 15 psi.

7.5.1.5　Lubrication

Ensure that all points are lubricated with specified grease, oil, etc.

7.5.1.6 Fluid Levels

All fluid levels should be checked and topped off as necessary.

7.5.1.7 Wheel Bearings

Wheel bearings should be repacked.

7.5.1.8 Battery

Disconnect battery terminals.

7.5.2 Indefinite Period

7.5.2.1 Engine

1. Gasoline

(1) Run the engine at 1,500 r/min and treat the upper cylinder by spraying an engine preservative oil into the carburetor (if equipped) air intake for about two minutes. This oil should be SAE #10, formulated for antirust and anticorrosion protection, along with being a high detergent type that meets requirements for most severe (M. S.) service and also Ford Specification M−2C35. Open throttle for a short burst of speed, then shut off the ignition and allow the engine to come to a stop while continuing to spray the oil into the carburetor air intake.

(2) Leave the spark plugs installed and cover all engine openings with dust-proof caps or shields.

(3) Drain oil from crank case.

(4) Drain water from radiator and engine block.

(5) Drain fuel from tank, carburetor and fuel lines.

(6) Attach a tag on steering wheel stating.

Caution: Oil and water has been removed from engine.

2. Diesel

(1) Run the engine until it reaches normal operating temperature and then shut the engine down.

Caution: Oil will be hot.

(2) Drain oil from crank case and refill with required amount.

(3) Drain water from radiator and engine block.

(4) Attach a tag on steering wheel stating.

Caution: Water has been removed from engine.

7.5.2.2 Transmission

1. Drain transmission fluid

2. Refill transmission

Note: Prolonged storage may be detrimental to the seals in the transmission.

7.5.2.3 Axle and Reduction Gear Box

(1) Drain drive axle and reduction gear box by removing drain plug located on the underside of the differential carrier housing. Reinstall plug after draining. The drive axle and

reduction gear box are connected internally and share the same oil.

(2) Attach a tag on steering wheel stating.

Caution: Oil has been removed from axle and reduction gear box.

7.5.2.4   Tires

The tractor should be raised and axles chocked to prevent tire contact with the ground. Tire pressure should be reduced to 15 psi and tires should be sprayed with a rubber preservative.

7.5.2.5   Lubrication

Ensure that all points are lubricated with specified grease, oil, etc.

7.5.2.6   Levels

All fluid levels should be checked and topped off as necessary.

7.5.2.7   Wheel Bearings

Wheel bearings should be repacked.

7.5.2.8   Battery

The battery should be removed and stored separately. The battery must be stored in a cool dry place and must not be exposed to direct sunlight. A recommended storeroom temperature of 32°F (0°C) to 90°F (32°C) should be maintained. If the battery is stored in the open, it must be covered for protection against dirt and moisture. A slow charge should be given to the battery every one to two months.

Notes to the Text

1. Run the engine at 1,500 r/min and treat the upper cylinder by spraying an engine preservative oil into the carburetor (if equipped) air intake for about two minutes. This oil should be SAE #10, formulated for antirust and anticorrosion protection, along with being a high detergent type that meets requirements for most severe (M. S.) service and also Ford Specification M - 2C35. Open throttle for a short burst of speed, then shut off the ignition and allow the engine to come to a stop while continuing to spray the oil into the carburetor air intake.

以 1 500 r/min 的速度运行发动机,通过向化油器(如果装有)进气管喷入发动机防护油约 2 min 来处理上汽缸。这种油的黏度应为 SAE♯10 级,为防锈和防腐蚀保护而配制,同时也是一种满足最高级别维修要求和福特 M - 2C35 标准的高效去垢剂。加油门让发动机短时间内急加速,然后关闭发动机,在发动机停下来的过程中继续把油喷到化油器的进气口。

Section 7.6   Servicing

7.6.1   Preparation for Use

1. Fuel system

(1) Check all fuel line connections to make sure they are properly connected.

(2) Check fuel level, and add fuel if needed.

2. Battery

(1) Check the battery fluid level. Top off with distilled or approved drinking water, as needed.

(2) Connect the battery cables, if disconnected.

3. Brake system

(1) Check master cylinder fluid level.

(2) If low, check for leaks in brake lines.

4. Engine and transmission

(1) Check engine oil level, add oil if needed.

(2) Check radiator coolant/water mixture level. If coolant is needed, make sure radiator drain valve is closed, then add required coolant/water mixture (deutz diesel units excluded).

(3) Check transmission fluid level.

Note: This check is to determine if there is fluid in the transmission. If any fluid shows on the dipstick, do not add transmission fluid until it is checked with the engine running, after the engine has reached normal operating temperature.

(4) Check transmission cooling lines for proper connections.

5. Rear (drive) axle

Check oil level in the rear axle and add oil, if needed.

6. Tire inflation

Check air pressure in tires (45 psi) and add air, if needed.

7. Units equipped with cummins B3. 3 engine

In cold weather, ensure that the engine preheat toggle switch, located on the engine side of the firewall, is in the ON position. The switch may be turned off in warm weather climates.

7.6.2　Periodic Maintenance Schedule

The tractor should be maintained in accordance with the following schedules:

1. Daily

(1) Check that all controls, including belt controls, are in good working order. Report a faulty unit and do not use until repaired.

(2) Check fuel level.

(3) Check engine oil level.

(4) Check hydraulic oil level. If low, check for leaks.

(5) Check radiator coolant/water level (deutz diesel units excluded).

Caution: The recommended tire pressure listed is for the tires that are standard on the 660 unit. Other optional tires should be inflated according to the tire manufacturer specifications.

(6) Check that the pressure of the standard pneumatic tires is 45 psi.

Warning: Do not change a pneumatic tire for a cushion tire or vice versa without first consulting tug.

(7) Check tire treads for damage. Remove any stones, etc.

(8) Check voltmeter (gasoline units) or ammeter (diesel units) for charging (Ford ESG – 642 4. 2L units excluded).

(9) Check governor operation (Ford 300 units only).

(10) Check operation of the headlights and if aimed correctly.

(11) Check operation of backup, tail, stop and turn signal lights.

(12) Check operation of the cab marker lights, the heater, the wiper and the cab light, if equipped.

(13) Sound horn.

(14) Durst Transmissions: Visually inspect transmission to determine if the oil cooler fins (if liquid to air) are clear of dirt, and to ensure that hoses and fittings are not damaged or leaking.

(15) Durst Transmissions: Check transmission oil for signs of water or other contaminants. Check (smell) oil for signs of burnt oil (overheating). The fluid must be changed, the transmission flushed and the torque converter changed if the oil is contaminated with water (fluid is pink), has gone rancid with a foreign substance (smell), has clutch debris (the fluid has a dark color) or burned (fluid smells burned). If these conditions exist, it may be necessary to remove and dismantle the transmission to flush and replace the torque converter.

2. Weekly or at 50 hours

(1) Check conveyor belt for proper tension and alignment.

(2) Check belt drive chain adjustment.

(3) Lubricate belt drive chain.

(4) Check rear finger guard adjustment.

(5) Check that the wheel lug nuts are tightened to 125 ft-lb (135. 6 N · m).

Warning: Lug nuts must be retightened to 125 ft-lb (135. 6 N · m) after any time the lug nuts have been loosened for any reason, and at the intervals specified in this periodic maintenance section.

(6) Check the level of the brake fluid in the master cylinder. If low, top off as necessary.

(7) Check battery charge and fluid level.

(8) Check drive axle mounting bolts and tighten if required.

(9) Check transmission fluid level (with engine running and at normal operating

temperature).

(10) Check cylinder head bolt torque and all nuts and bolts for tightness. If required, torque cylinder head bolts. See engine manufacturer specifications for torque ratings.

(11) Check drive axle oil level.

(12) Check underneath for oil leaks and damage.

(13) Check ball joints for looseness.

(14) Check exhaust system for leaks and holes.

(15) Check crankcase ventilation (PCV) system (gasoline only).

(16) Check/drain water separator (diesel only).

(17) Check neutral safety switch adjustment.

(18) Change engine oil and filter.

Note: Only the intial change is required at the 50-hour maintenance level, change monthly or every 200 h thereafter.

3. Bi-weekly or 100 h

(1) Durst Transmission: Inspect control lever operation for looseness or binding.

(2) Durst Transmission: Inspect cooler for signs of leakage, damage or loose mounting bolts.

(3) Durst Transmission: Inspect for damage or signs of leakage around bell housing and bolts.

4. Monthly or 200 h

(1) Check the service brake pedal for proper operation. Adjust if required.

(2) Check the parking brake. With the parking brake handle in "lock" position, the unit should not move on a 8% or less grade. If the tractor does not move, no adjustment is necessary, but if the tractor rolls, then adjustment is advised. Follow the recommended adjustment instructions in this section.

(3) Check the fan and alternator belts; tighten if necessary.

(4) Check air cleaner temperature control (Ford 300 units only).

(5) Check EGR system (gasoline only).

(6) Check hoses and clean radiator exterior.

(7) Check brake pads and shoes; replace if pads are less than 1/8 in.

(8) Change engine oil and filter.

Note: The intial oil change is required at the 50-hour maintenance level, change monthly or every 200 h thereafter.

(9) Check air filter. Clean or replace.

(10) Lubricate service brake pedal linkage pivots.

(11) Lubricate parking brake lever pivots.

(12) Lubricate throttle linkage.

(13) Lubricate conveyor side and handrail pivots.

(14) Lubricate motor access cover hinges.

(15) Lubricate belt reduction gear box using appropriate high quality gear oil.

(16) Lubricate seat slide.

(17) Lubricate drive shaft joints.

(18) Lubricate steering axle links and king pins.

(19) Lubricate exhaust control valve and free-up. Use penetrating oil and rust inhibitor (Ford 300 units only).

(20) Lubricate all grease fittings. (see paragraph 7. 6. 3 for list of zerk fitting locations.)

5. Every two months or 400 h

(1) Change hydraulic oil filter.

(2) Check cooling system for dirt or rust in coolant.

(3) Check ignition timing, adjust if necessary (Ford 300 units only).

(4) Check injector pump timing, adjust if necessary (diesel only).

(5) Check idle rpm and mixture, adjust if necessary.

(6) Check thermostat in cooling system.

(7) Check spark plugs; clean, adjust and test (gasoline only).

(8) Inspect front wheel bearings; clean and repack.

(9) Check tie rod ends looseness.

(10) Clean distributor cap (Ford 300 units only).

(11) Adjust governor (Ford 300 units only).

(12) Adjust carburetor (Ford 300 units only).

6. Every five months or 1,000 h

(1) Durst Transmission: Change transmission fluid and clean filter and screen within pan.

(2) Durst Transmission: Check transmission mounting bolts for tightness. Refer to the Durst Transmission Service Manual in Chapter 5 for torque specifications.

7. Every six months or 1,200 h

(1) Check cylinder compression.

(2) Check spark plug wire resistance (gasoline only).

(3) Check intake manifold bolts and tighten, if necessary.

(4) Change fuel filter.

(5) Change spark plugs (gasoline only).

(6) Change PCV valve (gasoline only).

8. Yearly or 2,400 h

(1) Adjust automatic transmission bands.

(2) Change transmission fluid and filter.

(3) Change drive axle oil, use appropriate high quality gear oil.

(4) Drain, flush, refill and bleed the hydraulic brake system.

7.6.3 Lubrication

(1) Periodic lubrication should be performed according to the Periodic Maintenance Schedule in paragraph 7.6.2 of this section (see also Table 7.3).

(2) Use grease at pressure zerks in the following locations as equipped:

1) Ball joints, upper and lower.

2) Tie rod ends.

3) Hydraulic lift cylinders.

4) Steering cylinder tie rod ends.

5) Conveyor roller bearings.

(3) See Figure 7.5, 7.6 and Table 7.3 for more lubrication information.

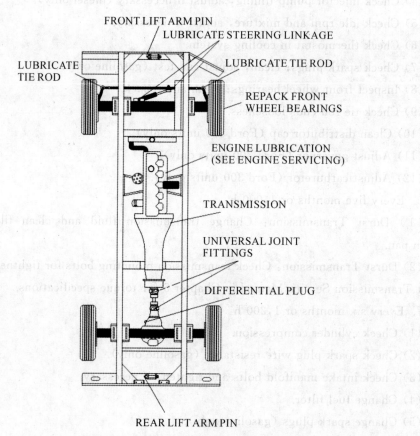

Figure 7.5 Chassis lubrication diagram

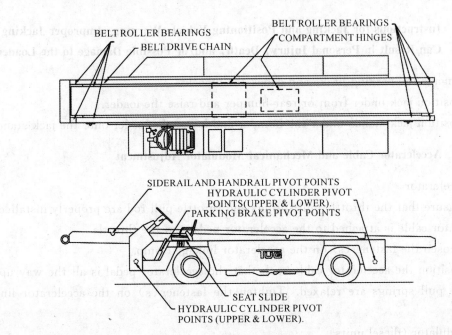

Figure 7.6　Body lubrication diagrams

Table 7.3　Lubrication instructions
Hydraulic oil（check level daily）

Ambient Temperature	Hydraulic Fluid
− 60°F to ＋ 50°F	Mobil DTE Ⅱ
− 15°F to ＋ 140°F	Mobil DTE 13

Engine oil（check level daily）

Ambient Temperature	Oil（Must Meet API Service SE/CC）	
− 10°F to ＋ 90°F and above	Gasoline engine：10W − 40	Diesel engine：15W − 40

Chassis components

Component	Interval	Lubricant
Transmission	Check level weekly	Dexron Ⅱ ATF
Brake cylinder	Check level weekly	DOT 3 heavy duty brake fluid
Conveyor drive chain	Lube weekly	Same as engine oil
Rear axle	Check at 6 months	85W − 90 limited slip
Zerk fittings	Lube at 6 months	Chassis grease
Wheel bearings	Repack at 24 months	Wheel bearing grease

7.6.4 Instructions for Jacking and Positioning Vehicle Warning: Improper Jacking Can Result in Personal Injury, Death, and/or Possible Damage to the Loader

(1) Engage the parking brake and chock the wheels.

(2) Position jack under front or rear bumper and raise the loader.

(3) Position jack stands under the bumper and lower the loader onto the jack stands.

7.6.5 Accelerator Cable and Mechanical Modulator Adjustment

1. Accelerator

(1) Ensure that the throttle pull springs and throttle pull rod are properly installed and the accelerator cable is attached to the accelerator pedal and the throttle.

(2) Loosen the fastener(s) on the accelerator linkage clamp.

(3) Position the accelerator cable such that the accelerator pedal is all the way up and the throttle pull springs are relaxed. Tighten the fastener(s) on the accelerator linkage clamp.

2. Modulator (diesel units)

(1) With the accelerator cable properly adjusted, fully depress the accelerator pedal.

(2) Measure the cable travel. Travel should be 1.5 in.

(3) Adjust as necessary by loosening the fasteners on the cable clamp, positioning the cable and tightening the fasteners.

7.6.6 Park Brake Adjustment

1. Discussion

When properly adjusted, the parking brake will prevent the loader from rolling when parked on a 8% grade. Since the parking brake is on the drive shaft and brakes through the reduction in the differential, engaging a properly adjusted parking brake should not require an excessive amount of force; however, some slight resistance will be present when engaging.

2. Procedure

(1) If parking brake hand lever is equipped with a locking set screw, loosen the set screw.

(2) To tighten parking brake, turn knurled knob at the end of the hand lever clockwise.

(3) To loosen parking brake, turn knurled knob at the end of the hand lever counterclockwise.

(4) If parking brake hand lever is equipped with a locking set screw, tighten the set screw.

7.6.7 Transmission Linkage Adjustment

Note: The transmission linkage must be properly adjusted to prevent damage and maintain safe operation.

(1) Place shifter in neutral position.

(2) Remove swivel from transmission shift lever and back off lockout from swivel.

(3) Place transmission in neutral.

1) Rotate shift lever fully counter clockwise.

2) Rotate shift lever clockwise to second position from full counter clockwise for 2-speed transmission (gasoline engine) or third position from full counter clockwise for 3-speed transmission (diesel engine).

(4) Tighten or loosen swivel one turn at a time until it is perfectly aligned with the hole in the shift lever, then reinstall swivel in shift lever.

(5) Tighten lockout until it is firmly engaged with swivel.

7.6.8   Hoof Governor Adjustment (Ford 300 Gasoline Engine)

1. Adjusting screw adjustment

(1) Remove welch plug covering the adjusting screw by drilling the plug with a 1/16 in drill. Insert a 1/16 in rod in drilled hole and pry off welch plug.

(2) Remove brass lock washer.

(3) Use an engine tachometer to read engine speed at full throttle. Adjust speed to 2,400 to 2,600 r/min.

Note: Use a screwdriver to turn the adjusting screw as needed on the housing to increase or decrease engine speed.

(4) Seal adjusting screw by replacing the brass lock washer in the screw slot and reinstall the welch plug. Tap lightly using a flat punch.

2. Secondary screw adjustment

The secondary adjusting screw in this governor has been factory set to cover a wide range of engine speeds. In setting governor to desired road or engine speed, use the main adjusting screw only. If governor control is too sharp or not sharp enough, adjust using instructions below.

Note: Only in rare instances will the secondary adjustment need to be changed.

(1) Remove welch plug covering the secondary adjusting screw by drilling the plug with a 1/16 in drill. Insert a 1/16 in rod in drilled hole and pry off welch plug.

(2) If governor control is too sharp, causing surging or hunting, turn the secondary adjusting screw clockwise 1/4 turn at a time.

Note: It will be necessary to turn the main adjusting screw counterclockwise approximately one turn to compensate for every 1/4 turn of clockwise secondary screw adjustment.

(3) If governor control is not sharp enough, causing too great a variation in speed

between load and no-load conditions, turn the secondary adjusting screw counterclockwise 1/4 turn at a time.

Note: It will be necessary to turn the main adjusting screw clockwise approximately one turn to compensate for every 1/4 turn of counterclockwise secondary screw adjustment.

(4) Seal secondary adjusting screw by replacing the brass lock washer in the screw slot and reinstall the welch plug. Tap lightly using a flat punch.

Caution: Too sharp a blow to the welch plug may upset the secondary adjustment screw adjustment.

New Words

1. fastener ['fɑːsnə(r)] *n.* 紧固件
2. bearing ['beəriŋ] *n.* 轴承，支座

Phrases and Expressions

1. pneumatic tire 充气轮胎
2. cushion tire 软心轮胎，缓冲轮胎
3. governor operation 调速器运行
4. ft-lb 英尺·磅
5. Bi-weekly 每两周
6. EGR system 排气再循环系统
7. drive shaft joints 传动轴接头
8. king pin 转向销
9. rust inhibitor 防锈剂
10. zerk fitting 油枪加油嘴
11. intake manifold 进气歧管
12. tie rod end 转向横拉杆球铰接头
13. accelerator cable 油门线
14. governor control 调速器控制
15. welch plug 拱形张紧堵片
16. lock washer 防松垫圈
17. grease fittings 油脂注入器，黄油嘴

Notes to the Text

1. Check the level of the brake fluid in the master cylinder. If low, top off as necessary. 检查主缸制动液的液位。如果很低，按要求加满。

2. In cold weather, ensure that the engine preheat toggle switch, located on the engine side of the firewall, is in the ON position. The switch may be turned off in warm weather

climates.

天气寒冷时,确保位于防火墙发动机一侧的发动机预热开关在"ON"位置上。这个开关可能在温暖的天气下被关闭。

3. Warning：Do not change a pneumatic tire for a cushion tire or vice versa without first consulting tug.

警告：不要在没有事先咨询 TUG 公司的情况下把充气轮胎换成软心轮胎,反之亦然。

4. Ensure that the throttle pull springs and throttle pull rod are properly installed and the accelerator cable is attached to the accelerator pedal and the throttle.

确保油门拉杆弹簧和油门拉杆正确安装,油门线连接油门踏板和油门。

5. Remove welch plug covering the adjusting screw by drilling the plug with a 1/16 in drill. Insert a 1/16 in rod in drilled hole and pry off welch plug.

移除覆盖在调整螺钉上的盘塞时,用 1/16 英寸的钻头在盘塞上钻孔。在钻孔中插入一根 1/16 英寸的杆来撬开盘塞。

Section 7.7　Troubleshooting

7.7.1　Engine/Transmission (See Table 7.4)

Table 7.4　Troubleshooting for engin and transmission

Problem	Cause	Correction
1. Engine will not crank	(a) Looseor corroded battery cables; (b) Undercharged battery; (c) Defective starter relay; (d) Loose or broken cable to starter; (e) Loose or open wiring through neutral switch; (f) Defective starter motor	(a) Clean and tighten cable connections; (b) Charge or replace battery; (c) Replace starter relay; (d) Tighten or replace cables; (e) Repair, adjust or replace; (f) Repair or replace
2. Engine will not crank-starter spins	(a) Defective starter motor; (b) Defective flywheel ring gear	(a) Remove starter, inspect for broken or worn drive; (b) Inspect ring gear teeth. Replace fly wheel and ring gear if necessary
3. No start in neutral	(a) Defective neutral safety switch	(a) Adjust or replace
4. Transmission cannot be shifted into or out of forward, neutral, or reverse gear range	(a) Loose or disconnected cable end; (b) Improper cable or lever adjustment; (c) Snow or ice obstructing movement	(a) Tighten cable ends; (b) Adjust cable or lever; (c) Remove obstacle

Continued

Problem	Cause	Correction
5. Excess binding transmission selector or hand throttle	(a) Lack of lubrication; (b) Foreign object or excessive dirt or corrosion in lever housing; (c) Kinked or mashed cable housing; (d) Carburetor or transmission linkage binding	(a) Clean and lubricate; (b) Clean and lubricate; (c) Repair or replacel; (d) Inspect and repair
6. Engine speed too high with hand throttle in idle position	(a) Hand throttle linkage out of adjustment; (b) Choke or idle speed is out of adjustment	(a) Adjust or repair linkage; (b) Inspect and adjust
7. Engine speed too low at maximum speed position of hand throttle	(a) Linkage or cable loose or disconnected	(a) Adjust or repair linkage

7.7.2 Brakes (See Table 7.5)

Table 7.5 Troubleshooting for brakes

Problem	Cause	Correction
1. Excessive pedal travel or pedal goes to floor consistently	(a) Low fluid level; (b) Hydraulic system; (c) Drum brakes out of adjustment, worn; (d) Loose or improper attachment of pedal, pedal support booster and master cylinder; (e) Misaligned anchor plate	(a) Add fluid, bleed system, and check for leaks; (b) Refer to master cylinder; (c) Repair or replace as required; (d) Repair or replace as required; (e) Refer misaligned disc brake anchor plate diagnosis
2. Brake pedal feels spongy when fully applied	(a) Low fluid level; (b) Brakes out of adjustment; (c) Front wheel bearing out of adjustment; (d) Disc brake caliper attachment loose; (e) Worn or damaged self adjusters	(a) Add fluid, bleed system, check for leaks; (b) Adjust brakes; (c) Adjust front wheel bearings; (d) Replace or tighten as required; (e) Remove drum and check lining for proper adjustment

Continued

Problem	Cause	Correction
3. Noise at wheels when brakes are applied — squeaks or chatter	(a) Worn or scored brake drums and lining or rotors and pads; (b) On disc brakes — missing or damaged brake pad insulators; (c) Burrs or rust on caliper that would obstruct seating of shoe to caliper; (d) Dirty, greased or glazed linings	(a) Inspect, repair or replace as required; (b) Replace disc brake pads; (c) Clean or deburr caliper; (d) Clean or replace
4. Brakes pull to one side	(a) Unequal air pressure in tires; (b) Grease or fluid on lining; (c) Loose or missing disc brake caliper attaching bolts; (d) Restricted brake lines or hoses	(a) Inflate tires to correct pressure; (b) Clean, sand and/ or replace linings; (c) Replace missing bolts tighten to proper torque; (d) Repair or replace as required
5. Brake warning light on	(a) Low fluid level; (b) Worn or damaged brake warning switch; (c) Worn or damaged master cylinder	(a) Add fluid, bleed system and check for leaks; (b) Replace switch; (c) Perform master cylinder diagnosis test. Repair or replace as required

7.7.3   Park Brake (See Table 7.6)

Table 7.6   Troubleshooting for park brake

Problem	Cause	Correction
1. Parking brake will not hold	(a) Parking brake cable out of adjustment; (b) Rear brakes out of adjustment; (c) Parking brake linkage, release lever, clevis and ratchet binding	(a) Adjust parking brake cable; (b) Adjust rear brakes; (c) Repair or replace linkage as required
2. Parking brake will not release or fully return	(a) Manual release brake control components binding or damaged; (b) Parking brake linkage and cable binding; (c) Worn or damaged rear brake components	(a) Repair or replace manual parking brake control; (b) Repair or replace as required; (c) Check rear brake shoe retracting springs and parking brake levers. Repair as required

Continued

Problem	Cause	Correction
3. Parking brake lever hard to pull up	(a) Brake handle out of adjustment; (b) Frozen or damaged brake cable	(a) Adjust or repair; (b) Light taps with a mallet may free cable and brakes. Otherwise, apply warm air from a ground equipment heater

7.7.4　Steering (See Table 7.7)

Table 7.7　Troubleshooting for steering

Problem	Cause	Correction
1. Steering is difficult	(a) Lack of lubrication on axle; (b) Low fluid in hydraulic reservoir; (c) Steering column bent or bearings are damaged; (d) Low hydraulic pressure; (e) Damaged Char-Lynn valve	(a) Lubricate; (b) Fill as needed; (c) Check steer column. Replace if necessary; (d) Check pressure. Gasoline units: Replace power steering pump. Diesel units: Check pressure regulator. If no problems found, replace power steering pump; (e) Replace
2. Front wheel appears to wobble	(a) Loose/worn studs or kingpins; (b) Worn or damaged bearings and/or hubs	(a) Tighten or replace; (b) Repair or replace

7.7.5　Conveyor Hydraulics (See Table 7.8)

Table 7.8　Troubleshooting for conveyor hydraulics

Problem	Cause	Correction
1. The front and rear lift arms will not raise conveyor	(a) Low oil level due to leak in system; (b) Hydraulic pump drive belt slipping or broken; (c) Defective hydraulic pump; (d) Relief valve set too low; (e) Lift control valve defective	(a) Service hydraulic tank, locate and repair leak. Recheck system; (b) Replace and/or adjust; (c) Repair or replace; (d) Adjust or repair; (e) Replace or repair

Continued

Problem	Cause	Correction
2. Conveyor rises too slowly	(a) Pump belt slipping; (b) Internal valve leak; (c) Internal cylinder leak; (d) Low oil level in tank; (e) Worn hydraulic pump; (f) System relief valve too low; (g) Relief valve leaking internally	(a) Adjust; (b) Repair or replace; (c) Repair or replace; (d) Add oil; (e) Repair or replace; (f) Adjust; (g) Repair or replace
3. Conveyor creeps down with control valve in the "OFF" position	(a) Internal leak in cylinder or valve	(a) Repair or replace

7.7.6 Conveyor Belt (See Table 7.9)

Table 7.9 Troubleshooting for conveyor belt

Problem	Cause	Correction
1. Conveyor belt will not run in either direction when controls are actuated	(a) Defective control valve; (b) Belt drive chain broken or too loose; (c) Belt drive shaft key missing; (d) Defective hydraulic motor; (e) Optional shifter-brake interlock out of adjustment with parking brake; (f) Defective control box, circuit or solenoid control valve; (g) Defective neutral start switch or circuit breaker. Check by stopping and restarting engine; (h) Open wiring or stop button stuck in open circuit	(a) Repair or replace; (b) Adjust or replace; (c) Replace; (d) Repair or replace; (e) Adjust or replace; (f) Repair or replace; (g) Adjust or replace; (h) Inspect circuit and repair
2. Forward belt speed too low	(a) Low oil level in tank; (b) Pump drive belt slipping; (c) Loose conveyor belt; (d) Too much load on conveyor and/or conveyor angle too steep; (e) Binding in belt drive or rollers	(a) Add oil; (b) Adjust; (c) Adjust rear take-up; (d) Basic capacity is 2,000 lb. At conveyor angles exceeding 15°, the capacity must be reduced below 2,000 lb; (e) Adjust, lubricate, repair or replace

Continued

Problem	Cause	Correction
3. Reverse belt speed too fast	(a) Belt drive motor holding valve out of adjustment	(a) Repair or replace
4. Belt will only run forward	(a) Belt drive motor holding valve out of adjustment; (b) Relief valve set too low or defective; (c) Open electrical circuit or defective solenoid	(a) Adjust or replace; (b) Adjust, repair or replace; (d) Repair or replace
5. Drive chain is noisy	(a) Chain is out of adjustment	(a) Adjust and lubricate
6. Belt does not track properly	(a) Rollers are out of adjustment	(a) Adjust
7. Belt will not continue to run after push button is released	(a) Defective R1 or R2 belt control relay or open wiring	(a) Repair or replace
8. Belt will not stop when "STOP" button is pushed	(a) Short to ground by-passing "STOP" button; (b) Short circuit across ineffective "STOP" button	(a) Repair; (b) Repair or replace
9. Belt stops running although engine continues to run and "STOP" button has not been pressed	(a) Defective control relay, solenoid or 15 ampere circuit breaker	(a) Replace

7.7.7 Electrical (See Table 7.10)

Table 7.10 Troubleshooting for electrical

Problem	Cause	Correction
1. Battery does not stay charged — engine starts O. K.	(a) Worn or damaged battery; (b) Loose or worn alternator belt; (c) Worn or damaged wiring or cables; (d) Defective alternator or regulator; (e) Other vehicle electrical system malfunction	(a) Test battery, replace if necessary; (b) Adjust or replace belt; (c) Clean, repair as required; (d) Perform general charging systems test. Repair or replace as required; (e) Check other systems for current draw. Repair as required
2. Alternator noisy	(a) Loose or worn alternator belt; (b) Bent pulley flanges; (c) Defective alternator	(a) Adjust or replace belt; (b) Replace pulley; (c) Perform alternator tests, replace as required

Continued

Problem	Cause	Correction
3. Charge indicator gauge shows steady charge	(a) Worn or damaged battery; (b) Poor regulator ground; (c) Loose wiring connection; (d) Defective alternator or regulator	(a) Check battery, replace if necessary; (b) Ensure good ground; (c) Tighten all wiring connections; (d) Perform general charging systems test. Repair or replace as required
4. Charge indicator gauge shows discharge	(a) Loose or worn alternator belt; (b) Worn or damaged wiring; (c) Defective alternator or regulator; (d) Defective charge indicator gauge wiring and connections; (e) Worn or damaged gauge; (f) Other vehicle electrical system malfunction	(a) Adjust or replace belt; (b) Check battery-to-alternator wiring for ground or open. Repair if necessary; (c) Perform general charging system test. Repair or replace as required; (d) Repair as required; (e) Replace gauge; (f) Check and repair as required

Phrases and Expressions

1. worn drive 蜗杆传动装置

Notes to the Text

1. Engine will not crank — starter spins.

发动机不转——起动机转动。

2. No start in Neutral.

空挡不能启动。

3. Transmission cannot be shifted into or out of forward, neutral, or reverse gear range.

变速器不能挂入或退出前进挡、空挡或倒挡。

4. Excess binding transmission selector or hand throttle.

变速器选择器或手油门太紧。

5. Engine speed too high with hand throttle in idle position.

手油门在怠速位置时发动机转速过高。

6. Engine speed too low at maximum speed position of hand throttle.

手油门在最高速位置时发动机转速过低。

7. Excessive pedal travel or pedal goes to floor consistently.

踏板行程过大或踏板总是接触地板。

8. Brake pedal feels spongy when fully applied.

制动踏板在踩到底时感觉松软。

9. Noise at wheels when brakes are applied — squeaks or chatter.

刹车时车轮发出噪声——尖叫声或震动。

10. Brakes pull to one side.

刹车时跑偏。

11. Steering is difficult.

转向困难。

12. Front wheel appears to wobble.

前轮出现摆动。

Chapter 8   Ground Power Unit

Section 8.1   General Information Ⅰ

8.1.1   General Description

The GPU comprises an air cooled diesel engine driving a brushless, rotating field, salient pole, 400 Hz alternator, the engine and alternator being mounted on a steel monocoque chassis which is designed to be mounted on a truck chassis. Details of the truck (Leyland 45 – 130) will be found in the truck manufacturer's manual.

Weather protection is provided by four panels hinged to a steel, channel-section spine member mounted above the engine and alternator. The spine member is supported by three angle-section steel arches, one at either end of the chassis and one in the middle.

Sheet steel box compartments located at the side of the chassis provide stowage for the output cables.

All the controls and most instrumentation are located at the conveniently positioned control console, which is mounted on the chassis but isolated from it by anti-vibration mounts. The engine and power output control sections comprise pull-out modules incorporating solid state devices in detachable printed-circuit board form. The contactors, current transformers and engine relays are mounted on panels located on a frame which is positioned astride the alternator.

The location of the major components is shown in Figure 8.1.

8.1.2   Chassis

The whole assembly is mounted onto a rugged sheet steel chassis fabricated from 3 mm thick sheet steel to give a monocoque structure forming a bedplate on which the engine, alternator and other associated components are mounted. The engine and alternator are coupled together by the use of flexible membranes.

The control panel, canopy and batteries are also supported by the chassis structure and on either side stowages are provided for the output cables with arches bolted to each end to support the central spine and canopy.

8.1.3 Canopy

Four weatherproof panels are hinged to a steel channel-section spine member which is supported by the end angle-section arches. The end arches are bolted in position on the chassis together with the centre arch and spine member thus simplifying removal of the engine and alternator. The weather panels are secured in the closed position by rubber toggles and supported in the open position by gas spring struts. The panel which covers the control console has an aperture on its vertical face through which the controls may be operated while the panel is closed. The indicators can be monitored through a transparent window in the upper vertical face of the panel. The ends of the GPU are enclosed by panels screwed to the end arches, the rear panel being equipped with air intake apertures while the front panel has apertures which accommodate the fuel tank gauge.

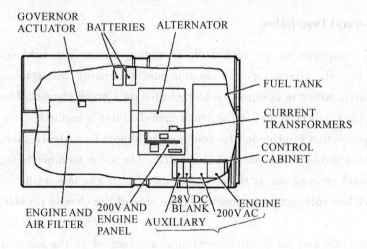

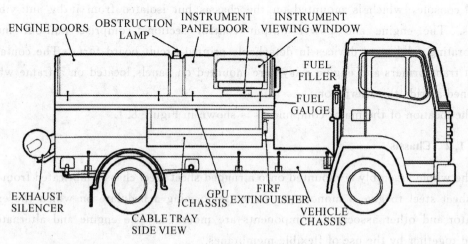

Figure 8.1   Overall view of ground power unit

The standard form of construction uses double skinned glass reinforced plastic (grp) material for the hinged panels, and steel for the end panels.

An amber hazard warning light is mounted on the spine member, the amber lamp being arranged to flash when the lights are on.

8.1.4   Engine

The alternator is driven by a Deutz air-cooled, four stroke, direct injection diesel engine with in-line configuration. The turbocharger is powered by the engine exhaust.

The fuel system includes a fuel lift pump, incorporating a hand priming lever and a gauze pre-filter, feeding fuel to the injector pump, surplus fuel being returned to the fuel tank. The full-flow fuel filter has a disposable cartridge. The lub-oil filter also uses disposable cartridges.

Air for combustion passes through the air inlet filter, having an element which can be cleaned a number of times before replacement. A Service Indicator fitted to the filter shows when cleaning is required. Cooling air is blown through a cowling over the cylinder heads and exhaust manifold by an axial fan, driven by a twin vee-belt. The belt tensioner incorporates a belt-failure warning switch.

A small alternator, fitted to the engine and driven by a vee-belt, incorporates a built-in rectifier and voltage regulator, and provides charging current for the 24 V battery which powers the engine electrical services, contactor circuits and lighting.

Fitted to the engine, when used in a GPU, are magnetic sensors for the over-speed trip, oil pressure transmitter and low oil pressure switch.

For further information on the engine, refer to the manufacturer's Engine Operation Manual, supplied with the GPU, which covers all regular engine servicing procedures. An Engine Workshop Manual is available from the engine manufacturers to users who have the facilities for more extensive engine maintenance.

The engine and alternator are bolted together to form an assembly which if required can be removed from the GPU as a unit. The assembly is mounted in the chassis on four anti-vibration mountings.

8.1.5   Control Systems

The 200 V A C output from the alternator is taken to an output contactor and the voltage is controlled by a fully transistorized automatic voltage regulator to within $+/-1.0\%$. The output contactor is controlled by pushbuttons on the panel fascia.

When a fault occurs on the 200 V output, the contactor will trip and the relevant lamp light. To reset the fault circuit depressing the RESET pushbutton. The circuitry is arranged such that following a fault trip is not possible to close the contactor unless the RESET pushbutton has been depressed. The alternator is fully protected by an overload unit with IDMT characteristics. The 200 V output is sensed at the aircraft connector and therefore gives automatic volt drop compensation.

8.1.6 Alternator

8.1.6.1 Construction

The ADE-HML alternator is manufactured in fabricated construction form, and generates a 200 V, 3-phase and neutral, 400 Hz supply. An exciter, mounted within the same frame and running on the rotor shaft, has its static field energized by a solid state automatic voltage regulator (AVR). The alternating current generated in the rotor of the exciter is rectified to DC by the three-phase diode assembly mounted on the rotor, and the DC output from this assembly is fed directly to the main field windings on the rotor, thus producing the required AC output from the three-phase windings of the stator. The arrangement eliminates the need for slip-rings and brush gear.

8.1.6.2 Stator Assembly

The stator core, which carries the main output winding, is arranged to form axial ventilating ducts to promote cooling of the core and windings. The laminated stator core is built up of low loss electrical steel, and the windings are protected by high-grade insulating material. The stator is impregnated with tropical grade insulating varnish to protect the surface winding from contaminants. The stator frame is provided with welded feet into which are let fixing holes for the alternator "bedding".

Flexible connections from the stator pass through the stator frame into a side mounted terminal box.

8.1.6.3 Rotor Assembly

The rotor assembly is carried on a high-grade steel shaft, to which the main field laminations are attached. A centrifugal steel fan is attached to the shaft at the drive end of the main core; the exciter rotor assembly and rotating rectifier are mounted at the non-drive end. The rotor and fan assembly are dynamically balanced.

The salient pole field windings comprise four coils which are wound directly on to the insulated pole body, fitted with winding end-supports, and then impregnated to consolidate the winding sections. Damper bars, carried on slots let into the pole face, are connected to a damper plate to each end of the rotor core to form the electrical connections between poles. After assembly, impregnation, and fitting of the ancillary parts, the complete rotor assembly is dynamically balanced to a high commercial standard.

8.1.6.4 Endplates

The non-drive end plate comprises a high grade cast aluminium assembly bolted to the main stator frame and housing the spherical bearing collar and the exciter stator assembly. Cooling air enters the machine at the non-drive end and is drawn up through the stator and rotor airways by the centrifugal fan fitted to the rotor.

At the drive end, the stator frame terminates in a flange enabling twelve bolts to secure the alternator frame to the Deutz diesel engine flywheel housing. Eight bolts and spring washers secure the rotor drive plate to the engine flywheel.

8.1.6.5　Exciter

The exciter has a six-pole rotating-armature mounted at the non-drive end of the alternator shaft. The three-phase output from the rotor winding is applied to the rotating rectifier mounted on the rotor shaft. The exciter stator core is fully laminated to provide quick response. Its poles are wound in a coil arrangement and are DC energized from the exciting supply (derived from the alternator output but supported for initial excitation by the 24 V battery) via the automatic voltage regulator. Cooling is provided by the main alternator air flow.

8.1.6.6　Rotating Rectifier

The rotating rectifier converts the three-phase output of the exciter rotor to DC necessary for energizing the main alternator field winding. It comprises a three-phase bridge arrangement using silicon diodes having a high reverse voltage capability and a liberal current rating to handle transient currents that can flow in the field winding during load changes.

8.1.6.7　Single Bearing Drive

For the ADE-HML single-bearing arrangement, the drive end of the alternator shaft is secured to the diesel engine flywheel with an accurately aligned rotor drive plate, and secured in position by eight bolts. Positive location between the alternator frame and the engine flywheel housing is achieved with twelve flange mounted bolts. The arrangement provides an accurate location for the axis of rotation of the alternator rotor relative to the centre-line of the crankshaft.

In the single-bearing coupling arrangement described above, small variations in manufacturing tolerances may occur, tending to displace the rotor from its true axis of rotation and creating excessive loading of the bearing at the non-drive end. This possibility is avoided by using a self-aligning spherical bearing, which also has the further advantage of high load bearing capability giving long running life.

8.1.6.8　Excitation System

The excitation system comprises a source of DC power for energizing the exciter field, a means of continuously sensing the alternator output voltage, and an automatic voltage regulator (AVR). The AVR compares the alternator output voltage sensing signal with a constant reference signal and uses the difference (error) to adjust the exciter field current so as to restore the alternator output voltage to normal limits. The excitation system includes circuitry whereby additional excitation, proportional to the alternator output current, can be applied to the exciter field so as to provide a quick response to load changes, thus improving voltage recovery time.

8.1.7　Control Cabinet

With the exception of the fuel gauge, all the controls and instruments are localised at the control console. The sheet steel cabinet has been designed on a modular basis so that control and instrumentation for each system are contained within their own removable module. Each module may be electrically disconnected-from its associated power by either one or two MS type multi-pin connectors with threaded locking rings.

The control cabinet is shown in Figure 8.2.

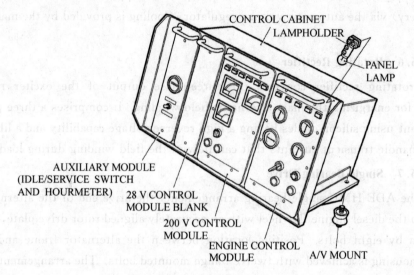

Figure 8.2　Control cabinet

The cabinet is mounted on support brackets by means of anti-vibration mountings. From right to left, the control module locations are:

(1) Engine control (2 connectors).

(2) 200 V AC control (2 connectors).

(3) Spare/blank.

(4) Auxiliary control module housing the Idle/Service switch and hours run counter.

8.1.8　Control Modules

Each control module is retained in the control cabinet by quick-release fasteners. The interior of each module contains the printed circuit boards of the control systems, the protection circuits and the switching devices.

The engine control module and 200 V AC control module are respectively identified in Figure 8.3, 8.4.

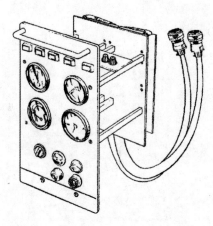
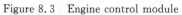

Figure 8.3　Engine control module

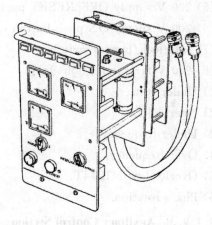

Figure 8.4　200 V AC control module

An extended lead is supplied as standard; this enables any module to be connected to the GPU after it has been withdrawn from the cabinet, so that adjustments can be made while the GPU is running.

8.1.9　Control Panel

A sheet steel, totally enclosed control panel contains engine 200 V controls and protection sections:

8.1.9.1　Engine Control Section

(1) Ammeter battery charge — moving coil type range 30 – 0.– 30.

(2) Oil pressure gauge — transducer operated.

(3) Engine temperature gauge — transducer operated.

(4) Fuel gauge — electrically operated.

(5) Engine start switch — OFF/RUN/START.

(6) Control panel illumination switch ON/OFF.

(7) Cold start push-button.

(8) Engine fault indications for:

1) Over speed.

2) Low oil pressure.

3) High engine temperature.

4) Coolant failure.

5) Battery charge alternator.

8.1.9.2　200 V AC Section

(1) Voltmeter — moving iron type.

(2) Voltmeter — select switch.

(3) Ammeter — moving iron type (0 – 400 A).

(4) Frequency meter.

(5) 200 V supply OFF/RESET push-button.

(6) 200 V supply ON push-button (illuminated).

(7) Interlock/bypass switch.

(8) 200 V fault indications for:

1) Under volts.

2) Over volts.

3) Under frequency.

4) Over frequency.

5) Overload with IDMT.

6) Phase rotation.

8.1.9.3 Auxiliary Control Section

(1) Hours run recorder.

(2) Idle/service select switch.

8.1.9.4 Control Panel Description

The above mentioned instrumentation and protection indications described above are mounted on the fascia of the control panel.

The control sections are held in the control panel with quick release fasteners and are connected via plug and socket connections. Each individual section has adjustable trip levels and an LED to indicated status.

For field service the entire section can be removed as the wiring is fed through a plug and socket connections. A replacement section can then be fitted and therefore the unit can be returned to service in a minimum amount of time. The section can then be taken for repair without the unit being out of service.

New Words

1. monocoque ['mɒnə(ʊ)kɒk] n. (汽车等的)无大梁结构

2. console [kən'səul] n. 控制台

3. contactor ['kɒntæktə] n. 接触器,触点

4. toggle ['tɒg(ə)l] n. 单向螺杆

5. strut [strʌt] n. 支柱

6. aperture ['æpətʃə] n. 孔,洞

7. amber ['æmbə] adj. 琥珀色的,黄色的

8. turbocharger ['tɜːbəutʃɑːdʒə] n. 涡轮增压器

9. gauze [gɔːz] n. (金属、塑料的)丝网

10. rectifier ['rektifaiə] n. 整流器(管)

11. transistorise [træn'zistəraiz] vt. 晶体管化

12. exciter [ik'saitə] n. 励磁机

13. core [kɔ:] *n.* 铁芯

14. insulating ['insəletiŋ] *adj.* 绝缘的

15. tropical ['trɒpik(ə)l] *adj.* 热带的

16. varnish ['vɑ:niʃ] *n.* 清漆，亮漆

17. centrifugal [sentri'fju:gl] *adj.* 离心（式）的

18. armature ['ɑ:mətʃə] *n.* 衔铁

Phrases and Expressions

1. salient pole 凸极

2. current transformer 电流变压器

3. gas spring 气弹簧

4. glass reinforced plastic 玻璃钢

5. fuel lift pump 加油泵，燃油泵

6. hand priming lever 手摇杆

7. injector pump 喷射泵

8. voltage regulator 稳压器，调压器

9. field winding 励磁绕组

10. cast aluminium 铸铝，生铝

11. spring washer 弹簧垫圈

12. silicon diode 硅二极管

Notes to the Text

1. The GPU comprises an air cooled diesel engine driving a brushless, rotating field, salient pole, 400 Hz alternator, the engine and alternator being mounted on a steel monocoque chassis which is designed to be mounted on a truck chassis.

地面电源机组由风冷的柴油发动机和柴油机驱动的无刷、旋转磁场、凸极、400 Hz 的交流发电机及安装发动机和发电机的桁架式底架组成，地面电源机组再安装在载重汽车底盘上。

2. A small alternator, fitted to the engine and driven by a vee-belt, incorporates a built-in rectifier and voltage regulator, and provides charging current for the 24 V battery which powers the engine electrical services, contactor circuits and lighting.

由 V 形带驱动，与发动机相配，内置整流器和稳压器的小交流发动机负责为发动机电气服务、接触器电路和照明提供电源的 24 V 的蓄电池充电。

3. An exciter, mounted within the same frame and running on the rotor shaft, has its static field energized by a solid state automatic voltage regulator (AVR). The alternating current generated in the rotor of the exciter is rectified to DC by the three-phase diode assembly mounted on the rotor, and the DC output from this assembly is fed directly to the main field windings on the rotor, thus producing the required AC output from the three-

phase windings of the stator. The arrangement eliminates the need for slip-rings and brush gear.

励磁绕组装在同一结构上并绕转子轴转动,固态自动电压调节器激活它的静电场。励磁绕组转子产生的交流电流由安装在转子上的三相二极管整流为直流电流,并直接输送给转子的主磁场绕组,从而从转子的三相绕组产生所需的交流电流。这种布置消除了集电环和电刷装置。

4. The rotating rectifier converts the three-phase output of the exciter rotor to DC necessary for energizing the main alternator field winding. It comprises a three-phase bridge arrangement using silicon diodes having a high reverse voltage capability and a liberal current rating to handle transient currents that can flow in the field winding during load changes.

旋转整流器将励磁绕组转子的三相输出转变为发电机主绕组所需的直流电。旋转整流器由三相桥式连接的硅二极管组成。硅二极管具有高逆向电压容量和宽范围额定电流使负载改变时,瞬时电流能在励磁绕组中流动。

5. The AVR compares the alternator output voltage sensing signal with a constant reference signal and uses the difference (error) to adjust the exciter field current so as to restore the alternator output voltage to normal limits.

自动电压调节器将发电机的输出信号与恒定的参考信号进行比较,并用两者的差值调节励磁电流,如此使发电机的输出电压恢复到额定值。

Section 8.2　General Information Ⅱ

8.2.1　Power Circuits

The circuits of the GPU are subdivided, in the description which follows, into two sections: Engine and 200 V AC. In each section, the power circuits (the major power-handling components and associated wiring), which are mounted on the 200 V and engine panel, are subdivided from the control circuits (including the protective trip circuits), most of which are housed in the removable control modules. The power and control circuit diagrams are placed on opposite pages and need to be read in conjunction with each other and with the functional descriptions which follow.

The 200 V and engine panel (see Figure 8.5(a)), which is carried on a support frame standing astride the alternator, houses the AVR, engine relays, 200 V output contactor, associated terminal blocks, and the fixed connectors which mate with the connectors of the control modules.

The current transformers for metering and overload sensing are carried on a smaller panel (see Figure 8.5(b)) mounted behind the 200 V and engine panel.

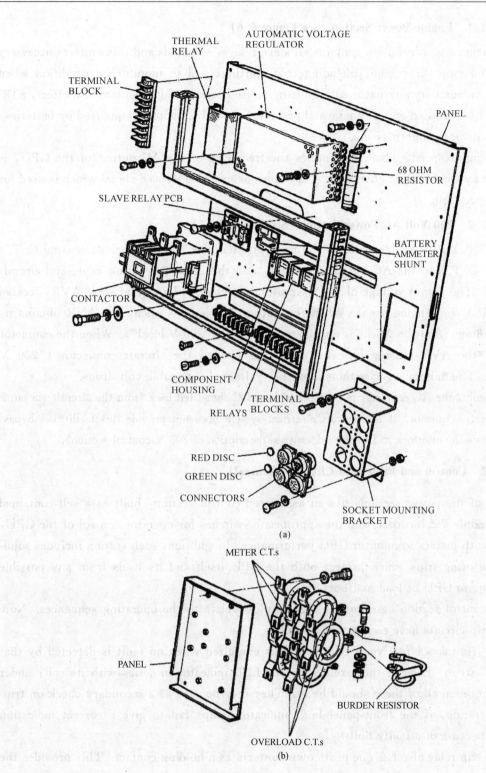

THERMAL
RELAY

AUTOMATIC VOLTAGE
REGULATOR

TERMINAL
BLOCK

PANEL

68 OHM
RESISTOR

SLAVE RELAY PCB

BATTERY
AMMETER
SHUNT

CONTACTOR

COMPONENT
HOUSING

TERMINAL
BLOCKS

RELAYS

RED DISC

GREEN DISC

CONNECTORS

SOCKET MOUNTING
BRACKET

(a)

METER C.T.s

PANEL

BURDEN RESISTOR

OVERLOAD C.T.s

(b)

Figure 8.5 The power circuits

(a) a 200 V and engine panel; (b) overload and meter C. T. panel

8.2.1.1 Engine Power Section (See Figure 8.6)

The engine power section contains all slave relays, solenoids and transmitters necessary to start and stop the engine, including cold starting, and to monitor its condition when running. An auxiliary alternator with built-in rectifiers and voltage regulator (in effect, a DC generator) is mounted on the engine and provides a 24 V DC supply, supported by batteries, for the contactors, lighting, etc.

The engine speed, which determines the frequency of the AC output of the GPU, is controlled by an electronic governor. A speed probe provides a speed signal which is used for the over speed trip.

8.2.1.2 200 Volt AC Power Section (See Figure 8.11)

The 200/115 V AC power from the main alternator is fed via overload sensing C.T.s and meter C.T.s to the AC output contactor, and thence to the output cable and aircraft connector. The output voltage of the alternator is controlled by a solid state AVR, located on the 200 V and engine panel. When the output contactor is open, the AVR obtains its sensing voltage from the AC lines out of the alternator ("200 V local"). When the contactor is closed, the AVR sensing is switched to the pins of the aircraft connector ("200 V remote"). This minimises variations in output voltage due to cable volt drops.

Normally, the AC contactor is interlocked by a DC signal fed back from the aircraft via pin F of the aircraft connector. If the aircraft electrical system does not provide this facility, a bypass switch allows the interlock to be over-ridden (see description of 200 V control section).

8.2.2 Control and Protection Circuits (General)

Each of the power sections has an associated control section, built as a self-contained control module and incorporating the appropriate switches for operator control of the GPU, together with meters to monitor GPU performance. In addition, each section includes solid-state monitoring trips which protect both the GPU itself and its loads from any possible damage due to GPU or load malfunction.

The control section descriptions which follow relate to the operating sequences. Note that the trip circuits have certain similarities.

Each trip takes effect via a relay which is energized when no fault is detected by that particular circuit. Each of these relays has an LED indicator in series with its coil; under normal operation all of these should be lit. They may be used as a secondary check on trip-circuit operation, if the front-panel fault indicator lamps fail to give a correct indication (perhaps because of a faulty bulb).

Each trip relay also has one of its own contacts as a holding contact. This provides the latching function; once the relay has de-energized and the holding contact has broken, disappearance of the fault condition cannot re-energize the relay.

To energize the trip relays initially, the holding contacts must be bypassed by an operator push-button; for engine control this is marked INHIBIT, and for AC and DC control it is marked OFF/RESET.

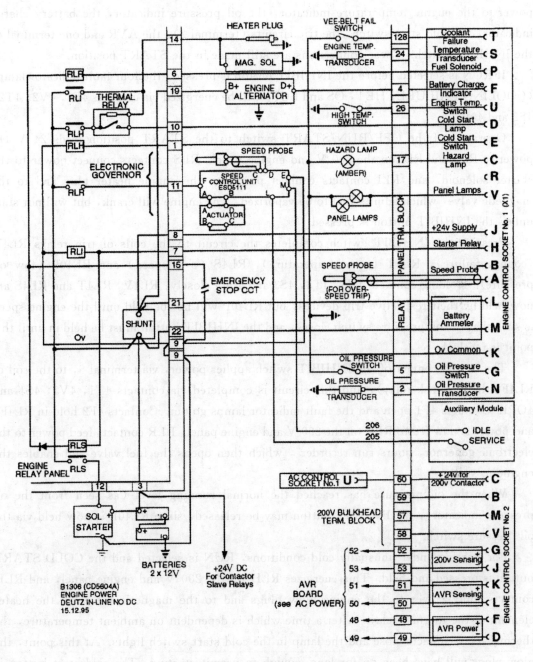

Figure 8.6　Engine power circuit

8.2.3　Engine Control and Protection Circuits

8.2.3.1　Engine Control Module (See Figure 8.7)

The module circuits operate from the engine battery 24 V DC. The sequence of operations for starting the engine is as follows:

Operation of the OFF/RUN/START switch to the RUN position applies 24 V DC power to the engine temperature indicator, the oil pressure indicator, the battery charge indicator, the cold start switch, the trip circuits (terminal 2), the AVR and one terminal of the INHIBIT switch. These connections are still made in the START position.

If RUN is selected before the INHIBIT button is pressed, the four fault indicator lamps (COOLING FAILURE, HET, OS and LOP) will be energized via contacts 4F1, 4V2, 4T2, 4S2 and 40P2.

Operation of the OFF/RUN/START switch to the START position feeds 24 V DC power to RLS and RLI on the 200 V and engine panel. RLS contacts connect power to the starter solenoid, and RLI contacts connect power to the engine heater plug and to the magnetic valve, which admits fuel to be vaporized. The engine will crank, but will not start unless the INHIBIT button is pressed.

Pressing the INHIBIT switch completes the circuit to the coils of trip relays RL4V (cooling failure), RL4T (over-temperature), RL4S (over-speed) and RL40P (low oil pressure). Holding contacts 4V3, 4T3, 4S3 and 40P3 closely; RL4V, RL4T and RL4S are now held by their respective trip devices, but RL40P will not be held until the engine speed is such that the oil pressure switch closes, and the INHIBIT button must be held in until this point is reached.

The second contact of the INHIBIT switch applies power, via terminal 6, to the coil of RL4F (the fuel-rack relay), and this circuit is completed via contacts 4T1, 4V1, 4S1 and 4OP1. Contacts 4F1 open and the fault indicator lamps go out. Contacts 4F2 hold in RL4F, and power is now fed to RLR on the 200 V and engine panel. RLR contacts feed power to the electronic governor, hours run recorder , which then opens the fuel valve and enables the engine to start.

When the oil pressure has reached the normal running level (as seen from the oil pressure indicator), the INHIBIT button may be released, since RL40P is now held via the oil pressure switch.

For starting under unusually cold conditions, RUN is selected and the COLD START button is pressed and held. This energizes RLH on the 200 V and engine panel, and RLH contacts apply power to the engine glow plugs and to the magnetic valve via the heater element of the thermal relay. After a time which is dependent on ambient temperature, the thermal relay contacts close and the lamp in the cold start switch lights. At this point, the glow plugs will have been on for long enough to permit starting. The cold start button is released and the normal starting procedure is followed.

The V-Belt switch and engine temperature switch have contacts which close under fault conditions, turning off the appropriate trip transistor and de-energizing RL4V or RL4T. Similarly an over-speed is sensed by trip circuit OS4, de-energizing RL4S, and the oil pressure switch contacts open under low oil pressure conditions, de-energizing RL40P. Any of these fault conditions opens the circuit to RL4F coil; 4F2 contacts open and power to the fuel solenoid is cut off, so that the engine is shut down. Contacts 4F1 apply power to the appropriate fault indicator lamp via contacts 4V2, 4T2, 4S2 or 40P2.

The LIGHTS ON/OFF switch applies 24 V DC power to the panel lamps and (via the flasher unit) to the amber hazard lamp.

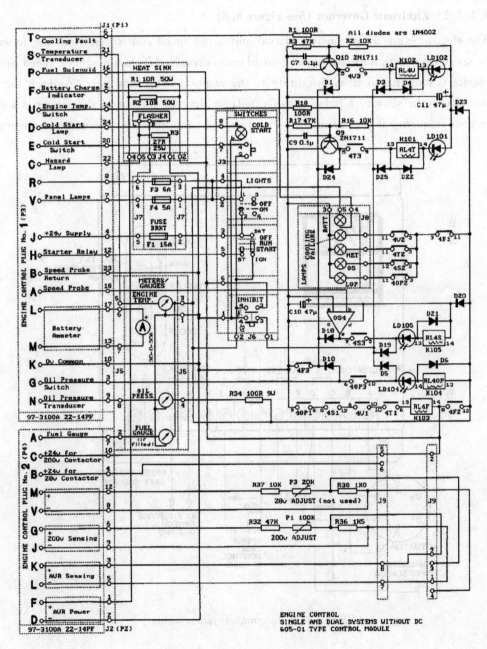

Figure 8.7 Engine control circuit

The potentiometer necessary for adjustment of output voltage via the AVR is located in the engine control module for convenience of access. The 200 V Sensing input (nominal 90 V DC) is derived from 3-phase transformer/rectifier circuits in the AVR, these being fed from the alternator output ("200 V Local" when the output contactor is open and "200V Remote" when it is closed). The potential dividers are normally adjusted so that 1. 6 V DC is fed out on pin K of engine control plug 2.

8.2.3.2 Electronic Governor (See Figure 8. 8)

The electronic governor consists of two units: the speed control unit and the actuator. Engine speed information for the speed control unit is received from a magnetic speed sensor. This device is mounted in close proximity to the engine starter ring gear. As each tooth of the gear passes the sensor, a pulse is generated, so that the output of the sensor is a pulse train whose frequency is proportional to engine speed (and hence to alternator frequency).

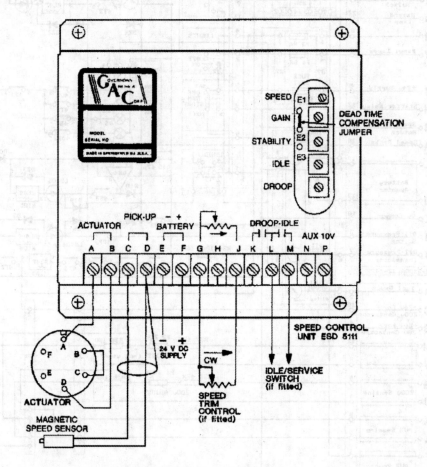

Figure 8. 8 Electronic governor wiring

Within the speed control unit, the pulse train is conditioned to give an analog speed signal which is then compared with the speed adjust set point input (set by the SPEED potentiometer in the speed control unit and the SPEED TRIM, if fitted) and fed via the dynamic control section to the switch-mode output circuit. This is capable of providing up to 10 A output to drive the actuator. The dynamic control section of the speed control unit has adjustments for gain and stability, which are set to match the control characteristics of the engine and so give the best response to the overall frequency control system.

When the IDLE/SERVICE switch in the auxiliary control module is set to IDLE, terminals L and M are linked. This overrides the normal governor action and limits the engine to idling speed. It is used both for starting and before engine shut-down. The DROOP facility is not used in this GPU.

The actuator is a rotary output, linear torque, proportional servo, controlled by the current output of the speed control unit (2 A approx. under normal conditions). An internal spring provides fail-safe operation by forcing the actuator to the fuel shut-off position when the actuator is de-energized. The unit is totally sealed and is maintenance-free.

During engine cranking, with IDLE selected, the actuator moves to the idling fuel position, minimising the surge of speed when the engine fires. When SERVICE is selected, the switching action of the output stage of the speed control unit increases the average current to the actuator, and the engine gains speed. As the correct governed speed is approached, the fuel flow is reduced until the balance point is reached. A speed anticipation circuit minimises speed overshoot when switching large increments of load.

8.2.3.3　Engine Over Speed Trip OS4 (See Figure 8.9)

The input pulses, derived from the magnetic pick-up mounted in the engine flywheel housing, are amplified and squared by the first stage transistors Q2, Q3 and Q4, transistor Q1 together with zener diode D8 limits the square wave amplitude to 5.1 volts.

The next stage is basically a diode pump circuit. The positive edge of the square wave passes through diode D15 to charge capacitor C2. The resistor and diode chain across capacitor C2 (R25, R26, R27, D11, D13, D14) discharges capacitor C2 between pulses such that the resultant DC voltage across the capacitor C2 is proportional to the frequency of the input pulses. Diode D12 conducts the negative edge of the pulses and sets the DC voltage level on capacitor C6 to give a linear circuit. Diodes D11, D13 and D14 are for temperature compensation.

The potentiometer P2 is used to set the frequency at which the output of the integrated circuit U1 passes 0 V and after further amplification by Q5 and Q6 puts a voltage of approximately 1 V on to the base of transistor Q7 to de-energize relay RL4S via the trigger circuit transistors Q7 and Q8. If the engine speed is above the setting, relay RL4S will de-energize and stop the engine.

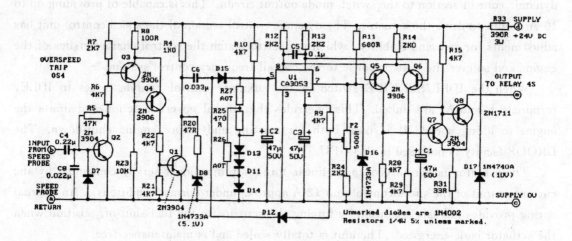

Figure 8.9 Circuit diagram: engine over speed trip OS4

8.2.4 Automatic Voltage Regulator (See Figure 8.10)

The AVR is the main controlling element in the closed-loop system which maintains the selected output of the GPU constant. It receives a sensing input, nominally 1.6 V DC, derived from the GPU output and controls the excitation current of the alternator in such a way as to maintain this sensing input constant.

The excitation current is controlled by a switching technique, to minimise the dissipation in the AVR. When the semiconductor switch is ON, 24 V DC is applied to the exciter winding, but because this winding has a high inductance, the current rises comparatively slowly. When the switch is OFF, the inductance causes the current to continue to flow through the "free-wheel diode" connected across the winding. Varying the duty cycle (the ratio of ON to OFF times) of the switch varies the average current in the winding, and hence controls the output of the alternator.

The switching cycle takes place at 2,400 Hz, and is in fact derived from the 400 Hz 3 - phase voltage being fed into the AVR, so that the switching is in synchronism with the output.

8.2.4.1 Power Supplies

The main power supply to feed the exciter winding is derived from the 3-phase output of the alternator ("200 V local" when the output contactor is open and "200 V remote" when it is closed), connected to terminals 45, 46 and 47 of the AVR. This is in turn connected via line fuses and radio frequency interference filter to step down transformers, the secondary output of the transformers being full-wave rectified to give a nominal 28 volt DC at 200 V AC line voltage. To provide initial excitation at start-up, and to maintain excitation under short-circuit and soft start conditions, a feed to the exciter winding is also provided from the 24 V battery supply. This is taken through dropping resistors (see engine control circuit) and a diode, so that excitation current is only drawn from the battery when the alternator output voltage is low.

The supply for the sensing and control circuitry is fed from the 28 V rail and stabilized by a shunt regulator, the negative line of this supply being common with the 24 V negative. From this stabilized rail, a precision reference diode provides the reference voltage for the comparison circuit (C).

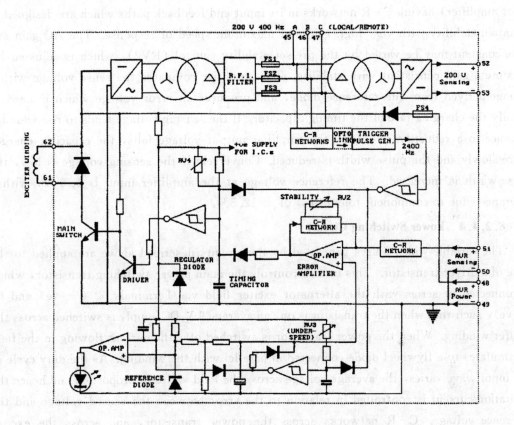

Figure 8.10　AVR circuit diagram(simplified)

8.2.4.2　Sensing Circuits

A 3-phase Delta-star arrangement of sensing transformers and 3-phase bridge rectifier provide the "200 V sensing" voltage of approximately 90 V DC. This is fed out via terminals 52 (+ ve) and 53 (- ve) to the 200 V sensing potential divider (incorporating the voltage trimmer potentiometer) located in the engine control module.

The AVR receives an "AVR Sensing" input of 1.6 V DC nominal on terminals 51 and 50, from whichever of the potential dividers has been selected by the VOLTS SELECT switch in the engine control module.

8.2.4.3　Variable Width Pulse Generator

A triple C – R network across the three-phase input is linked via opto-couplers to a trigger circuit which generates a 2,400 Hz pulse train. A timing capacitor is charged through a resistor network from the positive rail, and discharged by each of these pulses, to give a 2,400 Hz sawtooth waveform. This is fed to a Schmitt comparator; when the sawtooth waveform is above the threshold of the comparator, its output goes low. Hence the circuit generates a train of negative-going variable-width pulses.

The input sense voltage is applied to the inverting input of an operational amplifier (the error amplifier) having C – R networks in its input and feedback paths which are designed to maintain stability with very high gain and maximum speed of response. The AC gain and time constant may be varied by the pre-set stability control (RV2), which is adjusted for maximum gain consistent with stability. The amplifier compares the sense voltage with a voltage derived from the reference diode, and produces an error voltage which is used to modify the charging rate of the timing capacitor. If the sensing voltage tends to rise (due for instance to a reduction in load), the amplifier output voltage falls, the capacitor charges more slowly and the pulse width is reduced. Conversely if the sensing voltage is low, the pulse width is increased. The reference voltage at the amplifier input is 1.6 V, with a tolerance (due to component tolerances) of ± 12.5%.

8.2.4.4 Power Switching Circuit

The variable-width pulses generated by the circuits described above are applied to the base of a driver transistor. This in turn controls the main power switching transistor, which is connected in series with the alternator exciter field via Terminals 62 (+ ve) and 61 (−ve), such that when the transistor is on, an average 9 V DC supply is switched across the exciter winding. When the power transistor is switched off, the current flowing in the field commutates to a flywheel diode connected in parallel with the winding. As the duty cycle of the input wave varies, the average voltage across the field varies in proportion and hence the excitation current is controlled in relation to the error between the sensed voltage and the reference voltage. C – R networks across the power transistor, and across the exciter winding, protect the transistor from spikes due to fast switching.

8.2.4.5 Under-Speed Protection

If the alternator is driven at significantly below its normal speed, it will not be capable of generating its full output voltage. In attempting to overcome this, the AVR may force excessive excitation current to flow, damaging the winding. Under-speed protection is incorporated to guard against this risk.

A second timing capacitor and charging resistor network, incorporating a pre-set control (RV3), generates a second sawtooth waveform, again at 6 × output frequency, and this is fed to another Schmitt comparator. Under normal conditions, the sawtooth does not reach the threshold of the comparator, and no output results. However, if the frequency is below a value set by adjusting RV3, this comparator begins to produce negative output pulses. These are used to pull down the reference voltage fed to the error amplifier, and hence the alternator output is regulated at an appropriately lower voltage. When this circuit is in operation, a LED on the AVR board is lit, and this is used when setting up the circuit.

New Words

1. transformer [trænsˈfɔːmə] n. 变压器
2. probe [prəʊb] n. 探针
3. malfunction [mælˈfʌŋ(k)ʃ(ə)n] n. 故障,失灵
4. coil [kɒil] n. 线圈
5. latch [lætʃ] n. 门插销
6. transistor [tranˈzistə] n. 晶体管,三极管
7. solenoid [ˈsəʊlənɒid] n. 电磁线圈
8. potentiometer [pə(ʊ),tenʃiˈɒmitə] n. 电位计
9. capacitor [kəˈpæsitə] n. 电容器
10. amplification [æmpləfiˈkeiʃən] n. 放大,扩大,增大
11. semiconductor [semikənˈdʌktə] n. 半导体
12. inductance [inˈdʌkt(ə)ns] n. 电感
13. threshold [ˈθreʃəuld] n. 上限或下限
14. comparator [kəmˈpærətə] n. 比较器

Phrases and Expressions

1. slave relay 中间继电器
2. over speed trip 超速跳闸
3. glow plug 电热塞
4. thermal relay 热继电器
5. speed overshoot 速度过调量
6. zener diode 稳压二极管
7. shunt regulator 并联调节器(稳压器)
8. opto-coupler 光耦合器
9. trigger circuit 触发电路
10. sawtooth waveform 锯齿波形
11. error amplifier 误差信号放大器

Notes to the Text

1. The engine speed, which determines the frequency of the AC output of the GPU, is controlled by an electronic governor. A speed probe provides a speed signal which is used for the over speed trip.

发动机的转速,决定地面电源装置交流电的输出频率,由电子监控器控制。速度探测器的发动机转速信号用来超速跳闸。

2. Each of the power sections has an associated control section, built as a self-contained

control module and incorporating the appropriate switches for operator control of the GPU, together with meters to monitor GPU performance. In addition, each section includes solid-state monitoring trips which protect both the GPU itself and its loads from any possible damage due to GPU or load malfunction.

每一个动力部件都有与之相连的控制部件,设有配套齐全的控制模块和适当的开关以及监控 GPU 性能的仪表,便于操作者控制 GPU。此外,每一部件配有电子监控闸保护 GPU 及其负载,避免任何来自 GPU 或其负载故障产生的危害。

3. Engine speed information for the speed control unit is received from a magnetic speed sensor. This device is mounted in close proximity to the engine starter ring gear. As each tooth of the gear passes the sensor, a pulse is generated, so that the output of the sensor is a pulse train whose frequency is proportional to engine speed.

发动机转速控制信号来自于一电磁速度传感器。电磁速度传感器安装在发动机起动机的齿圈附近。当齿圈的每一个齿通过传感器时,将产生一个脉冲,因此,传感器的输出是一脉冲序列,且它的频率与发动机的转速成比例。

4. The AVR is the main controlling element in the closed-loop system which maintains the selected output of the GPU constant. It receives a sensing input, nominally 1.6 V DC, derived from the GPU output and controls the excitation current of the alternator in such a way as to maintain this sensing input constant.

自动电压调节器(AVR)是维持 GPU 所选的输出为常数的闭环系统的主要元件。AVR 接收 GPU 的输出信号作为输入信号,通常为 1.6 V 直流电,并控制交流发电机的励磁绕组,通过这种方式保持输入信号恒定。

5. The supply for the sensing and control circuitry is fed from the 28 V rail and stabilized by a shunt regulator, the negative line of this supply being common with the 24 V negative. From this stabilized rail, a precision reference diode provides the reference voltage for the comparison circuit (C).

传感和控制电路的供电经一个并联调节器由 28 V 稳态电源提供,供电线的负极与 24 V 的负极相同。28 V 稳态电源还有一个精密恒压二极管为比较电路提供参考电压。

6. Under normal conditions, the sawtooth does not reach the threshold of the comparator, and no output results. However, if the frequency is below a value set by adjusting RV3, this comparator begins to produce negative output pulses. These are used to pull down the reference voltage fed to the error amplifier, and hence the alternator output is regulated at an appropriately lower voltage. When this circuit is in operation, a LED on the AVR board is lit, and this is used when setting up the circuit.

正常情况下,锯齿形信号不会达到比较器的阈值,比较器不产生输出信号。但如果频率低于由 RV3 调定的值时,比较器将产生负的输出脉冲信号。这些负的输出脉冲信号将降低误差信号放大器的参考电压,从而将发电机的输出限制在一适当的低电压。当这个电路运行时,AVR 面板上的 LED 灯点亮,并且可用来检测电路。

Section 8.3　200 V AC Control and Protection Circuits

8.3.1　General

200 V AC control and protection circuits, shown in Figure 8.11, 8.12. The 200 V AC control module provides the means of applying, disconnecting and monitoring the condition of the 200/115 V, 3-phase, 400 Hz power circuit of the GPU. Meters are incorporated for the operator to monitor the voltage and current in any phase, and the frequency. The trip circuits open the AC contactor in case of overload, high and low frequency, high and low volt, and incorrect phase rotation. The control circuits for the AC contactor operate from the 24 V battery supply.

The control circuits, and the trip circuits with their associated relays, are carried by the Logic Board; in the description which follows they are subdivided for ease of reference. The trip circuits have a power supply derived from the alternator and transformed and rectified to 20 V DC on the power supply board. The power supply board also carries the interconnections between the logic board and the switches and controls on the front panel.

Aircraft electrical systems provide a 28 V DC feed to pin E or F of the aircraft connector to indicate that the aircraft circuits have detected a valid 200 V AC supply as present. The INTERLOCK/BYPASS switch is normally set to INTERLOCK, and this signal is used to lock the appropriate AC contactor. If the facility is not available (e.g. when testing using a load bank), the switch is set to BYPASS, and the contactor is then locked by circuits within the GPU.

8.3.2　Contactor Control

Assuming that the GPU engine is running, the sequence of events for closing the AC contactor is as follows.

The OFF/RESET button must first be pressed, energizing relays R3A and R3AB so that the contacts 3A and 3AB close. The fault indicator lamps OL (overload), HF (high frequency), LF (low frequency), LV (low voltage), HV (high voltage) and PR (phase rotation) all light momentarily, but assuming no faults in the parameter monitored, relays R3OL, R3HF, R3LF, R3HV, R3LV and 3PR energize, and thus change over their respective contacts. One of the NC contacts of each relay, which closes, holds in the relay after the OFF/RESET button has been released, whilst one of the NC contacts of each relay, which opens, turns off the appropriate fault indicator lamp.

Closure of contacts 3A also energizes R3H, which is held in by contacts 3H. If the INTERLOCK/BYPASS switch is in the BYPASS position, R3F and R3FB are also energized at this point.

After release of the OFF/RESET button, the +24 V supply is connected via contacts

3H and one contact of each of the trip relays (in series) to the ON button.

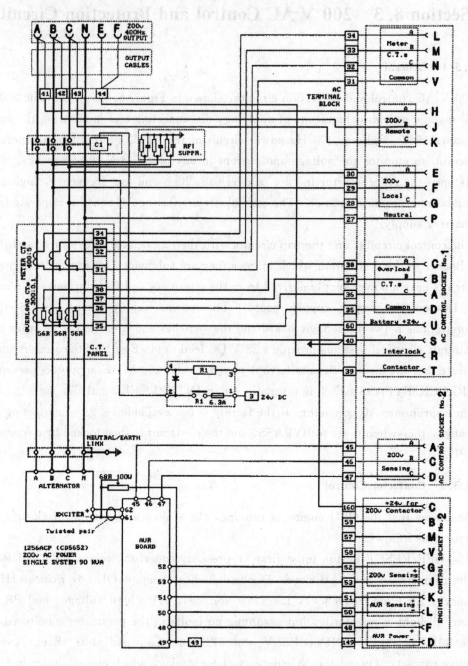

Figure 8.11   200V AC power circuit

When the contactor ON button is depressed, relay R3B is energized and +24 V supply is fed to the contactor slave relay via contacts 3B and AC control plug pin 1T. If the INTERLOCK/BYPASS switch is in the INTERLOCK position, and the DC interlock voltage is being received back from the aircraft, relays RL3F and FB will already be energized. A holding circuit is now established round the ON button via contacts 3B and 3F so that the contactor remains to be locked on. The lamp in the ON pushbutton switch is also lit.

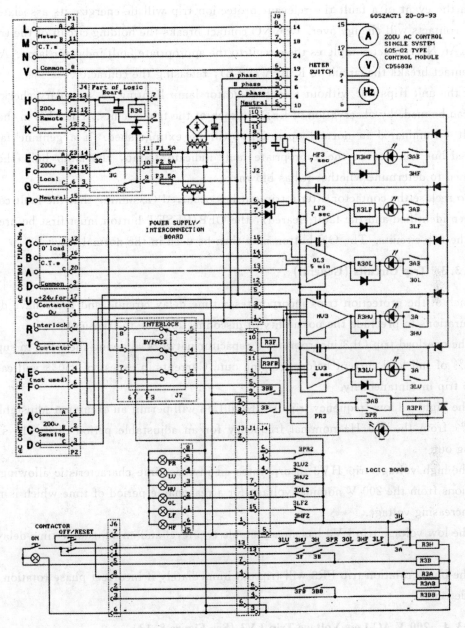

Figure 8.12 200V AC control circuit

Relay RG provides for remote AC sensing. When the contactor is closed, a 200 V AC supply is present at pins H, J and K of AC control plug 1, and relay RG is energized. The changeover of RG contacts transfers the 200 V sensing for the AVR and the trip circuits from Local (the alternator terminals) to Remote (the aircraft connector).

Pressing the OFF/RESET button breaks the circuit to R3B, opening contacts 3B and releasing the contactor, and turns off the lamp in the ON switch. The trip relays remain energized.

In the event of a fault the relevant protection trip will de-energize its associated relay whose contacts will change over. One NO contact breaks the holding circuit of the trip relay itself. An NC contact completes the circuit to the appropriate fault indicator lamp. A second NO contact breaks the circuit to R3B(A and B), releasing the contactor.

If the unit trips out without a fault indicator lamp lighting, a secondary check on the fault can be made by observation of LEDs in each of the trip circuits. The LED in the circuit at fault (illuminated in normal operation) will be extinguished. If a genuine fault has occurred but has not lit the appropriate fault indicator lamp, the lamp bulb should be examined to determine whether it has blown.

To reclose the contactor after a trip-out has occurred (assuming that the fault condition has been identified and the fault cleared), the OFF/RESET button must first be pressed to reset the trip circuits. The ON button may then be used in the normal way.

8.3.3 Trip Circuits (General)

Each of the protection trips incorporates a time delay circuit, and these have differing configurations to preclude the possibility of disconnection due to transients.

The overload trip OL3 incorporates a capacitor network allowing the GPU to supply up to 125% of full load current before tripping out. Currents in excess of 250% will cause the unit to trip instantaneously.

The high and low frequency trips HF3 and LF3 will permit an excursion adjustable up to $\pm 7.5\%$ from the 400 Hz nominal frequency for an adjustable period of 0 – 10 s before tripping out.

The high voltage trip HV3 incorporates an inverse time characteristic allowing minor excursions from the 200 V nominal voltage for a reasonable period of time which is reduced with increasing voltage.

The low voltage trip LV3 has an adjustable trip level and an adjustable time delay of 0 – 10 s.

The phase rotation trip PR3 will trip out immediately if incorrect phase rotation (C,B, A) is detected.

8.3.4 200 V AC Low Voltage Trip LV3 (See Figure 8.13)

The input to the trip circuit, derived from the three-phase output of the alternator, is rectified by the diodes D51, D53 and D54. The DC voltage appearing across resistor R111 is proportional to the average three-phase voltage which is compared to the voltage appearing on the slider of potentiometer P7. The difference in voltage is amplified by the integrated circuit U2 and transistors Q37 and Q38 to switch "ON", via the trigger circuit transistors Q35 and Q36, the timing circuit.

When the average voltage of the three-phase input is below the voltage appearing on the slider of potentiometer P7, the timing capacitor C35 is allowed to charge to approximately 6 V to de-energize relay RL3LV via the trigger circuit transistors Q31, Q32, Q33 and Q34. The charging current to capacitor C35 is adjusted by means of potentiometer P6, which thus sets the time delay.

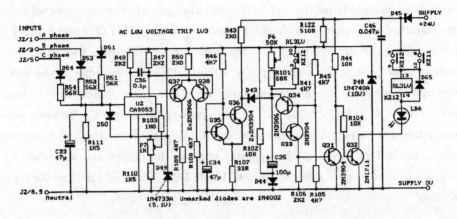

Figure 8.13　Circuit diagram: 200 V low voltage trip LV3

If the average voltage is below the trip level setting for longer than the time delay, adjustable between 0 – 10 s by potentiometer P6, the relay RL3LV will de-energize and open the AC contactor(s).

8.3.5　200 V AC High Voltage Trip HV3 (See Figure 8.14)

The three-phase input to the trip circuit, derived from the output of the alternator, is rectified by diodes D52, D55 and D56, smoothed and applied to the gating circuit of transistors Q39, Q40, Q41 and Q42. The high voltage trip level is set by potentiometer P8. If any one of the three input voltages exceeds the setting, relay RL3H V will be de-energized via the trigger circuit transistors Q43 and Q44, and open the AC contactor(s).

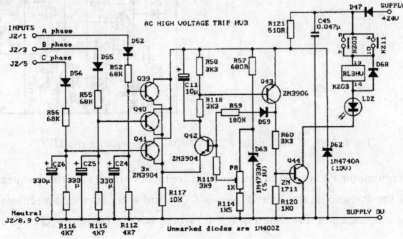

Figure 8.14　Circuit diagram: 200 V high voltage trip HV3

The charging of capacitors C24, C25 and C26 via the resistors R52, R55 and R56 respectively ensures an approximate inverse time characteristic to the gating circuit of transistors Q39, Q40 and Q41.

8.3.6　200 V AC High Frequency Trip HF3 (See Figure 8.15)

Input pulses, derived from the 200 V AC output of the alternator via signal transformer T202, are amplified and squared by the first stage transistors Q13, Q14 and Q15. Transistor Q16 together with zener diode D19, limits the square wave amplitude to 5.1 V.

The next stage is basically a diode pump circuit. The positive edge of the square wave passes through diode D21 to charge capacitor C6. The resistor and diode chain across capacitor C6 (R21, R22, R23, D22, D23, D24) discharge capacitor C6 between pulses such that the resultant DC voltage across the capacitor C6 is proportional to the frequency of the input pulses. Diode D20 conducts the negative edge of the pulses and sets the DC voltage level on capacitor C13 to give a linear circuit. Diode D22, D23 and D24 are for temperature compensation.

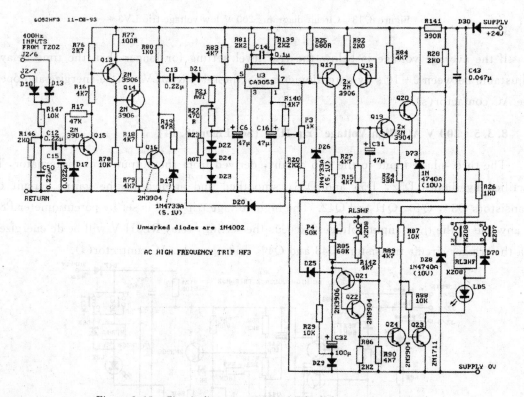

Figure 8.15　Circuit diagram: 200 V AC high frequency trip HF3

The potentiometer P3 is used to set the high frequency trip level (the frequency at which the output of the integrated circuit IC1 passes 0 V) and after further amplification by Q17 and Q18 puts a voltage of approximately 1 V on to the base of transistor Q19 to allow timing capacitor C32 to charge to approximately 6 V to de-energize relay RL3HF via the trigger circuit transistors Q21 and Q22. The charging current to capacitor C8 is adjusted by means of potentiometer P4, thus setting the time delay. If the frequency is above the trip level setting for longer than the time delay, the relay RL3HF will de-energize and open the AC contactor(s).

8.3.7　200 V AC Low Frequency Trip LF3 (See Figure 8.16)

Input pulses, derived from the 200 volt AC output of the alternator via signal transformer T202, are amplified and squared by the first stage transistors Q1, Q2 and Q3. Transistor Q4 together with zener diode D5, limits the square wave amplitude to 5.1 V.

The next stage is basically a diode pump circuit. The positive edge of the square wave passes through diode D4 to charge capacitor C3. The resistor and diode chain across capacitor C3 (R4, R6, R7, D6, D7, D8) discharge capacitor C3 between pulses such that the resultant DC voltage across the capacitor C3 is proportional to the frequency of the input pulses. Diode D20 conducts the negative edge of the pulses and sets the DC voltage level on capacitor C19 to give a linear circuit. Diode D6, D7 and D8 are for temperature compensation.

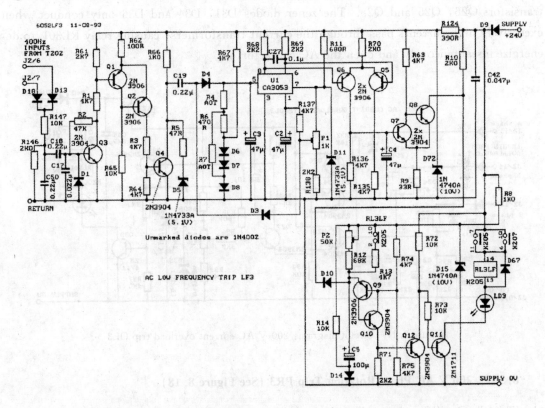

Figure 8.16　Circuit diagram: 200 V AC low frequency trip LF3

The potentiometer P1 is used to set the low frequency trip level (the frequency at which the output of the integrated circuit IC1 passes zero volt) and after further amplification by

Q5 and Q6 puts a voltage of approximately 1 V on to the base of transistor Q7 to allow timing capacitor C5 to charge to approximately 6 V to de-energize relay RL3LF via the trigger circuit transistors Q9 and Q10. The charging current to capacitor C8 is adjusted by means of potentiometer P2, thus setting the time delay. If the frequency is below the trip level setting for longer than the time delay, the relay RL3LF will de-energize and open the AC contactor(s).

8.3.8　200 V AC Current Overload Trip OL3 (See Figure 8.17)

The input, derived from burden resistors across the output of current transformers located in each phase of the 200 V AC supply, is rectified and smoothed and applied to the gating circuit of transistors Q25, Q26, Q28 and Q27. The overload trip level is set by potentiometer P5. If any one of these three inputs exceeds the setting, relay RL30L will be de-energized, via the trigger circuit transistors Q29 and Q30. Diode D37 and resistor R36 provide some positive feedback to ensure that relay RL30L is switched cleanly.

The charging of capacitors C22, C21 and C30 via resistors R32, R33 and R35 respectively ensures an approximate inverse time characteristic to the gating circuit of transistors Q26, Q25 and Q28. The zener diodes D31, D34 and D36 only conduct when excessively high currents pass through the current transformers, causing relay RL30L to de-energize instantaneously and open the AC contactor(s).

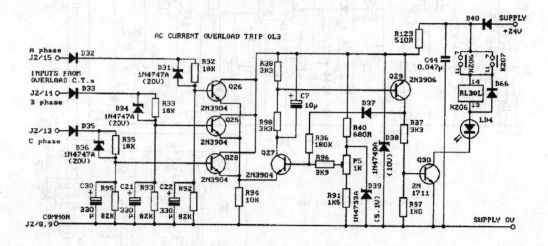

Figure 8.17　Circuit diagram: 200 V AC current overload trip OL3

8.3.9　200 V AC Phase Rotation Trip PR3 (See Figure 8.18)

This circuit differs from the other trip circuits in that the power to energize the trip relay is derived from the input lines, rather than from the common DC supply rail. The input, derived from the 200 V AC output of the alternator, is rectified by the diode bridge D57, D58, D60 and D61. The resultant DC volt will energize relay RL3PR when the phase rotation is correct. When the phase rotation is incorrect the DC voltage of the diode bridge is too low to energize relay RL3PR and thus prevents closing of the AC contactor.

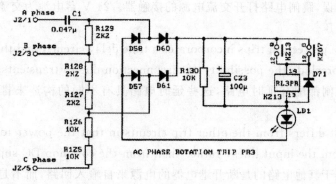

Figure 8.18　Circuit diagram：200 V AC phase rotation trip PR3

New Words

1. pin [pin] *n.* 插头，电子管管脚
2. configuration [kənˌfigəˈreiʃ(ə)n] *n.* 结构
3. preclude [priˈkluːd] *vt.* 排（消）除，预防
4. instantaneously [instənˈteinjəsli] *adv.* 立即，立刻地
5. excursion [ikˈskɜːʃ(ə)n] *n.* 飘移，偏移
6. feedback[ˈfiːdbæk] *n.* 反馈，反应

Phrases and Expressions

1. trip circuit 跳闸电路
2. phase rotation 相序
3. trip relay 跳闸继电器
4. time delay circuit 延时（迟）电路
5. nominal frequency 额定频率，标称频率
6. timing circuit 计时（定时）电路
7. supply rail 供电轨，供应轨

Notes to the Text

1. The 200 V AC control module provides the means of applying, disconnecting and monitoring the condition of the 200/115 V, 3-phase, 400 Hz power circuit of the GPU.

200 V 交流电源控制模块为 GPU 的 200/115 V,3 相,400 Hz 的电路准备应用方式,监控使用情况并控制通断。

2. Meters are incorporated for the operator to monitor the voltage and current in any phase, and the frequency. The trip circuits open the AC contactor in case of overload, high and low frequency, high and low volt, and incorrect phase rotation. The control circuits for

the AC contactor operate from the 24 V battery supply.

相应仪表帮助操作者监控各相的电压、电流和频率。万一出现过载、过频（或欠频），过压（或欠压）和相序错误，跳闸电路打开交流电源的接触器。24 V 蓄电池为交流电源接触器的控制电路运行提供电压。

3. Each of the protection trips incorporates a time delay circuit, and these have differing configurations to preclude the possibility of disconnection due to transients.

每一个保护跳闸都配有延时电路，这些延时电路具有不同的构造来排除瞬时断开的可能性。

4. This circuit differs from the other trip circuits in that the power to energize the trip relay is derived from the input lines, rather than from the common DC supply rail.

这种电路不同于其他电路的是断开继电器的电源来自输入回路，而不是来自普通的直流供电轨。

Chapter 9　　Aircraft Airstarter

Section 9.1　　General Description

9.1.1　General

The model LPA6/300/TR low pressure air starter comprises a Cummins type KTA19 – C600 water-cooled diesel engine driving an Atlas Copco ZA6C single-stage compressor through a step-up gearbox, to provide a supply of 300 lb/min (136 kg/min) of completely oil-free air at a delivery pressure of 40 psi nominal (30 psi for air conditioning), shown in Figure 9.1.

The diesel engine can be started without additional external aids at ambient temperatures down to − 20℃. The assistance of the Ether Start Kit is required for temperatures below 10℃. An inlet air filter and an exhaust silencer are fitted to both the engine and the compressor. Although essentially designed for starting aircraft gas turbine engines, the unit is also used for air conditioning the aircraft while it is parked.

The engine, step-up gearbox and compressor constitute an integral assembly which is mounted on a fabricated steel bedplate and enclosed in a weather protecting steel canopy. The engine exhaust gas is piped to a high level within the canopy and passed through an exhaust silencer. Lockable hinged doors in the canopy provide generous access to the control panel for routine operations and normal maintenance. A perspex viewing window and a slot beneath, give the unit the ventilation required for operation with the access doors closed. A semi-rotary hand pump for filling the radiator is fitted.

The bedplate carrying the engine and compressor assembly is mounted on a steel subframe, which has outriggers to carry the canopy. The subframe, which also carries the 640 litre fuel tank, is mounted on an Leyla and 45 – 130 truck chassis with left hand driving. The fuel contents gauge is of the direct reading type, fitted to the tank. An aluminium access ladder is stowed on the L. H. side of the chassis.

The unit is equipped with a 24 V DC negative-earthed battery system which starts and subsequently monitors the unit. The 2 × 12 V batteries are mounted inside the canopy.

Temperature and pressure switches, associated with both the engine and the compressor, control an engine run solenoid so that, in the event of overheating or oil pressure failure, the engine is automatically shut down. A fault lamp on the control panel and another on the roof of the canopy will illuminate in the event of a fault shut down.

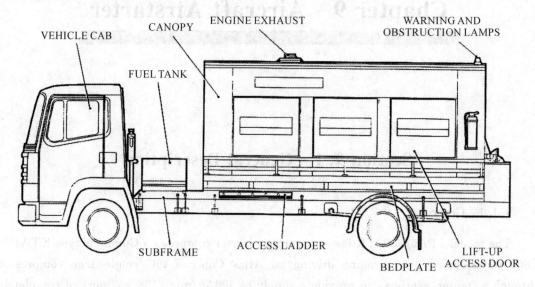

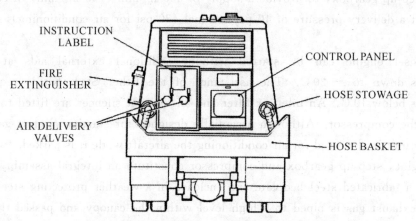

Figure 9.1   Overall view of airstarter

All operating controls, switches, gauges and indicators are grouped on a single control panel mounted at the rear of the starter. A perspex viewing window with a slot beneath allow operation of the unit without opening any access doors. Also located at the rear of the canopy, on the left-hand side, there are a pair of air shut-off valves.

Standard equipment supplied with the starter includes a pair of 15.2 m (50 ft) long air hoses, each with anti-scuffing jacket and a Roylan AB1040 aircraft coupling, and a 7.2 kg B. C. F. fire extinguisher. The air hoses are stowed on either side of the unit.

Safety switches fitted to the air shut-off valves prevent the airstarter engine being started unless both valves are closed.

9.1.2　Compressor Description

9.1.2.1　General

The engine unit is directly flanged to the housing of the step-up gearbox by means of an adapter housing, and drives the step-up gear shaft via a flexible coupling. The compressor unit is bolted to the step-up gear housing. The overall assembly arrangement of engine-gearbox-compressor ensures vibration-free running, rigidity and correct alignment at all times.

The Cummins KTA19 – C600 engine is fully described in a supplementary handbook. The following paragraphs describe the compressor unit.

9.1.2.2　Compressor Element

The compressor unit consists of a housing in which are assembled the compressor driving components and the compressor element. The element, shown in Figure 9.2, comprises two precision ground Archimedean-type screw rotors, each of which is suspended by ball-and-roller bearings and arranged to intermesh with each other along their longitudinal axes within the compression space of the housing.

The male rotor 8 is driven via a gear wheel 6 from the stepped-up gearbox output, and drives the female rotor 4 via a pair of timing gears 1 which ensure that a slight clearance is maintained between the lobes of each rotor. The male rotor has four lobes, the female rotor has six. In addition to allowing air to be forced along, the clearance between the two rotor lobes practically eliminates wear and power loss.

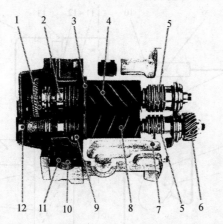

Figure 9.2　Compressor element; cut-away view

1—timing gears；　2—rotor bearings；　3—end plate；　4—female rotor；　5—sealing packages；

6—drive gear wheel；　7—cooling jacket；　8—male rotor；　9—vent hole, air leak；

10—oil outlet port, element-to-oil sump；　11—drain hole, bearing oil return；　12—balancing piston

The compressor housing around the rotors incorporates a jacket which allows cooling oil, pumped from the step-up gearbox, to circulate. The oil system also lubricates the rotor bearings and timing gears.

In order to prevent air and oil leakages, sealing rings held in special retainer packages are fitted at the ends of each rotor shaft. The sealing packages located next to compression space prevent air leakage and those located next to the rotor bearings prevent lubricating oil from entering the compression space. To avoid penetration of oil into the compression space, drains evacuate any oil that may gather between the oil and air sealing packages.

An air-operated balancing piston 12 is fitted behind the rear thrust bearing of the male rotor of the compressor element. The purpose of this balancing piston is to equalize the load on the rear and front thrust bearings of the male rotor during the load periods of the compressor. During the load periods, the balancing piston is actuated by air at working pressure; during off-load periods, the piston is at atmospheric pressure.

The air discharge from the compressor comprises a main delivery pipe via an outlet silencer which incorporates a safety relief valve. This valve is set to operate at a pressure of 3.5 bar (50 psi). Figure 9.3 shows the overall compressor air circuit.

9.1.3　Compressor Lubrication and Cooling-oil System

9.1.3.1　General

The compressor element is cooled — and its driving components and bearings are lubricated — by a common oil system contained in and pumped from the step-up gearbox via an oil cooler which is mounted on the engine casing. The cooling/lubrication flow circuit is shown in Figure 9.3.

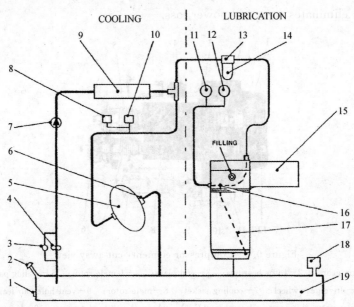

Figure 9.3　Compressor oil cooling and lubrication circuit

1—drain plug; 2—oil level dipstick; 3—by-pass valve (oil pump); 4—oil pump; 5—compressor element; 6—restrictor; 7—non-return valve; 8—temperature switch (high sensing) combined with (10); 9—oil cooler; 10—pressure switch (low sensing); 11—pressure gauge; 12—temperature gauge; 13—by-pass valve (oil filter); 14—oil filter; 15—step-up gearbox housing; 16—oil port to front bearings; 17—oil pipe to rear bearings and timing gears; 18—breather; 19—oil sump

The system is monitored for oil pressure and oil temperature by a pair of gauges on the control panel at the rear of the starter. In addition, low oil pressure and high oil temperature protection switches are located at strategic points in the oil flow circuit. Operation of either switch, both of which are in series with other switches forming part of the engine protection circuit, will therefore result in engine shut down and a general (flashing lamp) warning.

9.1.3.2   Description

(1) Oil from the oil sump 19, located in the bottom of the step-up gearbox housing 15, is pumped by a gear-type oil pump 4 which is driven by the gearbox main drive shaft. The oil is pumped via a non-return valve 7 to oil cooler 9. The non-return valve is included to prevent oil from draining back into the sump and thus un-priming the pump. From the oil cooler, the oil is passed via a two-way connection to:

1) The cooling jacket of the compressor element 5. From the cooling oil outlet port on the compressor, the oil is returned to the sump via a restrictor 6.

2) The full-flow oil filter 14. From the filter the oil is passed to the gearbox casing 15 to lubricate the step-up gearbox and, via a duct in the gearbox, to lubricate the front bearings of the compressor 16. A pipe 17 takes the oil to the rear of the compressor to lubricate the rear bearings and timing gears. After passing through the bearings and over the gears, the oil is returned to the sump.

(2) Two by-pass valves protect the oil system against over pressure:

1) One is located in the oil pump housing 3, and allows oil to by-pass back to the sump whenever the maximum permissible oil pressure is exceeded on the delivery side of the pump.

2) The other 13 is located in the the header of the filter. It opens when the pressure drop over the filter is above normal due to any clogging of the filter element. In such an event, the oil is passed unfiltered to the lubrication points.

9.1.4   Compressor Air Circuit

9.1.4.1   General

The airstarter unit is started while selected to operate in a passive (unloaded) condition. It is subsequently selected to either one of two active operating modes — jet start or air conditioning. These modes are selected at the air control selector valve located on the control panel. In addition to its mechanical operation as a valve, the selector also serves as a switch. This subsection describes the porting functions carried out by the valve, and also describes the operation of the pressure operating valve and the regulating valve in the air system. The overall compressor air circuit is shown in Figure 9.4, which also shows details of the pressure operating valve and the regulating valve.

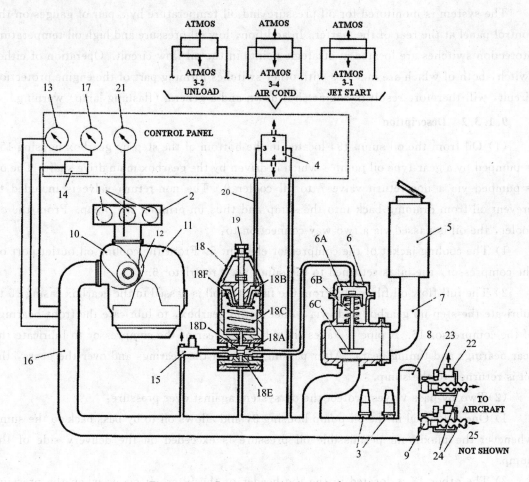

Figure 9.4　Compressor air circuit(schematic)

1－air filters;　2－air filter housing;　3－safety relief valves;　4－air control selector valve/switch;　5－dump silencer;

6－regulating valve;　7－duct;　8－air delivery pipe assembly;　9－air shut-off valves;　10－compressor element;

11－step-up gear housing;　12－compressor element balancing piston;　13－delivery air temperature gauge;

14－delivery air temperature switch;　15－three-port solenoid valve;　16－cross piece adaptor;

17－delivery air pressure gauge;　18－pressure operating valve;　19－adjusting screw;　20－fine duct;

21－air filter restriction indicator;　22－pressure relief solenoid valve;　23－micro switch;

24－aircraft delivery hose;　25－aircraft hose connector

9.1.4.2　Compressor Air Flow

All indrawn air to the compressor passes through two replaceable paper cartridge type filter elements 1, fitted behind the compressor air inlet plenum chamber 2. These filter elements ensure that dust and other foreign matter are removed from the air before they enter the compressor elements. A vacuum gauge (compressor air filter) on the control panel will indicate in the red area when the filters are due for servicing.

The indrawn air is compressed in the compressor elements 10 and discharged through the cross piece adaptor 16 to the air delivery pipe assembly 8. The cross piece adaptor incorporates the regulating valve 6 which relieves the air pressure to atmosphere via the dump silencer 5 whenever the selected low or high pressure is exceeded. The two air hoses are connected to the air delivery manifold via air shut-off valves 9 with associated couplings.

An air-operated balancing piston 12 is fitted behind the rear thrust bearing of each male rotor of the compressor elements. During the start-up and idle periods of the airstarter the balancing pistons are at atmospheric pressure; during on-load periods the balancing pistons are air-operated at working pressure to equalize the load between the rear and front thrust bearings of the male rotors.

The pressure operating valve 18, in conjunction with the air control selector valve (4) and a three-port solenoid valve 15, enable a low or high discharge pressure to be provided for air conditioning or jet start respectively.

Also incorporated in the compressor air circuit there are two spring-loaded safety relief valves 3, which are set to operate at a pressure of 3.51 bar (50 psi), and an air temperature safety switch 14. A delivery air pressure gauge 17 and a delivery air temperature gauge 13, with sensors in the air circuit, are fitted on the control panel.

9.1.4.3　Start and Unload

In order that the airstarter unit can be started, the air control selector valve 4 must be set to the UNLOAD position. The switch function of the selector while in this position ensures that the throttle change solenoid is energized, thus dictating an engine idling speed of 950 r/min, while the porting action via ports 3 - 2 vents to atmosphere chamber 18B, thus relieving the spring pressure on valve 18D. At idling speed, the engine speed sensing switch is closed so that three-port solenoid valve 15 is energized, allowing chamber 6A to be vented to atmosphere through the three-port solenoid valve via pressure operating valve chamber 18A. The compressor air pressure is therefore able to open valve 6C and discharge to atmosphere via blow-off silencer 5.

Note: Whenever the air control selector is in the UNLOAD position, a supply is made to energize both the left-and right-hand pressure relief solenoid valves.

9.1.4.4　Air Conditioning

The switch function of the air control selector, when it is in the AIR CONDITIONING position, breaks the supply to the throttle change solenoid so that the engine throttle is allowed to move to its maximum speed setting. The porting action via ports 3 - 4 of the selector vents to atmosphere chamber 18B, thus relieving the spring pressure on valve 18D. Because the engine speed is beyond 1,800 r/min, the engine speed sensing switch will be open to keep the three-way solenoid valve 15 in the de-energized condition. Part of the compressor air flow is now able to pass via chamber 18A to chamber 6A, thus exerting a pressure on spring-loaded piston 6B, which is pushed towards its valve seat. Pressure now starts to build up in the air delivery manifold and, simultaneously, in chamber 18E of the pressure operating valve.

Valve 6C continues to move towards its seat until the pressure in chamber 18E is positive enough to open valve 18D and allow part of the air flow to chamber 6A to be vented to atmosphere. At this "balance" point, the air pressure is steady at the required air conditioning pressure of 2.1 bar (30 psi).

9.1.4.5　Jet Start

The switch function of the air control selector, when it is in the JET START position, breaks the supply to the throttle change solenoid so that the engine throttle is allowed to move to its maximum speed setting; the porting action is via ports $3-1$, admitting part of the compressor air flow to chamber 18B so that piston 18F increases the tension of spring 18C on valve 18D, which is pushed harder down into its seat. Because the engine speed is beyond 1,800 r/min, the engine speed sensing switch will be open to keep the three-way solenoid valve 15 in the de-energized condition. Part of the compressor air flow is thus able to pass via chamber 18A to chamber 6A, exerting a pressure on spring-loaded piston 6B which is pushed towards its valve seat. Pressure now starts to build up in the air delivery manifold and, simultaneously, in chamber 18E of the pressure operating valve.

Valve 6C moves harder into its seat until the pressure in chamber 18E is positive enough to open valve 18D against the force of spring 18C imposed by the pressure on piston 18F. When this occurs, part of the air flow to chamber 6A is vented to atmosphere. At this "balance" point, the air pressure is steady at the required jet starting pressure of 2.81 kgf/cm^2 (41 lbf/in^2) (1 kgf/cm^2=0.1 MPa=14.5 lbf/in^2).

9.1.4.6　Air Shut-Off Valves

Air is delivered from the air delivery manifold 8 via two air shut-off valves 9 each complete with a female coupling, which accepts the delivery air hose, a relief solenoid valve and a micro-switch.

One or both of the air shut-off valves (depending on aircraft requirements) are opened to commence the jet start or air conditioning operation. In the closed position of each valve, the micro-switch is actuated to energize the associated pressure relief solenoid valve and vent to atmosphere all pressure in the delivery hose side, allowing safe uncoupling of the hose from the aircraft. Both air shut-off valves must be closed (micro-switches operated) to allow the airstarter engine to be started.

9.1.5　Electrical System and Safety Devices

9.1.5.1　General

The 24 V DC electrical system for the airstarter unit is physically divided into two areas: one located in the engine area and the other on the control panel.

An overall circuit of the electrical system is shown in Figure 9.5. For the purpose of description, the overall circuit can be split into three aspects.

(1) Ancillary circuits (those supplying the control panel lamps and the roof obstruction

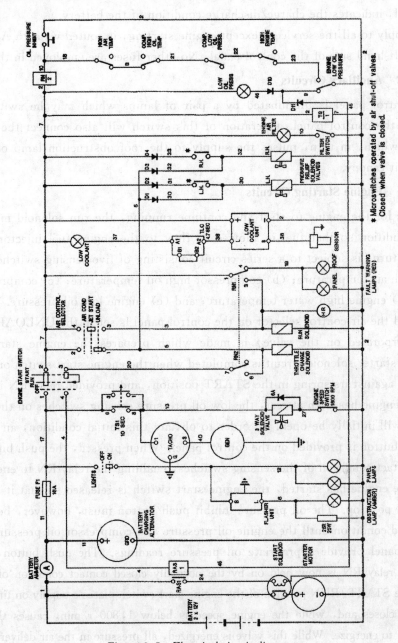

Figure 9.5　Electrical circuit diagram

lamp).

（2）Engine starting and automatic shut down circuits (shut-down occurring in the event of a fault condition affecting the engine or compressor).

（3）Control and safety circuits.

The electrical system depends on two 12 V, series-connected, lead-acid batteries. Once the engine is started and running, an engine-driven alternator and associated voltage regulator maintain the battery in a charged condition. A battery ammeter, located on the control panel, indicates the charge/discharge condition of the battery.

The supply to all the services, except engine starting, is routed via a 16 A fuse located at the top left-hand side of the control panel. No other fuses are employed in the system.

9.1.5.2　Ancillary Circuits

The control panel is illuminated by a pair of lamps which may be switched on, as required, at the control panel. Operation of this switch will also connect the supply to a flasher unit which, in turn, pulses the supply to the roof obstruction lamp on top of the airstarter unit canopy.

9.1.5.3　Engine Starting Circuits

In order for the engine to start and continue running, the run solenoid must be in an energized condition so that fuel is allowed to flow to the engine fuel injectors. The fuel solenoid, in turn, is subject to a series circuit consisting of five sensing switches associated with (a) high air temperature; (b) compressor high oil temperature; (c) compressor low oil pressure; (d) engine high water temperature; and (e) engine low oil pressure.

Provided the air control selector on the control panel is set to the UNLOAD position, a switch (incorporated on the valve) is made which prepares the engine starter solenoid circuit. The starter solenoid circuit is completed when the engine start switch on the control panel is held against its spring in the START position, and provided relay RN is energized. Because the engine has been at rest, the low oil pressure sensing switches on the engine and compressor will initially be open. In order to obviate this initial condition, an oil pressure inhibit push button is provided on the control panel. When pressed, the push button bridges the open contact condition of the sensing switches, enabling the relay RN to energize.

Once the engine has started, the engine start switch is released so that it assumes the normal RUN position. The oil pressure inhibit push button must, however, be maintained in the pressed condition until the engine oil pressure and compressor oil pressure gauges on the control panel provide appropriate oil pressure readings. The push button can then be released since relay RN is now held on by the normally closed contact condition of the sensing switches. The START/RUN position of the engine start switch places a supply on timer TS. The timer contact closes and, while the engine speed is below 1,800 r/min, causes the three-port solenoid valve to energize. While this valve is energized, all pressure in the air delivery manifold is relieved. The engine speed sensing switch opens to de-energize the three-port solenoid valve at speed above 1,800 r/min (assuming that JET START or AIR COND has been selected at the air control selector valve).

Should any fault occur affecting one of the sensing switches listed above, the relevant switch will open to de-energize relay RN. Contact RN 1 will change over, causing the run solenoid to de-energize and therefore starve the engine of fuel, and placing the supply on a general fault warning lamp on the control panel and a warning lamp on the roof of the unit.

9.1.5.4　Control and Safety Circuits

While the air control selector is in the UNLOAD position，the switch on the valve also makes the supply to the throttle change solenoid. While this solenoid is energized the engine throttle is maintained at its minimum speed setting. In addition，the supply is made to the pressure relief solenoid valve，so that no build-up of pressure in the manifold is possible.

When the air control selector valve is set to the JET START position，the switch on the valve removes the supply from the pressure relief solenoid valve，which closes，and from the throttle change solenoid，which de-energizes and causes the engine to speed up to maximum rotate speed. At the same time，the switch places a hold-on supply to the run solenoid. The purpose of this hold-on supply is to ensure that the engine will not shut down（and thus interfere with the stringent jet start cycle）. Any one of the fault sensing switches shouldn't be operated. .

Caution：If the general fault warning lamps light during the jet start cycle，select unload as soon as the cycle is complete. The unit will then shut down. Do not shut down the unit by selecting off at the engine start switch.

When the air control selector is set to AIR COND，the circuit conditions are the same as described for JET START except that the bypass supply to the run relay RN is removed. The engine is therefore subject to automatic shut down in the event of any of the faults.

When either of the two air shut-off valves at the rear of the unit is closed，the appropriate pressure relief valve solenoid is energized to relieve pressure in the air hose，irrespective of the setting of the air control selector valve.

9.1.5.5　Shut-down

When the engine start switch is set to OFF for shut-down，relay RN and timer TS are immediately de-energized. Contact RN 1 removes the supply from the run solenoid，thus starving the engine of fuel. Timer contact TS 1，however，holds on for a period of 10 s so that the three-port solenoid valve is kept energized for a period sufficient to ensure that all pressure in the air delivery manifold is relieved. The DC supply to the tacho generator is also maintained during this period.

New Words

1. airstarter ['ɛəstɑːtə] *n.* 气源车
2. silencer ['sailənsə(r)] *n.* 消声装置，消声器
3. canopy ['kænəpi] *n.* 天蓬，遮蓬
4. perspex ['pəspeks] *n.* 塑胶（有机、防风）玻璃
5. stow [stəʊ] *v.* 安装
6. flange [flændʒ] *n.* 法兰
7. intermesh [intə'meʃ] *v.* 使互相结合，使互相啮合
8. lobe [ləʊb] *n.* 凸角叶片

9. jacket ['dʒækit] *n.* 水套,油套

10. penetration [peni'treiʃn] *n.* 穿透,渗透

11. dipstick ['dipstik] *n.* 量杆,量油尺

12. clogging ['klɒɡiŋ] *n.* 堵塞

13. indrawn [,in'drɔːn] *adj.* 吸入的,吸进的

14. cartridge ['kɑːtridʒ] *n.* 滤芯,滤筒

15. ammeter ['æmitə(r)] *n.* 电流表,安培计

Phrases and Expressions

1. ambient temperature 周围温度,环境温度

2. exhaust silencer 排气消声器

3. gas turbine engine 燃气涡轮发动机

4. fire extinguisher 灭火器

5. ball-and-roller bearing 滚珠轴承

6. timing gear 正时齿轮,定时齿轮

7. sealing ring 密封环

8. pressure gauge 压力计,压力表

9. temperature gauge 温度计,温度表

10. vacuum gauge 真空计,真空表

11. series-connected 串联的

Notes to the Text

1. Although essentially designed for starting aircraft gas turbine engines, the unit is also used for air conditioning the aircraft while it is parked.

气源车本来是设计用来起动飞机燃气涡轮发动机的,但飞机发动机停车时,也可以用作飞机的空气调节装置。

2. The engine, step-up gearbox and compressor constitute an integral assembly which is mounted on a fabricated steel bedplate and enclosed in a weather protecting steel canopy.

发动机、升级变速箱和压缩机构成一个完整的装置,安装在钢铁制成的底盘上,并附有钢铁制成的顶棚以防风雨的侵蚀。

3. Temperature and pressure switches, associated with both the engine and the compressor, control an engine run solenoid so that, in the event of overheating or oil pressure failure, the engine is automatically shut down.

发动机和压缩机配有温度和压力开关,控制发动机运转电磁线圈,一旦过热或压力过低,发动机会自动停车。

4. The compressor unit consists of a housing in which are assembled the compressor driving components and the compressor element. The element comprises two precision ground Archimedean-type screw rotors, each of which is suspended by ball-and-roller bearings and arranged to intermesh with each other along their longitudinal axes within the

compression space of the housing.

压缩机由装在其箱体内的驱动装置和转子组成。转子部分是两个在箱体压缩空间内沿着纵轴相互啮合的、精密磨光的阿基米德型螺杆,并由滚珠球轴承支撑。

5. The compressor element is cooled — and its driving components and bearings are lubricated — by a common oil system contained in and pumped from the step-up gearbox via an oil cooler which is mounted on the engine casing.

压缩机转子的冷却,驱动装置和轴承的润滑由一个公共的润滑系统(包含升级变速箱)经安装在发动机箱体上的冷却器泵油完成。

6. The indrawn air is compressed in the compressor elements 10 and discharged through the cross piece adaptor 16 to the air delivery pipe assembly 8. The cross piece adaptor incorporates the regulating valve 6 which relieves the air pressure to atmosphere via the dump silencer 5 whenever the selected low or high pressure is exceeded. The two air hoses are connected to the air delivery manifold via air shut-off valves 9 with associated couplings.

吸入的空气在转子 10 内被压缩,经十字块适配器 16 输送到空气输送管 8。十字块适配器与调压阀 6 制成一体。调压阀只要选择低压(或高压)超出范围,便经排气消声器 5 释放压缩空气的压力到大气。两个空气输送管经关断阀 9 及相应的接头连接到空气输送总管上。

7. The electrical system depends on two 12 V, series-connected, lead-acid batteries. Once the engine is started and running, an engine-driven alternator and associated voltage regulator maintain the battery in a charged condition. A battery ammeter, located on the control panel, indicates the charge/discharge condition of the battery.

电气系统由两个 12 V、串联的铅-酸蓄电池提供电源。发动机启动并运转时,发动机驱动的交流发电机和与之相连稳压器给蓄电池充电。安装在控制面板的蓄电池电流表显示蓄电池是充电还是放电。

8. When either of the two air shut-off valves at the rear of the unit is closed, the appropriate pressure relief valve solenoid is energized to relieve pressure in the air hose, irrespective of the setting of the air control selector valve.

当气源车后部的两个空气关断阀任意一个关闭时,不论空气控制选择阀处于什么状态,相应的压力释放阀的电磁线圈通电,释放空气输送管内的压力。

Section 9.2　Operation

9.2.1　Operating Controls and Indicators

9.2.1.1　General

All the controls, indicators and gauges necessary for operating the unit are grouped together on a control panel fitted within the canopy (see Figure 9.6), and are located at eye level on the rear of the unit. That part of the panel containing the various gauges is fronted by a perspex cover, all the "handed" controls and switches are at the bottom of the control panel.

Two manually-operated air shut-off valves are located at the rear of the airstarter unit on the left-hand side.

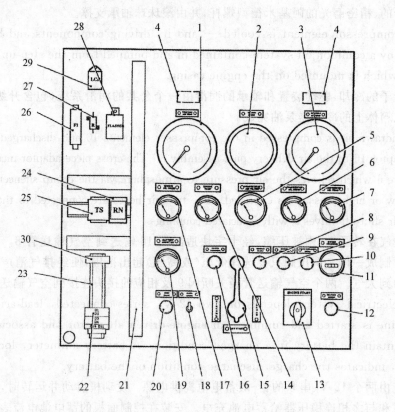

Figure 9.6   Operating controls and indicators

1—deliver air temperature gauge;   2—tachometer（×100）;   3—delivery air pressure gauge;   4—panel lamps;

5—compressor air filter;   6—compressor oil pressure gauge;   7—compressor oil temperature gauge;

8—lamp indicator（general fault warning）;   9—lamp indicator（engine air filter restriction）;

10—lamp indicator（low coolant）;   11—hours run recorder;   12—oil pressure inhibit button;

13—engine start switch;   14—engine water temperature gauge;   15—air control selector valve（and switch）;

16—lamp indicator（engine oil pressure warning）;   17—engine oil pressure gauge;

18—roof and panel light switch;   19—battery ammeter;   20—cold start control;

Mounted on the extreme left-hand side of the panel, but not in view, are the following components of the electrical circuit:

21—cable trunking;   22—diode module;   23—terminal block;   24—engine run relay RN;

25—timer TS;   26—control fuse,16 A;   27—resistor 25 ohm 25 W（used with the flasher unit）;

28—flasher unit;   29—low coolant relay;   30—timer module

9.2.1.2 Gauge Readings

The following is a list of readings that should be obtained at the various gauges under normal operating conditions. The numbers in parenthesis relate to those shown in Figure 9.6.

Ammeter（19）   Charge 0 – 40 A

Engine oil pressure（17）   Minimum 30 psi（2.1 bar）

Engine water temperature（14）   150 – 210℉（65.5 – 99℃）

Compressor oil pressure (6)	Average 20 psi (1. 4 bar)
Compressor oil temperature (7)	100 – 180°F (38 – 82°C)
Engine tachometer (2)	Unload: 1,100 r/min
	Air conditioning: 2,100 r/min
	Jet start: 2,100 r/min
Delivery air temperature (1)	Not to exceed 250°C (480°F)
Deliver air pressure (3)	Should not exceed 50 psi (3. 5 bar)
	Air conditioning: 30 psi (2. 1 bar)*
	Jet start: 40 psi (2. 8 bar)*

* At normal temperature 60°F (15. 6°C) and barometric pressure 29. 5 inHg (750 mmHg).

9. 2. 1. 3　Controls

(1)A control board fitted within the canopy at the rear of the unit comprises:

Air pressure gauge (direct reading 0 – 60 psi).

Tachometer (electric operated 0 – 3,000 r/min).

Air temperature gauge (transducer operated 0 – 300°C).

Engine oil pressure gauge (transducer operated 0 – 100 psi).

Engine water gauge (transducer operated 40 – 120°C).

Hours run recorder (electric).

Compressor oil pressure gauge (transducer operated 0 – 100 psi).

Compressor oil temperature gauge (transducer operated 40 – 120°C).

Battery charge ammeter (40 – 0 – 40).

Compressor air filter vacuum gauge (0 – 30 in).

Cold start kit (ether-manual operation).

OFF/RUN/START engine switch (with safety interlock).

Oil pressure inhibit pushbutton.

3 Position air supply valve allowing idling, air conditioning, jet start.

Engine low coolant indication.

(2)Protection circuits for:

Low engine oil pressure.

Low compressor oil pressure.

High engine water temperature.

High compressor coolant temperature.

High air outlet temperature.

Note: These protection circuits are not operable in "JET START".

(3)(Pressure relief valves: Spring loaded variety set release at 50 psi.

(4)Pressure operating valve: Dual mode valve set at 30 psi for air conditioning and 42 psi for jet engine starting and control operation.

9.2.2 Operating Instructions

9.2.2.1 Post Delivery Checks

On receipt of the airstarter a careful check should be made to ensure that the consignment is complete and undamaged. All visible components should be checked for security. Should breakages, damage or shortfalls be noted please inform ADE-HML immediately as shown on the instructions on the advice note. Failure to inform the company of this may invalidate the warranty.

9.2.2.2 Commissioning before Use

Proceed as follows:

Check that the tyre pressures are correct.

Depending upon transit distance and/or time, the unit may be despatched with its batteries dry and uncharged. The condition of the batteries should therefore be checked and initial filling and charging carried out as necessary. A label attached to the batteries gives filling and charging instructions. Note that the specific gravity of a fully charged lead acid battery should be 1.265/1.280 in a temperate climate and 1.235/1.250 in a tropical climate.

Check the engine as follows (refer to the engine handbook for specific details):

(1) Water level. Fill with water, and add inhibitor and anti-freeze solution if in a temperate climate.

(2) Oil level. Fill with the correct grade and type of oil.

(3) Fuel level. Fill with the correct grade of diesel fuel. The tank capacity is 640 litres (140 gallons).

Check the compressor gear box oil level using the gear box dip stick. Top up as necessary to the indicated level with the correct grade of oil.

Check that the fire extinguisher is full.

Ensure that the side panels of the canopy are secure.

9.2.2.3 Positioning the Unit

Caution: The airstarter must be started before it is connected to an aircraft.

Proceed as follows:

Manoeuvre the unit into position adjacent to the aircraft air hose connection and apply the parking brake.

Connect the delivery air hose(s) to the delivery air coupling(s) on the airstarter unit. Ensure that the hose(s) is/are completely uncoiled and stretched out with no kinks.

9.2.2.4 Starting the Unit

Proceed as follows:

Turn the air control selector to the UNLOAD position.

Turn the engine start switch to the START position and hold it in this position and press and hold oil pressure inhibit button. When the engine fires release the switch, which will assume the RUN position, maintain oil pressure inhibit until oil pressure reaches approx. 20 psi and release. (While the air control selector is in the UNLOAD position, the engine throttle is automatically set so that the engine idles at 1,100 r/min.)

Turn the air control selector to the JET START or AIR CONDITIONING position, as required.

Check all readings at the gauges on the control panel while the unit is warming up and ensure that they are as required for jet start or air conditioning operation.

When entirely satisfied that pressures are as required, connect the delivery hose(s) to the air starter coupling(s) on the aircraft. Commence delivery, as commanded, by opening the air shut-off valve(s).

9.2.2.5　Cold Start

Caution:The cold start facility must only be used when the engine is being cranked during starting.

If the engine fails to start, or if the engine is to be started during cold conditions, another operator will be required to assist in the starting sequence. The sequence is exactly described as above in 9.2.2.4. Except that, while the engine is cranking, the other operator pulls out the COLD START knob on the control panel and then releases it, thus causing starting fluid (ethyl ether) to be injected into the engine intake manifold. Repeat the sequence if the engine still fails to start.

9.2.2.6　Shut Down

Proceed as follows:

Turn the air control selector to the UNLOAD position.

Allow the unit to run in the unload mode for two/three minutes, enabling the engine to run in the idling condition and thus permit normal cool-down.

Turn the engine start switch to OFF.

Switch the panel and roof lamps OFF.

9.2.2.7　Moving the Airstarter

Proceed as follow:

Disconnect the air hoses from the delivery air couplings.

Stow the hoses correctly in their trays.

Drive the airstarter to its new location.

9.2.3　Output

The unit has a maximum output of 136 kg/min (300 lb/min) at 2.9 bar (40 psi) for jet starting, or 2.1 bar (30 psi) for air conditioning, under normal conditions of barometric pressure, temperature arid humidity. Because air is used as the operating medium, performance depends on the pertaining conditions of temperature and barometric pressure, and these conditions can vary any time, any day. Detailed performance figures of the unit for such varying conditions are shown in graph Figure 9. 7, which also gives a table of the maximum discharge pressure that can be obtained for various airfield heights (see Table 9. 1).

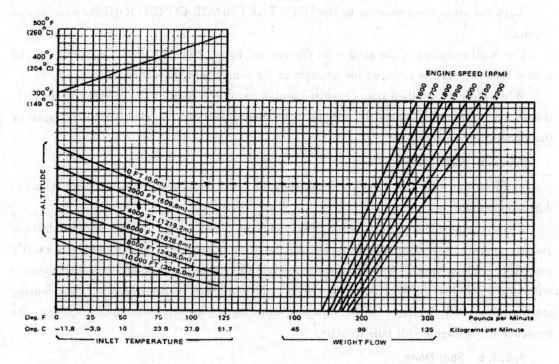

Figure 9. 7   Performance versus altitude and temperature

The most important operating consideration is airfield height, and the appropriate adjustments — for air starting and air conditioning applications.

Table 9. 1   The maximum discharge pressure for airfield heights

Altitude		Maximum Discharge Pressure
(Feet)	(Meters)	
0	0. 0	43 lbf/in² (2. 95 kgf/cm²)
2,000	609. 8	40 lbf/ in² (2. 74 kgf/cm²)
4,000	1,219. 2	37 lbf/ in² (2. 53 kgf/cm²)
6,000	1,828. 8	35 lbf/ in² (2. 39 kgf/cm²)
8,000	2,436. 4	32 lbf/in² (2. 18 kgf/cm²)
10,000	3,048. 0	29 lbf/in² (2. 03 kgf/cm²)

New Words

1. tachometer [tæˈkɒmitə] *n.* 转速计(表)
2. barometric [bærəˈmetrik] *adj.* 大气压力的
3. transducer [trænzˈdjuːsə] *n.* 传感器
4. trunking [ˈtrʌŋkiŋ] *n.* 线槽
5. commissioning [kəˈmiʃəniŋ] *n.* 试运转,试车
6. inhibitor [inˈhibitə] *n.* 抑制剂,防锈剂
7. manoeuvre [məˈnuːvə] *vt.* 操纵
8. uncoil [ʌnˈkɔil] *vt.* 解开
9. kink [kiŋk] *n.* 纽结,绞缠
10. tray [trei] *n.* 绞盘

Phrases and Expressions

1. eye level 平视
2. specific gravity 相对密度
3. temperature climate 温暖气候
4. tropical climate 热带气候
5. anti-freeze solution 防冻剂
6. airfield height 机场海拔高度

Notes to the Text

1. All the controls, indicators and gauges necessary for operating the unit are grouped together on a control panel fitted within the canopy, and are located at eye level on the rear of the unit.

操作气源车必需的所有控制器、显示器和量表都集中在遮蓬的控制面板上,并安装在气源车后部的视平线高度上。

2. On receipt of the airstarter a careful check should be made to ensure that the consignment is complete and undamaged. All visible components should be checked for security. Should breakages, damage or shortfalls be noted please inform ADE-HML immediately as shown on the instructions on the advice note. Failure to inform the company of this may invalidate the warranty.

收到气源车后应立即进行仔细检查以确保托运的货物完整、无损坏。出于安全考虑,所有明显的部件都应该检查。如果有破损、毁坏或短缺请立即按照通知书的说明书面告知 ADE – HML。若未告知,ADE – HML 公司将不承担责任。

3. When entirely satisfied that pressures are as required, connect the delivery hose(s) to the air starter coupling(s) on the aircraft. Commence delivery, as commanded, by opening the air shut-off valve(s).

当压力完全满足要求时,将输送软管连接到飞机的接口上,按照要求,打开空气关断阀,开

始输送。

4. The unit has a maximum output of 136 kg/min (300 lb/min) at 2.9 bar (40 psi) for jet starting, or 2.1 bar (30 psi) for air conditioning, under normal conditions of barometric pressure, temperature arid humidity.

在正常的大气压力、温度和湿度条件下,气源车在最大输出流量为 136 kg/min (300 lb/min),压力为2.9 bar(40 psi)时启动飞机发动机;压力为 2.1 bar (30 psi)进行空气调节。

Section 9.3　Maintenance

9.3.1　Servicing

9.3.1.1　General

Normally the regulating system governing the engine speed requires no servicing other than minor lubrication of the linkage connecting the speed control solenoids to the engine governor. The air delivery pressure of the unit, however, will require adjustment to suit different aircraft jet starting requirements, and to be compatible with the height above sea level at which the unit is being operated.

9.3.1.2　Engine Speed Adjustment

The engine speed at full load is controlled by the high speed adjustment on the engine governor. Removing the cover and turning the allen head screw clockwise lowers the full load speed.

Note: The full-load speed must be set to 2,100 r/min.

The engine idle speed is controlled by the low speed adjustment on the engine governor. Removing the cover and turning the allen head screw clockwise increases the idle speed.

Caution: Do not allow the engine to idle below 1,100 r/min.

9.3.1.3　Delivery Pressure Adjustment for Jet Start

The delivery pressure can be raised or lowered to meet any aircraft requirements for jet starting provided the pressure does not exceed the values listed in Table 9.2 for a given altitude. Adjustment is carried out on the pressure operating valve (see Figure 9.8).

Table 9.2　Maximum discharge pressure for a given altitude

Altitude		Maximum Discharge Pressure at Control Panel for Jet Start
0	feet (0 meters)	42 psi (3.0 bar)
2,000	feet (600 meters)	39 psi (2.7 bar)
4,000	feet (1,200 meters)	36 psi (2.5 bar)
6,000	feet (1,800 meters)	34 psi (2.4 bar)
8,000	feet (2,400 meters)	31 psi (2.2 bar)
10,000	feet (3,000 meters)	29 psi (2.0 bar)

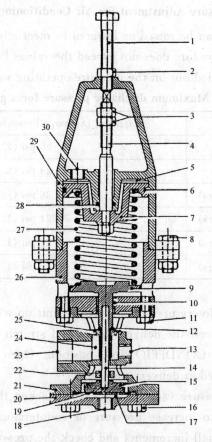

Figure 9. 8   Pressure operating valve

1—adjusting screw (air conditioning);   2—check nut;   3—adjusting nuts (jet start);   4—piston rod;   5—chamber;

6—O-ring;   7—piston;   8—bolt;   9—spring seat;   10—guide bush;   11—setscrew;   12—ring;

13—connection to 3-port solenoid;   14—stem;   15—plunger;   16—setscrew;   17—chamber;

18—connection to air delivery;   19—membrane;   20—cover;   21—valve housing;   22—guide bush;

23—connection to regulating valve;   24—chamber;   25—oblique bracket;   26—spring housing;

27—spring O-ring;   28—trapezoidal seal ring;   29—cylinder;   30—connection to air control selector

Before adjusting the delivery pressure, start the unit up with the air control selector in the UNLOAD position. Uncoil the delivery hoses and stretch them out, then turn the air control selector to the JET START position, then open the air delivery valves. The hoses will become pressurised at delivery pressure.

To set the jet start delivery pressure (see Figure 9. 8), slacken the lock nuts 3 while holding the squared end of piston rod 4. To increase pressure, hold the piston rod and adjust the nuts upwards; to decrease the pressure, adjust the nuts downwards. Make the adjustments in small increments and check the pressure frequently so as to avoid over pressure of the compressor.

Caution: Ensure that the adjusting screw lock nuts are tight after making the pressure adjustment.

9.3.1.4 Delivery Pressure Adjustment for Air Conditioning

The delivery pressure can be raised or lowered to meet any aircraft requirement for air conditioning provided the pressure does not exceed the values listed in Table 9.3 for a given altitude. Adjustment is carried out on the pressure operating valve (see Figure 9.8).

Table 9.3   Maximum discharge pressure for a given altitude

Altitude		Maximum Discharge Pressure at Control Panel for AC
0	feet (0 meters)	30 psi (2.1 bar)
2,000	feet (600 meters)	28 psi (2.0 bar)
4,000	feet (1,200 meters)	26 psi (1.8 bar)
6,000	feet (1,800 meters)	24 psi (1.7 bar)
8,000	feet (2,400 meters)	22 psi (1.6 bar)
10,000	feet (3,000 meters)	20 psi (1.4 bar)

Before adjusting the delivery pressure, start the unit up with the air control selector in the UNLOAD position. Uncoil the delivery hoses and stretch them out, then turn the air control selector to the AIR CONDITIONING position. Open the air delivery valves; the hoses will become pressurized at delivery pressure.

To set the delivery pressure (see Figure 9.8), slacken the check nut 2 and turn the adjusting screw clockwise to increase pressure or counterclockwise to decrease pressure. Make the adjustments in small increments and check the pressure frequently so as to avoid over pressurizing the compressor.

Caution: Ensure that the adjusting screw check nut is tight after making the pressure adjustment.

9.3.2 Maintenance

Note: See the vehicle manufacturer's operating manual for the servicing requirements of the truck chassis.

9.3.2.1 General

The maintenance intervals given below are only a guide. They can be adapted to coincide with the maintenance schedule for the engine and/or the maintenance routine usually practised by the user.

The inspections are cumulative, i.e. items given against the 10-hour inspection should be included in the 50 – 60 hour inspection schedule, and so on.

Reference should be made to the engine operator's manual for more detailed particulars of engine maintenance.

9.3.2.2 Daily

Proceed as follows:

Check the coolant level in the overflow tank of the engine radiator. Maintain the level between the minimum and maximum marks.

Check the oil level in the engine and compressor sumps. Top up as necessary with the recommended oil.

Fill the fuel tank at the end of the day's run. This helps to prevent moist air from condensing on the inside walls of the tank and contaminating the fuel with water.

9.3.2.3   Every 10 Hours or Weekly

Continue from paragraph 9.3.2.2 as follows:

Clean out the unit. Inspect for fuel, oil or coolant leaks.

Check the electrolyte level in the batteries. Also check that the battery terminals are clean and tight.

Check that the pressure operating valve and regulating valve together maintain the pre-set operating pressures for air conditioning and air start.

9.3.2.4   Every 50 – 60 Hours or Monthly

Continue from paragraph 9.3.2.2, 9.3.2.3 as follows:

Clean and examine the compressor air intake filter cartridges; replace them as necessary.

Note: If the unit is operated in a dust-laden atmosphere, the filter elements should be changed more frequently.

Drain any water or sediment which may have been collected in the fuel tanks. Drain until clean fuel flows from the drain cock of each tank.

Lubricate the ball-and-socket joints of the engine throttle operating linkage. Also lubricate all hinges and locks, etc.

Check the condition and tension of the engine fan belts; adjust the tension if necessary. Replace the belts if they are worn or frayed.

Check the condition of the air delivery hoses.

Check the fire extinguisher and see that the seal is intact. Follow the instructions given on the data plate attached to the extinguisher.

Check that the engine air filter restriction and the low oil pressure warning lamps on the control panel are serviceable. Replace any defective lamps.

Carry out the first renewal of the compressor lubricating oil and the compressor oil filter element. Subsequent oil changes and oil filter element renewals are to be made after every 500 h of operation or yearly.

9.3.2.5   Every 150 – 180 Hours or Half Yearly

Continue from paragraph 9.3.2.2 – paragraph 9.3.2.4 as follows:

(1) Inspect the exterior of the radiator and compressor oil cooler cores, and clean as follows:

Remove any accumulated dust and dirt with a stiff-fibre brush. Do not use a wire brush or any metal objects.

If the dirt is of an oily nature and cannot be removed by air jet, or if compressed air is not available, clean the cores with a quality grease solvent. A spray gun should preferably be used to apply the solvent to the cores. Rinse with a powerful water jet and blow dry with compressed air, if available. Repeat the process as many times as necessary (Another method of cleaning the cores is the use of steam or a steam cleaning device, if available).

(2) Remove the compressor air intake filters and clean them with an air jet.

9.3.2.6 Every Year or 2,000 Hours of Operation

Continue from paragraph 9.3.2.2 – paragraph 9.3.2.5 as follows:

Renew the compressor air intake filter cartridges.

Drain the lubricating oil from the compressor sump while the unit is thoroughly warm. Securely tighten the drain plug after draining and fill up with the recommended oil.

Renew the compressor oil filter element.

Clean the breather of the step-up gear housing as follows:

Remove the breather.

Dismantle it and wash the two steel mesh pads in diesel fuel oil or some similar cleaning solvent. Wash the sintered bronze filter disc in trichlorethylene.

Dry the parts with compressed air.

Reassemble and reinstall the breather on the pipe.

Remove and inspect the rubber diaphragm of the balancing pistons on the compressor element.

Inspect the rolling diaphragm of the regulating valve (see Figure 9.9).

Dismantle, clean and inspect the parts of the pressure operating valve (see Figure 9.8).

Check the operation of the pressure relief valves.

Check the operation of the engine and compressor safety switches.

9.3.2.7 Engine Fuel Filters

Refer to the engine manufacturer's operation manual.

The absolute necessity of regular fuel filter maintenance is specially stressed in the engine operator's manual. A clogged fuel filter means fuel starvation and reduced engine performance.

It should be noted that the condition and the quality of the fuel determine the frequency of fuel filter renewal.

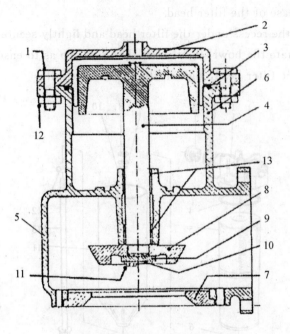

Figure 9.9　Regulating valve

1—screw；2—cover；3—rolling diaphragm；4—piston；5—valve housing；6—valve seat；7—valve；
8—disk；9—lock plate；10—screw；11—nut；12—bush (dry low-friction type)；13—sealing ring

9.3.2.8　Compressor Oil Filter

The compressor oil filter assembly comprises a filter element bolted to a filter head (see Figure 9.10). The filter element must be replaced each time the compressor lubricating oil is changed. Proceed as follows：

Place a drain pan under the filter.

Unscrew the centre bolt 13 then remove the bowl 11 from the filter head 1.

Withdraw the filter element 5 from the bowl; discard the element.

Remove circlip 6 from the centre bolt 13, then remove guide 7, centre bolt upper seal 8, washer 9 and spring 10.

Withdraw centre bolt 13 with its lower seal 12.

Rinse the bowl and all other filter parts in a suitable cleaning solvent and then blow-dry using an air jet.

Clean the filter head 1, paying particular attention to the recess that engages with the filter bowl.

Inspect all the seals for damage or hardening; renew as necessary.

Locate seal 12 on the centre bolt, then pass the bolt through the bottom of the bowl.

Locate spring 10, washer 9, seal 8 and guide 7 on the centre bolt and secure in position with circlip 6.

Insert a new element into the bowl.

Fit seal 4 to the base of the filter head.

Fit the bowl into the recess under the filter head and lightly secure in position by means of the centre bolt. Rotate the bowl slightly left then right so as to ensure that the bowl has seated correctly into the filter head, tighten the centre bolt fully.

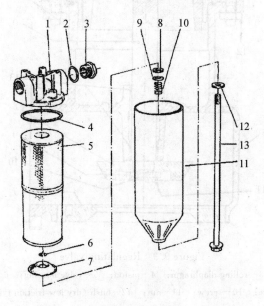

Figure 9.10   Compressor oil filter

1—filter head;   2—washer;   3—hollow screw;   4—seal;   5—filter element;   6—circlip;   7—guide;
8—centre bolt upper seal;   9—rubber washer;   10—spring;   11—bowl;   12—lower seal;   13—centre bolt

9.3.2.9   Cleaning the Compressor Air Filters

The compressor air filters comprise a set of three paper cartridges. They should be removed at the recommended periods and cleaned and inspected as follows:

Carefully knock the end faces of each cartridge on a flat, soft surface such as the palm of the hand or a rubber tyre so as to remove the majority of the heavy dry contaminant. Do not strike the side or bottom of the cartridge on a hard surface.

Blow dry air, at a pressure not exceeding 5 bar (73 psi), up and down the pleats on the inside of the cartridges, then blow up and down the length of the pleats on both sides of the cartridges. Maintain a reasonable distance between the air nozzle and the pleats.

Inspect each cartridge for any signs of damage by placing a bright light inside it. Thin (weak) spots, pin holes or the slightest rupture in the paper will render a cartridge unfit for further use. Each replacement cartridge must be examined for tears or punctures in the same way before being fitted.

Caution: If the compressor is separated from the engine for any reason, it is essential that the bolts securing the flywheel to the driving flange of the flexible coupling are tightened to the correct torque when re-assembling, before bending the tab washers.

The torque figure is (140 ±14) lb-ft, (190 ± 19) N · m.

Use new tab washers, ade part NO. D883 - AD41856.

9.3.3　Trouble Shooting

9.3.3.1　General

This section contains a trouble shooting chart which will aid the operating personnel by indicating possible troubles, the probable causes of the troubles, and the remedies for correcting the troubles. The operator must keep in mind that trouble-shooting is the organised investigation of an operating problem and its subsequent correction. The remedies for correcting the trouble include adjusting and repair/replacement of the component(s) causing the trouble.

Reference should be made to the "Trouble Shooting Section" of the engine operator's manual for trouble shooting of the engine and its accessories.

9.3.3.2　Trouble Shooting Chart (See Table 9.4)

Table 9.4　The troubleshooting chart

Fault/Symptom	Possible Cause	Remedy
A.　High compressor discharge temperature	Discharge pressure is too high for the altitude at which the unit is operating	The maximum discharge pressure settings for a given altitude
	Inaccurate delivery air pressure gauge reading	Check reading with a calibrated gauge
	Inaccurate delivery air temperature gauge reading	Check reading with a calibrated gauge
	Restricted compressor inlet air passage	Check for damaged inlet duct. Be certain no rags or foreign material are in the air duct
	Excess compressor rotor discharge end clearance	Call Atlas Copco engineer
	Compressor rotors out of line and contacting	Call Atlas Copco engineer
B. Low discharge pressure on JET START with hose (s) stretched out and not connected to aircraft	(1) JET START setting incorrect in pressure relief valve	Refer to fault/symptom a remedy NO. 1
	(2) Leak in control piping	With air control selector set to JET START, feel control piping with hands or check for leaks with soap suds. The only air leak should be the normal discharge from the JET START pilot valve
	(3) Inaccurate delivery air pressure gauge reading	Check reading with calibrated gauge
...	...	...

New Words

1. clockwise [ˈklɔːkwaiz] *adv.* 顺时针方向（地）
2. slacken [ˈslæk(ə)n] *vt.* 松弛
3. setscrew [ˈsetskruː] *n.* 调节螺钉（栓）
4. plunger [ˈplʌn(d)ʒə] *n.* 活塞，柱塞
5. membrane [ˈmembrein] *n.* 膜，隔膜
6. counterclockwise [kauntəˈklɔːkwaiz] *adv.* 反时针方向（地）
7. electrolyte [iˈlektrəlait] *n.* 电解液，电解质
8. sediment [ˈsedimənt] *n.* 沉淀（物），沉积
9. fray [frei] *v.* 磨损
10. grease [griːs] *n.* 滑脂
11. rinse [rins] *n.* 漂洗
12. dismantle [disˈmænt(ə)l] *vt.* 拆除，拆卸
13. diaphragm [ˈdaiəfræm] *n.* 隔膜，隔板

Phrases and Expressions

1. engine governor 发动机调速器
2. spring seat 弹簧座
3. guide bush 导向衬套
4. maintenance interval 维修间隔期
5. coincide with 与……相符，与……一致
6. top up 加满，充值
7. drain cock 排气（水）旋塞
8. ball and socket joint 球窝接头（关节）

Notes to the Text

1. The engine speed at full load is controlled by the high speed adjustment on the engine governor. Removing the cover and turning the alien head screw clockwise lowers the full load speed.

发动机全负荷转速由发动机调速器的高速调节装置控制。打开外盖，顺时针方向旋转六角头螺钉将降低全负荷转速。

2. To set the jet start delivery pressure, slacken the lock nuts while holding the squared end of piston rod. To increase pressure, hold the piston rod and adjust the nuts upwards; to decrease the pressure, adjust the nuts downwards. Make the adjustments in small increments and check the pressure frequently so as to avoid over pressure of the compressor.

为了设定启动飞机发动机的传输压力，固定活塞杆方头的同时，旋松锁紧螺母。活塞杆固定，向上旋转锁紧螺母，压力增加；反之，向下旋转锁紧螺母，压力降低。以微小增量调整并经常检查压力，以免压缩机超压。

3. The maintenance intervals given below are only a guide. They can be adapted to coincide with the maintenance schedule for the engine and/or the maintenance routine usually practised by the user.

下面给出的维护间隔仅供参考。维护间隔可以和发动机的维护进度或用户例行维护周期一致。

4. Dismantle it and wash the two steel mesh pads in diesel fuel oil or some similar cleaning solvent. Wash the sintered bronze filter disc in trichlorethylene.

拆下它并在柴油或类似的清洗剂中清洗两个金属网衬垫。在三氯乙烯中清洗烧结的青铜滤芯。

5. The absolute necessity of regular fuel filter maintenance is specially stressed in the Engine Operator's Manual. A clogged fuel filter means fuel starvation and reduced engine performance.

发动机操作手册特别强调了定期维护燃油滤清器的必要性。燃油滤清器的堵塞将导致燃油不足并降低发动机的性能。

6. Carefully knock the end faces of each cartridge on a flat, soft surface such as the palm of the hand or a rubber tyre so as to remove the majority of the heavy dry contaminant. Do not strike the side or bottom of the cartridge on a hard surface.

在平坦的、柔软的表面上,如手掌或橡胶轮胎,敲打滤芯的端面,除去大多数的干的、较重的沉积物。不能在坚硬的表面上击打滤芯的侧面或地面。

Chapter 10   Towbarless Aircraft Tractor

Section 10.1   General Description

10.1.1   General

Towbarless aircraft tractor is a narrow and low-profile diesel-powered, four-wheel drive tractor designed for aircraft push and tow. With an optional front cab that can be either stationary or elevated, the vehicle can hold a driver. The towbarless aircraft tractor is powered by an eight-cylinder Caterpillar (Cat) diesel engine; it can reach a top speed of approximately 17 mile/h (27.4 km/h) with no vehicle in tow. Towbarless aircraft tractor of the module AST-3 is shown in Figure 10.1.

Figure 10.1   Towbarless aircraft tractor

Mounted with the towbarless aircraft tractor of AST-3's structural frame is a diesel engine, a main hydraulic pump powered four speed transmission, a driveline assembly, a fuel system, an air intake system, a lubrication system, a cooling system, a steering system, a brake system, and an electrical system.

10.1.2 Major Components or System

10.1.2.1 Engine

The AST – 3 is powered by a four-stroke, eight-cylinder, turbocharged Caterpillar diesel engine. From inside the engine's cast iron block, a single camshaft drives all valves and the fuel injector pumps. At the engine's forward end is a vertically aligned gear train that drives the hydraulic pump, the lubricating oil pump and the fuel pump, and connects the entire assembly with the crankshaft and camshaft. For more detailed information, refer to the engine service manual.

10.1.2.2 Hydraulic System

The AST – 3's hydraulic system, powered by the diesel engine, provides for the driving force for four speed, steering, braking, and optional elevating front cab and internal self-jacking systems of AST – 3.

10.1.2.3 Main Hydraulic Pump

Hydraulic pressure for driving force, the steering, braking, front cab elevation and optional internal jacking system is provided by a single, variable displacement, axial piston, gear driven hydraulic pump mounted on the transmission. At an engine speed of 2,400 r/min, the pump will displace 45 gallons per minute (170.3 L/min) of fluid at 2,000 psi (13,789.5 kPa).

10.1.2.4 Emergency Steering Pump

An electrically driven hydraulic pump activated by the vehicle operator, in conjunction with low hydraulic pressure, provides emergency steering if the main hydraulic pump fails. The emergency steering pump is activated by pressing the momentary ON emergency steering pump switch when the hydraulic pressure in the steering system is below 1,800 psi.

10.1.2.5 Hydraulic Reservoir

The 60-gallon hydraulic reservoir contains the hydraulic fluid necessary to operate all subsystems and provide a reserve. The reservoir contains an integral filler cap and level gage, strainer, and vent.

10.1.2.6 Hydraulic Filter

A hydraulic filter assembly is located between the return manifold and the hydraulic reservoir. A filter service indicator is mounted on the filter housing to allow a visual indication of the relative accumulation of contaminants. A diverter/bypass valve in the filter head allows full-flow filtration until a bypass pressure is reached.

10.1.2.7 Dual Check/Distribution Valve

Hydraulic fluid leaving the main hydraulic pump is distributed to the major hydraulic subsystems by a dual check/distribution valve. The valve also routes fluid to the return manifold, provides a pick-off for the emergency steering pump switch, and incorporates check valves that block fluid flow in the wrong direction.

10. 1. 2. 8 Steering Subsystem

The steering subsystem is composed of the following components.

1. Single cab steering subsystem

(1) Char-lynn steering control unit. A 48-cubic inch displacement char-lynn steering control unit connected to the main hydraulic system receives steering wheel inputs and controls the porting of hydraulic power to the front axle steeling cylinders and to solenoid valve. The direction of turn determines the side of the steering cylinders and the solenoid valve that the unit ports power and fluid to. A left turn will send power directly to the left-turn side of the front steering cylinders. The return flow from these cylinders then drives solenoid valve which, in turn, ports power to the appropriate rear axle steering cylinders. For a right turn, the unit ports power directly to solenoid valve 1 and then to the right side of the front steering cylinders.

(2) Solenoid valve. Solenoid valve controls the rear axle steering cylinders. Its porting of hydraulic power is dependent upon which steering mode is selected. The valve incorporates an internal double-pilot check valve.

2. Dual-cab steering subsystem

(1) Orbital steering select valve. This valve selects the front or rear orbital steering valves. If the operator selects front cab operation, the valve directs pressure to the front orbital steering valve and relieves pressure from the rear orbital steering valve. Rear cab selection reverses that process. Shutting down the tractor will de-energize the valve's solenoids, moving it to the center position.

(2) Orbital steering valves, front & rear. Each valve is a manually operated, directional control valve and meter within a single valve. Rotating the steering wheel causes the attached valve to direct fluid to the appropriate steering cylinders on each axle.

(3) Steering manifold. The steering manifold coordinates the movements of the steering cylinders via two solenoid-operated directional control valves and their associated connections. It also provides a place for test gauge ports.

(4) Steering mode select valves. The valves are mounted on the steering manifold. When the tractor is being operated from the front cab, valve determines the tractor's steering mode. Another valve determines the steering mode when the tractor is being operated from the rear cab. When track steering mode is selected, the appropriate valve is de-energized or in its center position. Otherwise, the valves direct hydraulic fluid to obtain four-wheel coordinated or crab steering modes.

(5) Steering cylinders. The steering cylinders are double-acting hydraulic cylinders attached to the axles. When hydraulic pressure is directed into the cylinders by a steering valve, the cylinders extend or retract and turn the wheels in the required direction.

10. 1. 2. 9   Service Brake Subsystem

The service brake subsystem is composed of five components:

(1)Dual check relief valve. This valve splits the supply pressure into separate circuits for the front and rear brakes, checks the flow in one direction, and provides pressure relief for the service brake subsystem.

(2)Accumulators. The accumulators provide emergency service brake pressure if a loss of normal or emergency supply pressure occurs. Each service brake circuit contains one accumulator.

(3)Brake valve. The brake valve controls service brake application via inputs from the front or rear brakes and their slave cylinders. It receives main hydraulic system pressure (or brake system accumulator pressure if main system pressure disappears) and applies it to both front and rear service brakes when commanded.

(4)Front and rear brake master and slave cylinders. Service brake control pressure is supplied by the brake master and slave cylinders. There is one brake pedal assembly per cab, each equipped with dual master cylinders. When the operator applies pressure to the brake pedal, control pressure from the master cylinder acts against a slave cylinder. The slave cylinder is mechanically linked to the service brake valve, which diverts hydraulic system supply pressure to the front and rear brakes. The master/slave cylinder circuit uses DOT 3 brake fluid only.

(5)Service brake assemblies. Two service brake assemblies, a front and a rear, reside on each wheel. Each brake assembly consists of a caliper and its internal components and two brake pads surrounding a large brake disk. Hydraulic pressure applied to the caliper pushes the brake pads against the disk, providing the friction for the braking action.

10. 1. 2. 10   Parking Brake Subsystem

The AST – 3 has two caliper disk parking brakes. They are mounted on the front and rear axle shafts. Both parking brakes have a hydraulic parking brake release cylinder. The parking brake is spring-applied but hydraulically released. Hydraulic pressure is supplied by the main hydraulic pump and can be supplied by a remote hand pump. The parking brake subsystem is composed of three major components. These components are described as follows.

(1) Parking brake valve assembly. The parking brake valve assembly contains two solenoid valves that simultaneously change position to pressurize or relieve the parking brake calipers. In their de-energized state (parking brake on), the calipers are relieved, causing the caliper springs to apply the brakes. In their energized state (parking brake off), the calipers are pressurized, causing the brakes to release.

(2) Remote hand pump and shuttle valve. These components provide a means of releasing the parking brake should the main piston pump and the emergency pump fail. Once normal operation is resumed, a check valve and passages within the shuttle valve allow any

remaining oil pressure in the hand pump circuit to return to the reservoir.

（3）Parking brake calipers. When applied, the parking brake calipers prevent the rotation of the axle pinions, effectively locking the wheels.

10.1.2.11　Cab Lift Subsystem

The cab lift subsystem contains the cab lift valve assembly and the cab lift cylinder. The cab lift valve assembly is composed of two solenoid valves that alternately change positions to fill or, drain hydraulic oil from the cab cylinder. The single-acting cab lift cylinder provides 28 inches of cab lift.

10.1.2.12　Driveline Assembly

The front and rear drive shafts are coupled to the transmission via a transfer case mounted on the transmission's bottom. Turned by gears in the transfer case, the drive shafts transmit torque to the front and rear differentials which, in turn, transmit their torque to the axles, the planetary wheel assemblies, and the wheels.

10.1.2.13　Fuel System

Fuel from the fuel tank is pulled by the fuel transfer pump through a water separator (standard equipment) and a filter. From the filter, fuel flows to the housing for the injector pumps. Fuel enters this housing at the top and through an inside passage to the fuel transfer pump. From the fuel transfer pump, fuel under pressure fills the housing for the fuel injection pumps. Pressure of the fuel in the housing is controlled with a bypass valve. Fuel pressure at full load is 30 psi (205 kPa). If fuel pressure exceeds this limit, the bypass will open, allowing excess fuel to return to the inlet of the fuel transfer pump.

The engine mounted fuel filter is a single-unit spin-on cartridge.

10.1.2.14　Air Intake System

Air is drawn through a large single-element air filter mounted on the side of the vehicle into the intake side of an exhaust gas-driven turbocharger. The turbocharger pressurizes the intake air that is then cooled by an air-to-air heat exchanger to improve the combustion efficiency of the engine. The air is then ducted to the engine's intake manifold, which routes it to the individual cylinders. For more detailed information, refer to the engine service manual.

10.1.2.15　Exhaust System

The engine's hot exhaust gases are flushed out through its exhaust valves to the exhaust manifold. The exhaust manifold carries them into the turbocharger where they spin a single-stage turbowheel (The turbowheel is connected to an impeller on the air intake side of the turbocharger). After passing over the turbowheel, the exhaust gases travel through a muffler before being directed out to the rear of the vehicle.

10.1.2.16　Lubrication System

The engine lubrication system consists of an oil pump, oil cooler, a bypass and full-flow filter, a pressure regulator valve, and a pressure relief valve.

A gear-type engine lube, oil pump mounted on the vertical gear train pumps oil from the oil pan into a lube oil cooler mounted on the engine block. Oil is then pumped through an oil filter past a pressure relief valve. From the oil filter, oil progresses through the main oil gallery that lubricates the crankshaft, camshaft, connecting rod bearings, and, via a nozzle line, the pistons. Tappets pick up oil from the camshaft and route it through the pushrods to the rocker arms and valves. Some of the used engine oil is also routed to the turbocharger. Oil return lines from the turbocharger and the engine return the oil to the oil pan/sump. For more detailed information, refer to the engine service manual.

10.1.2.17   Cooling System

The vehicle's cooling system consists of the engine cooling system and the components used to cool the tractor's transmission system.

The engine cooling system consists of a coolant filter mounted atop the engine, a thermostat that controls the engine's temperature, an engine-driven coolant pump, a heat exchanger behind an engine-driven fan, a lube oil cooler, and paths within the engine to circulate coolant around the cylinder heads.

A radiator forward of the engine-mounted charge air cooler cools the hydraulic fluid from the transmission.

New Words

1. aircraft ['eəkrɑːft] *n.* 飞机
2. lube [luːb] *n.* 润滑油,润滑剂
3. lubricate['luːbrikeit] *vt.* 使润滑,加润滑油
4. contaminant[kən'tæminənt] *n.* 污染物
5. jack [dʒæk] *n.*(*v.*) 千斤顶
6. transmission [trænz'miʃ(ə)n] *n.* 传动装置,变速箱
7. camshaft ['kæmʃɑːft] *n.* 凸轮轴
8. mount [maunt] *n.*(*v.*) 安装,装配
9. torque [tɔːk] *n.* 扭矩,转矩
10. muffler ['mʌflə] *n.* 消声器
11. pushrod ['puʃrɔd] *n.* 阀门挺杆
12. strainer ['streinə] *n.* 网式过滤器
13. accumulator [ə'kjuːmjuleitə] *n.* 蓄电池

Phrases and Expressions

1. planetary wheel 行星齿轮,行星轮
2. fuel transfer pump 输油泵(低压油泵)
3. bypass valve 旁通阀
4. pressure regulator valve 压力调节阀

5. pressure relief valve 减压阀
6. level gage 液面指示器
7. check valve 单向阀
8. dual check valve 双向止回阀
9. crab steering mode 蟹行转向
10. slave cylinder 从动缸,辅助油缸
11. service brake 行车制动
12. parking brake 停车制动

Notes to the Text

1. The engine's hot exhaust gases are flushed out through its exhaust valves to the exhaust manifold. The exhaust manifold carries them into the turbocharger where they spin a single-stage turbowheel (The turbowheel is connected to an impeller on the air intake side of the turbocharger). After passing over the turbowheel, the exhaust gases travel through a muffler before being directed out to the rear of the vehicle.

发动机的高温废气经排气门到排气歧管。排气歧管将部分废气输入到涡轮增压器以驱动涡轮(涡轮与涡轮增压器进气端的压气机相连)。经过涡轮后,废气再经过牵引车后部的消声器排出。

2. The engine cooling system consists of a coolant filler mounted atop the engine, a thermostat that controls the engine's temperature, an engine-driven coolant pump, a heat exchanger behind an engine-driven fan, a lube oil cooler, and paths within the engine to circulate coolant around the cylinder heads.

发动机冷却系统由一个安装在发动机顶上的冷却液过滤器,一个控制发动机温度的温度调节装置,一个发动机驱动的冷却液泵,一个安装在发动机驱动风扇后面的热交换器,一个润滑油冷却器及许多使冷却液循环的冷却液通道组成。

3. Hydraulic pressure for the steering, braking, front cab elevation and optional internal jacking system is provided by a single, variable displacement, axial piston, gear driven hydraulic pump mounted on the transmission.

转向、制动、前驾驶室升降和选装的自身顶升系统的液压力由安装在变速箱上单级的、变位移的、柱塞的、齿轮驱动的液压泵提供。

4. The valves are mounted on the steering manifold. When the tractor is being operated from the front cab, valve SV2 determines the tractor's steering mode. Valve SV3 determines the steering mode when the tractor is being operated from the rear cab. When track steering mode is selected, the appropriate valve is de-energized or in its center position. Otherwise, the valves direct hydraulic fluid to obtain four-wheel coordinated or crab steering modes.

转向模式选择阀安装在转向管路中。从前驾驶室操作牵引车时,由 SV2 阀决定牵引车的转向模式;当从后驾驶室操作牵引车时,则由 SV3 阀决定转向模式。牵引车的转向模式一旦选定,相应的控制阀断电或处于中心位置。否则,模式选择阀将控制液压油使四轮同步或蟹行转向模式。

Section 10. 2　General Safety Description

10. 2. 1　Safety Instructions

The same basic safety regulations is applied to the maintenance of the AST – 3 as for the control and operation of the vehicle (see Operating Manual Primary Information "Safety Instructions").

During all maintenance work on the AST – 3, the local accident prevention regulations and other official directives must be observed. Any deviation from regulations and approvals carried out by the user must be clarified in each case with the manufacturer.

10. 2. 2　Accident Prevention

(1) To avoid accidents resulting in personal injury, the following regulations must always be observed:

Maintenance work may only be carried out by authorized and specially qualified personnel.

For all maintenance work, switch off engines, ensure non-energized state, main vehicle switch OFF, battery isolating switch OFF. The only exceptions to this ruling are formed by individual maintenance steps for adjustment and trial runs or test driving. In these cases, maintenance staff must take suitable precautions.

During load tests of the hydraulic system (high pressure up to 420 bar) leaks occurring in the high-pressure section can cause danger to life.

Never open the coolant circuit when at operating temperature. Danger of scalding!

When changing oil, avoid the hot oil coming into contact with metal components.

Never tighten or open pressurized pipes.

For cleaning work using compressed air, wear respiratory protection and goggles.

Soak up spilled oil immediately using a binding agent. Danger of slipping!

Gaskets and other components containing asbestos must be specially identified and disposed of separately.

(2) During test runs, particularly in confined rooms.

Demarcate the danger area around the vehicle.

Keep a safe distance from turning or moving parts.

Extract exhaust gases using an extraction system.

During load tests of the hydraulic system (high pressure up to 420 bar) leaks occurring in the high-pressure section can cause danger to life.

When testing the nitrogen accumulators, only use the stipulated special tools.

Fuels and hydraulic oils are combustible. Never smoke or approach the vehicle with a naked flame.

When working under the cabin or under the vehicle, the vehicle must be jacked up and supported and underlaid at the frame under the front axle and in the rear axle area.

Use jacks and devices with adequate load-bearing capacity.

When the diesel engine is running, never enter the wheel houses. Danger of crushing!

Secure the vehicle with wheel chocks against rolling.

Place a fire extinguisher ready.

10.2.3　Safety Instructions for Welding Work

Before carrying out major welding work, always first consult the manufacturer. The local accident prevention regulations must always be observed.

Before carrying out welding work on the AST – 3, the entire electrical system must be disconnected from the power supply (24 V battery, external supply both 24 V and 220/380V). To disconnect from the battery, use the battery isolating switch S01, shown in Figure 10.2.

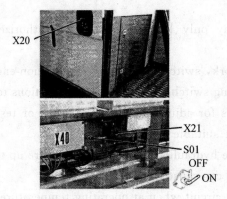

Figure 10.2　Battery isolating switch S01

10.2.4　Avoidance of Environmental Damage

Waste lubricants, fluids and oils of the vehicle following exchange of renewal must be disposed of in accordance with local regulations.

This affects in the main:

- Fuels;
- Engine oils;
- Hydraulic oil;
- Anti-freeze;
- Filter cartridges;
- Greases;
- Paints.

10.2.5   Hydraulic Pipes and Hoses

The hydraulic pipes and the hydraulic hoses must be checked for any damage and possible leaks at regular intervals.

Note: We recommend to have the pipes and hoses of your vehicle inspected by a special work ship after 5 years at the latest. Further inspections are then to be carried out every year.

10.2.6   Graphic Representation of AST – 3 / Name Plates

The position of the name plates may change in the event of technical modifications. For this reason, the serial number of the AST – 3 and the built-in units should always be referred to in the vehicle master sheet, shown in Figure 10.3.

1= Name Plate AST – 3;

2= Chassis Number;

3= Name Plate Engine / Engine Number.

Figure 10.3   The name plates of the AST – 3

The AST – 3 must only be supported or lifted at the points marked in the drawings. The support points "3" serve as a safeguard, so that the supported AST – 3 with advanced cabin does not tilt. If the tractor has an advanced cabin, do never work on and under the AST – 3 as long as the support points "3" are not sustained.

Warnning: First, the AST – 3 has to be supported at the support points "1". Then it has to be supported at the support points "2". The lifting load of the lifting tools has to be sufficiently dimensioned, shown in Figure 10.4.

1 = Support point   min 5 tons per side;

2 = Support point   min 3 tons per side;

3 = Support point   min 1 tons per side;

4 = Lifting point   min 5 tons per side;

5 = Lifting point   min 3 tons per side.

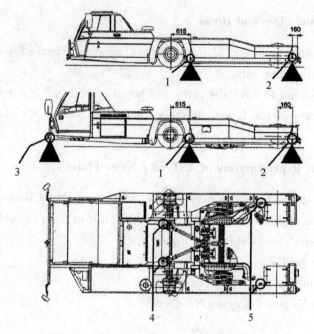

Figure 10.4   Lifting points / jacking points

New Words

1. energized [ˈenədʒaizd] *adj.* 激励的
2. goggles [ˈgɔːglz] *n.* 护目镜，防护眼镜
3. asbestos [æzˈbestɒs] *n.* 石棉
4. demarcate [ˈdiːmaːkeit] *vt.* 划分界线
5. combustible [kəmˈbʌstib(ə)l] *adj.* 易燃的
6. extinguisher [ikˈstiŋgwiʃə] *n.* 灭火器
7. Anti-freeze [ˈæntifriːz] *adj.* 防冻

Phrases and Expressions

1. safety regulations 安全守则
2. identification number 标识号，机器编号
3. naked flame 明火
4. serial number 序列号
5. accident prevention 事故预防，安全措施

Notes to the Text

1. Classification of the maintenance work referring to the assembly groups, is made analogous to the structure of the spare parts catalogue. Description of the maintenance work is found in the corresponding sections, with numerical structure.

维护工作参照装配组件进行分类,与备件目录的结构相似。维护工作的描述可以在相应数字部分查到。

2. Regular and proper maintenance of the vehicle are essential to a reliable and efficient operation for a long time. The maintenance procedures mentioned below represent the minimum work requirement and form part of our warranty cover.

车辆规范和正确的维护对其工作的可靠性、操作的有效性及使用寿命是至关重要的。下面论及的维护程序表示所要求的最低限度的工作量,形成我们保修期的一部分。

Section 10.3   General Maintenance

10.3.1   General Maintenance Steps

Maintenance must be carried out in the following steps:

• Cleaning;
• Visual inspections;
• Functional checks.

10.3.1.1   Cleaning

1. General information

Cleaning the vehicle regularly is required for an expert maintenance and makes operation of the vehicle and its controls easier.

After cleaning the vehicle, grease the vehicle according to the lubrication plan.

For handling and clearing away cleaning agents or auxiliary material the legal regulations have to be observed.

2. Cleaning with steam-jet cleaner

When cleaning the vehicle with a steam-jet cleaner, observe the following points:

You must not use cleaning agents that damage paint work, seals, plastics and chrome or the environment.

Do not clean electrical parts (plugs, sockets, distributor boxes, etc.) and their direct environment with a steam-jet cleaner.

Never apply steam-jet cleaner directly to bearing points, lubrication points and hydraulic switch units.

Do not clean chrome-plated parts with a steam-jet cleaner.

Do not clean cooling fins (high pressure) with a steam-jet cleaner.

10.3.1.2   Visual Inspections

Visual inspections refer to all components and assemblies listed in the following sections and must include appraisal of the external and internal condition, noise development and level of soiling.

Soiled appliances must be cleaned. Defective components, where these impair functional performance, must be immediately removed.

Visual inspections also include a check of fixtures, as well as appliance, cable and hose clamps, etc.

10.3.1.3 Functional Performance Checks

The individual functions of the vehicle are described in the Operating Manual. The necessary operating procedures must be carried out in accordance with the Manual.

10.3.2 Scope of Work Processes

The completion of the individual maintenance steps generally involves the following supplementary work processes:

Clean the vehicle before all maintenance work.

Make absolutely sure that the vehicle is horizontal.

Before all testing and setting work, run vehicle up to operating temperature (hydraulic oil temperature: minimum = 30℃, ideal = 45℃).

Tighten screwed connections with specified torque using a torque wrench.

Renew loosened seals and gaskets.

Always change the O-rings used in the high-pressure section of the drive system, e. g. after undoing screwed connections of flanges, valves, etc.

Clean detached attachments and components before replacing on the vehicle and check for damage. Exchange, if necessary.

Fit hydraulic components, pipes and hoses with dust caps after removing until replaced.

After all maintenance work, set all controls in driving position.

10.3.3 Test Drive

Check filling levels of all fuels and oils before and after a test drive.

Check tyre pressure before test driving.

Check driving and braking performance.

Check the function of the pick-up device.

Check the lighting, signaling and work lamps.

Check the window and mirror heating, windscreen wiper/washing system.

Check the function of the heating and fan.

If the test drive is not carried out by the workshop foreman, any defects occurring must be reported to him without delay.

10.3.4 Check Filling Level and Top Up

Top up to target level.

Check the firm fit and seal of the filling and draining cap.

Clean the area of the filler neck and drain screw, etc. before and after carrying out maintenance work.

Top up missing quantities of grease and fluids (fuel, cooling water, hydraulic oil, transmission oil, engine oil, etc.).

Check anti-freeze content of cooling fluids.

Check the tightness of the sealing caps.

Check vented sealing caps for air permeability.

Report unnaturally high consumption to the workshop foreman.

10.3.5　Check Condition

1. Battery

Check fluid level and acid density.

Electrical equipment in general.

Check cable connections, cable eyes with regard to tight fit.

Check cables with regard to scuff marks.

Check electrical contacts with regard to corrosion.

Check condition and fixing of fuses.

2. Mechanical assemblies

Check for external damage.

Firm fit of fixtures and screwed connections, crack formation, occurrence of corrosion.

Easy running of moving parts.

3. Lubrication status

Hydraulic, fuel and air pipes and hoses.

Check for abrasion on hoses, particularly on moving parts.

Report unnaturally high wear to the work ship foreman.

Check for leaks.

Check for the emergence of grease, fluids and compressed air at connections, screwed connections, tanks, etc.

10.3.6　Authorized Personnel

Testing and maintenance work requires specialized knowledge, and may therefore only be carried out by suitably qualified and specially trained personnel. The user is responsible for delegating authority to carry out maintenance work.

10.3.7　Maintenance

10.3.7.1　Pick-up Device (See Figure 10.5)

In Figure 10.5:

1 — Ball and socket joint (not shown);

2 — Lifting cylinder (not shown);

3 — Pull arms / slide rails;

4 — Stop pads (pull arms are completely open);

5 — Roller supports;

6 — Restrainer;

7 — Platform / rollers;

8 — Sensor rollers;

9 — Contact strips.

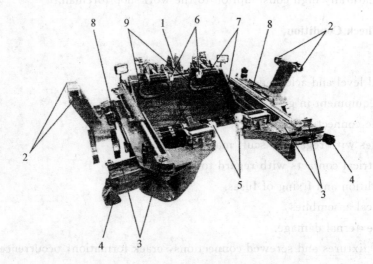

Figure 10.5   Position / overall view of construction parts (mechanical) — pick-up device

10.3.7.2   Maintenance Work

Clean the vehicle before starting any kind of maintenance work. Maintenance of this section is basically limited to checking condition, fixing, function and tightness.

Lubrication plan (see Figure 10.6).

Open the cap on the ball-and-socket joint of the triangular link. Check the grease filling and if necessary refill. Only a special "Long term 2/78 G" should be used for the grease filling, as shown in Figure 10.7.

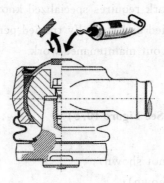

Figure 10.6   Lubrication from ball and socket joint of the triangular link

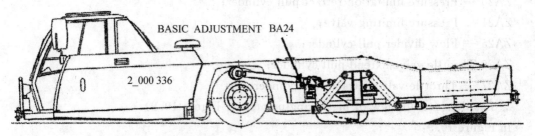

Figure 10.7　Sensor installation potentiometer lifting pick-up device BA24

10.3.7.3　Hydraulic System Pick-up Device

1. Position/overall view of construction parts (see Figure 10.8)

In Figure 10.8：

BA20 — Pressure switch pull cylinder level 1；

BA51 — Pressure sensor lifting cylinder；

YA01 — Lower platform；

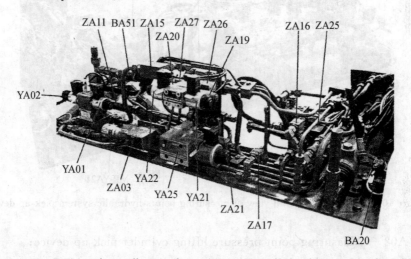

Figure 10.8　Position / overall view of construction parts-hydraulic system pick-up device

YA02 — Lift platform；

YA21 — Retract pull arms；

YA22 — Extend pull arms；

YA25 — Switch off pressure reduction；

ZA03 — Tension restrainer pressure reducing valve；

ZA11 — Flow regulating valve lower platform；

ZA15 — Pressure limiting valve lower platform；

ZA16 — Ball-type on/off valve emergency release pull cylinder；

ZA17 — Ball-type on/off valve emergency release lifting cylinder；

ZA19 — Pressure limitation extend pull cylinder；

ZA20 — Pressure limitation retract pull cylinder;

ZA21 — Pressure limiting valve;

ZA25 — Flow divider pull cylinder;

ZA26 Throttle valve extend pull cylinder;

ZA27 — Throttle valve retract pull cylinder.

2. Position/overall view of measuring points(see Figure 10. 9)

In Figure 10. 9:

MP – 3 — Measuring point pressure oil supply pull arms and restrainer;

MP – 4 — Measuring point pressure oil supply lifting cylinder;

MP – 6 — Measuring point pressure oil supply emergency operation with hand pump or hydraulic power pack;

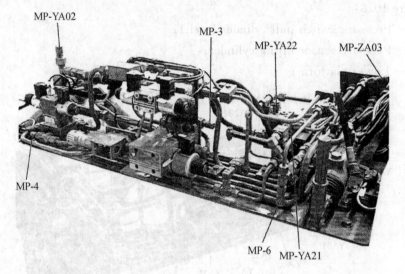

Figure 10. 9　Position / overall view of measuring points-hydraulic system pick-up device

MP – YA02 — Measuring point pressure lifting cylinder pick-up device;

MP – YA21 — Measuring point pull pressure pull arms;

MP – YA22 — Measuring point extension pressure pull arms;

MP – ZA03 — Measuring point pressure restrainer cylinder.

3. Pressure check pressure limitation/pressure reduction valves

(1) Pressure check pressure limitation valve ZA19.

Connect pressure gauge (400 bar) to measuring point MP – YA22.

Remove plug from sensor BA06.

Start engine and extend pull arms.

Pressure gauge has to show max. pressure ("control values" ZA19).

If the pressure differs from the control value ZA19 ("control values" ZA19, shown in Figure 10. 10), check hydraulic system with regard to errors and leaks. If necessary repair hydraulic system. For further trouble shooting contact the Goldhofer technical service.

After finishing the checks and adjustments connect plug of the sensors again.

Figure 10.10　Pressure check pressure limitation valve ZA19

(2) Pressure check pressure reduction valve ZA20(see Figure 10.11).

Figure 10.11　Pressure check pressure reduction valve ZA20

Connect pressure gauge (400 bar) to measuring point MP – YA21.

Start engine and retract pull arms completely.

Pressure gauge has to show max. pressure ("Control values" ZA20).

If the pressure differs from the control value ZA20 ("Control values" ZA20), check hydraulic system with regard to errors and leaks. If necessary repair hydraulic system. For further trouble shooting contact the Goldhofer technical service.

(3) Pressure check pressure limitation valve BA20(see Figure 10.12).

BA20 stops the retraction process, when exceeding the pressure reduction at the pull cylinder.

Connect pressure gauge (400 bar) to measuring point MP – YA 21.

Start engine and retract pull arms completely.

Switch on sensors BA26 / BA27 (actuate switch bracket).

By changing ZA21 increase pressure to 150 bar.

Switchbox X35: LED of clamp 306 has to be off.

By changing ZA21 reduce pressure slowly to control value BA20 ("control values" BA20).

Control value BA20: LED of clamp 306 light up.

After finishing the pressure check, readjust ZA21 to its control value.

Figure 10.12　Pressure check pressure limitation valve BA20

(4) Pressure check pressure limitation valve ZA21(see Figure 10.13).

Connect pressure gauge (400 bar) to measuring point MP – YA 21.

Start engine and retract pull arms completely.

Switch on sensors BA26 / BA27 (actuate switch bracket).

Pressure gauge has to show max. pressure with the switched on sensors (switch bracket actuated) ("control values" ZA21), if necessary adjust pressure of pressure limitation valve ZA21.

Figure 10.13　Pressure check pressure limitation valve ZA21

(5) Pressure check pressure reduction valve ZA03(see Figure 10. 14).

Figure 10. 14   Pressure check pressure reduction valve ZA03

Connect pressure gauge (400 bar) to measuring point MP – ZA03.

Remove plug from sensor BA03.

Start engine and lift pick-up device to final position (until stop). (For pressure check keep push-button of seat console pushed).

Pressure gauge has to show max. pressure ("control values" ZA03).

If the pressure differs from the control value ZA03 ("control values" ZA03), check hydraulic system with regard to errors and leaks. If necessary repair hydraulic system. For further trouble shooting contact the Goldhofer technical service.

(6) Pressure check pressure reduction valve ZA15(see Figure 10. 15).

Figure 10. 15   Pressure check pressure reduction valve ZA15

Start engine and lift pick-up device.

Connect pressure gauge (400 bar) to measuring point MP – 4.

Pressure gauge has to show max. pressure while lowering the pick-up device ("control values" ZA15).

If the pressure differs from the control value ZA15 ("control values" ZA15), check hydraulic system with regard to errors and leaks. If necessary repair hydraulic system. For further trouble shooting contact the Goldhofer technical service.

10. 3. 7. 4   Hydraulic System Drive

1. Position/overall view of construction parts(see Figure 10. 16)

In Figure 10. 16:

YF09 — G-switching, dependent from driving direction;

YF29 — Switch on solenoid valve drive rear axle;

YF33 — Driving direction valve ZF10 = forward;

YF34 — Driving direction valve ZF10 = backward;

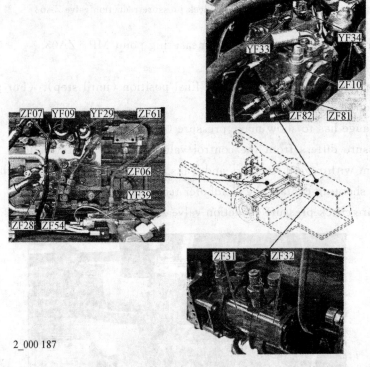

2_000 187

Figure 10. 16   Position / overall view of construction parts-hydraulic system drive

YF39 — Valve X - over steering rear axle;

ZF06 — Flushing valve;

ZF07 — Inch valve;

ZF10 — Drive pump 1;

ZF28 — Emergency steering switch over;

ZF31 — Gear wheel pump1. brake/2. pull arms pick-up device;

ZF32 — Gear wheel pump1. steering/2. lifting cylinder pick-up device;

ZF54 — Pressure limiting valve for flushing valve;

ZF61 — Shuttle valve G-switching rear axle;

ZF81 — DA-control valve;

ZF82 — Pressure limiting valve supply pressure.

2. Position/overall view of measuring points (see Figure 10.17)

In Figure 10.17:

MP – A — Measuring point for high or low pressure, dependent from driving direction;

MP – B — Measuring point for high or low pressure, dependent from driving direction;

MP – C — Measuring point for high pressure, independent from driving direction;

MP – GH — Measuring point for external setting pressure hydraulic engine front axle;

MP – GV — Measuring point for external setting pressure hydraulic engine front axle;

MP – P1 — Measuring point 1. brake / 2. pull arms pick-up device;

MP – P5 — Measuring point for automatic emergency supply steering;

MP – S — Measuring point reduction for emergency actuation;

MP – PST — Measuring point for control pressure driving pump;

MP – T1 — Point for input pressure lateral flushing hydr. pump / hydr. engine;

MP – Z — Point for brake pressure, driving pump swivels back.

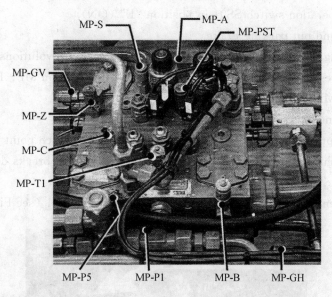

Figure 10.17   Position / overall view of measuring points-hydraulic system drive

3. Pressure checks drive

(1) Pressure check internal pressure cut-off.

Connect pressure gauge (600 bar) to measuring point MP – GV or MP – C.

Emergency operation switch SS01 into position "E" (ON).

Start engine and put parking brake.

Engage driving direction selector lever and increase engine revolutions to max.

Pressure gauge has to show max. pressure ("control values" ZF80), if necessary adjust pressure at ZF80.

Emergency operation switch SS01 into position "D" (OFF).

(2) Pressure check inch valve ZF07.

Connect pressure gauge (40 bar) to measuring points MP – PST and MP – Z.

Emergency operation switch SS01 in position "E" (ON).

Start engine and put parking brake.

Engage driving direction selector lever and increase engine revolutions to 1,200 r/min.

Actuate service brake slowly at increased revolutions.

If on pressure gauge MP – Z lies the control pressure of ZF07 ("Control values" ZF07), the pressure on pressure gauge MP – PST must break down, if necessary adjust pressure at inch valve ZF07.

Emergency operation switch SS01 in position "D" (OFF).

(3) Pressure check emergency steering switch over ZF28.

Connect pressure gauge (600 bar) to measuring point MP – P5, one pressure gauge (600 bar) to measuring point MP – C and one pressure gauge (40 bar) to measuring point MP – P2.

Emergency operation switch SS01 in position "E" (ON).

Start engine and put parking brake.

Engage driving direction selector lever and increase engine revolutions until on pressure gauge MP – C a pressure of 200 bar has been built up. (Pressure on MP – P5 has to be lower than pressure on MP – C).

Turn off engine.

The existing pressure on MP – P2 slowly decreases until switch point ("Control values" ZF28). After reaching the switch point, the pressure on MP – C breaks down. If necessary adjust pressure on ZF28.

(4) Adjustment of drive away start — DA-control valve ZF81 (see Figure 10.18).

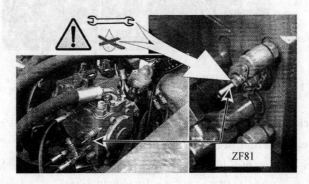

Figure 10.18   DA-control valve ZF81

Bring the vehicle to operating temperature (Hydraulic oil temperature minimum 30℃).

Put the driving direction lever at the requested position.

Increase the rpm of the diesel engine slowly until the vehicle starts rolling.

Compare the engine speed with the control value ZF81 at the beginning of motion (see maintenance manual, part 2, "control values" ZF81).

In the case of a variation, adjust the value at ZF81 until the control value has been reached.

(5) Pressure check of pressure limitation valve for supply pressure ZF82 (see Figure 10.19).
In Figure 10.19:

MP – A — Measuring point for high or low pressure, dependent from driving direction;

MP – B — Measuring point for high or low pressure, dependent from driving direction;

MP – G — Measuring point for supply pressure;

MP – X_1 — Point setting chamber;

MP – X_2 — Point setting chamber.

Connect pressure gauge (60 bar) to measuring point MP – G.

Start engine and put parking brake.

Read pressure.

If the pressure differs from the control value ZF82 ("control values" ZF82), check hydraulic system with regard to errors and leaks. If necessary repair hydraulic system. For further trouble shooting, contact the Goldhofer technical service.

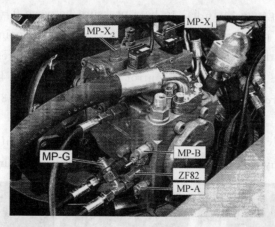

Figure 10.19　Pressure check pressure limitation valve supply pressure ZF82

(6) Control start hydro engines front axle ZF40 / ZF41 (see Figure 10.20).

Connect one pressure gauge (600 bar) to measuring point MP – GV.

Connect another pressure gauge (600 bar) to the test connector of the setting chamber of the wheel motor.

Find out and note down measure X of the Q_{min}-screw.

Clamp off clamp X30:320 (entrance parking brake switch) in the switch box X30 (cabin), so you can drive against the parking brake.

Start engine and put parking brake.

Release lock nut at the Q_{min}-screw (box wrench size 17), at the same time fix Q_{min}-screw with allen key (size 5) so that it cannot turn.

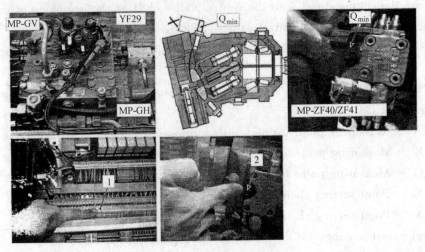

Figure 10. 20   Control start hydro engines front axle ZF40/ZF41

Increase engine revolutions slowly until Q_{min}-screw can be screwed in ("control start"). At the same time read the pressure at pressure gauge MP – GV. Pressure at pressure gauge has to correspond to control value ZF40 / ZF41 ("control values" ZF40 / ZF41). Up to now, at the second pressure gauge there has to be a pressure between 0 – 20 bar. This pressure remains until there is a pressure increase of 100 bar at measuring point MP – GV. From this pressure increase on both pressure gauges have to show the same pressure.

If the control value ZF40 / ZF41 ("control values" ZF40 / ZF41) is not correct, you must adjust the pressure at the adjusting screw (box wrench size 13, allen key size 3, see Figure 10. 20, Pos 2).

Readjust Q_{min}-screw to the determined measure X and secure with lock nut.

(7) Control start hydro engines rear axle ZF42 / ZF43 (see Figure 10. 21).

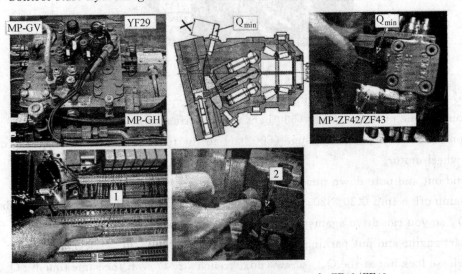

Figure 10. 21   Control start hydro engines rear axle ZF42/ZF43

Connect one pressure gauge (600 bar) to measuring point MP – GH.

Connect another pressure gauge (600 bar) to the test connector of the setting chamber of the wheel motor.

Find out and note down measure X of the Q_{min}-screw.

"Switch on solenoid valve drive HA" actuate YF29 manually with bayonet.

Clamp off clamp X30：320 (entrance parking brake switch) in the switchbox X30 (cabin), so you can drive against the parking brake (see Figure 10.21, Pos 1).

Start engine and put parking brake.

Release lock nut at the Q_{min}-screw (box wrench size 17), at the same time fix Q_{min}-screw with allen key (size 5) so that it cannot turn.

Increase engine revolutions slowly until Q_{min}-screw can be screwed in ("control start"). At the same time read the pressure at pressure gauge MP – GV. Pressure at pressure gauge has to correspond to control value ZF42 / ZF43 ("control values" ZF42 / ZF43). Up to now, at the second pressure gauge there has to be a pressure between 0 – 20 bar. This pressure remains until there is a pressure increase of 100 bar at measuring point MP – GV. From this pressure increase on both pressure gauges have to show the same pressure.

If the control value ZF42 / ZF43 ("control values" ZF42 / ZF43) is not correct, you must adjust the pressure at the adjusting screw (box wrench size 13, allen key size 3, see Figure 10.21, Pos 2).

Readjust Q_{min}-screw to the determined measure X and secure with lock nut.

10.3.8　Lubricants and Operating Fluids

10.3.8.1　General

The use of suitable lubricants and operating fluids is essential for the operating safety and a long service life of the vehicles. You should carry on using the lubricants and operating fluids which were used for initial filling. Lubricating oils, hydraulic oils and lubricating greases cannot be automatically mixed. If this is not possible for organizational reasons, you should only use products which meet the specifications and requirements as indicated below. No responsibility can be accepted for the indicated lubricants and operating fluids.

Each supplier of lubricants maintains customer service centers to give advice and information regarding lubricants. For the lubricants and operating fluids of engine, wheel gears and distributor gears, please see also the documentation from the sub-suppliers.

10.3.8.2　Disposal of Lubricants, Operating Fluids, Filters and Auxiliary Material

The disposal of lubricants, operating fluids, filters and auxiliary material must be effected according to the waste disposal and recycling law in force. You must also refer to the local regulations in the country of use.

10.3.8.3　Diesel Fuels

The specifications of the engine manufacturer must be observed. The use of high grade diesel fuel is essential to obtain the engine performance as noted. Use only diesel fuels which meet minimum specifications as indicated below. The sulfur content in the fuel should not exceed 0.5%. A higher sulfur content can affect the oil change intervals and the engine service life expectancy.

1. Winter diesel fuels

Special winter diesel fuels are available for engine operation at ambient temperatures below 0℃. These fuels have a lower viscosity and also limit the wax formation in the fuel at low temperatures. If wax formation occurs, this could stop the fuel flow through the filter and the fuel lines.

2. Additives for diesel fuels (flow improvers)

Commercially available flow improvers also enhance the diesel fuel ability to perform reliably in cold climate conditions. When using such flow improvers, follow the manufacturer's recommendations regarding use and dosage.

When adding a flow improver and/or other fuel additives, always add the additive first, due to its lighter specific weight (for example add petroleum first, then the diesel fuel).

10.3.8.4　Lubricating Oils (Lube Oils) for Diesel Engines

Follow the specifications of the engine manufacturer.

For modern diesel engines, only highly alloyed lube oils are used. They consist of base oils with an additive.

The lube oil specification is usually based on the following specifications and requirements:

CCMC specifications;

API — Grade classification;

And/or American military specifications.

Depending on the type of engine and load application, a specific minimum grade is required, whereas oils of a higher grade can extend the oil change intervals.

The selection of the lube oil viscosity is made according to SAE classification (Society of Automotive Engineers). The ambient temperature is a determining factor for the selection of the correct SAE classification. If the viscosity is too high, starting problems can occur and if the viscosity is too low, the lubrication effectiveness could be endangered. The temperature ranges on the following chart is guidelines and for a short time (shown in Figure 10.22), it is permitted to go above or below the values given.

10.3.8.5　Coolant Specification

The quality of the coolant which is used can have a great effect on the efficiency and life of the cooling system. To mix the coolant fluid, only clean, preferably soft water should be used. Do not use sea water, brackish water, salt water or industrial waste water.

To ensure a corrosion protection of the cooling system, it is advisable to use all-year round a mixture of 50% corrosion/antifreeze and 50% water. Differing information from the engine manufacturer is to be observed. Do not use more than 60% corrosion/antifreeze mixture, a higher percentage would reduce the cooling effectiveness and antifreeze protection. This ratio of antifreeze/water mixture must be checked during the maintenance work. The prescribed change intervals of 2 years must be observed.

In certain exceptional circumstances, and if ambient temperatures are constantly above the freezing point, for instance in tropical regions, and if there are no corrosion or antifreeze fluids available, then a mixture of 99 Vol −% water and 1 Vol. −% additive (corrosion protective oil) may be used. When using coolant refiners, the coolant must be changed once a year. When using coolant refiners, no other glycol based antifreeze fluids or other corrosion protectors should be used. When changing from corrosion/antifreeze fluids to refiners, the complete coolant and heating circuit must first be flushed with water.

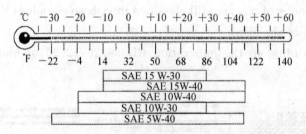

Figure 10.22　Temperature ranges, oil viscosity

10.3.8.6　Hydraulic Oils

Quality, cleanness and operating viscosity of the hydraulic oil are decisive for the operating safety, economical use and service life of the hydraulic equipment. Mineral, synthetic or biological hydraulic oils are offered on the market. Standard oils used in Goldhofer vehicles are mineral hydraulic oils of classification HLP 22 and HLP 32. Exceptions are made on specific customer request.

For the hydraulic oil filling of your vehicle, see "technical data" and/or "control values/ pressures". Depending on the requirement, mineral hydraulic oils are classified in various grades according to the DIN 51524. In order to improve their characteristics, different additives are added to the hydraulic oils. Basic types are as follows.

1. HL-oils

HL-hydraulic oils are pressure fluids of mineral oils with active substances to enhance the corrosion protection and the aging resistance. These oils are used in systems with a high thermal load applications and pressures up to about 200 bar. These agents do not contain any additives to reduce the wear in the mixing friction field.

2. HLP-oils

HLP-hydraulic oils are pressure fluids of mineral oils with active substances to enhance the corrosion protection, the ageing resistance and to reduce the wear in the mixing friction field. These oils are used in systems with a high wear due to high pressures (above 200 bar).

3. HLPD-oils

HLPD-hydraulic oils are pressure fluids of mineral oils with active substances to enhance the corrosion protection, the ageing resistance and to reduce the wear in the mixing friction field. Moreover, they contain finely divided additives which keep impurities and water drops floating.

4. HV-oils

HV-hydraulic oils are pressure fluids of mineral oils with active substances to enhance the corrosion protection, the aging resistance and to reduce the wear in the friction mixing zones. Moreover, they contain additives to increase the viscosity index for use in systems which are subject to high temperature fluctuations.

The individual groups of oils are still classified in viscosity grades (VG), which indicate the temperature range of the oils.

Classification according to the publication RD 90220/12.95 of Mannesmann Rexroth (see Figure 10.23).

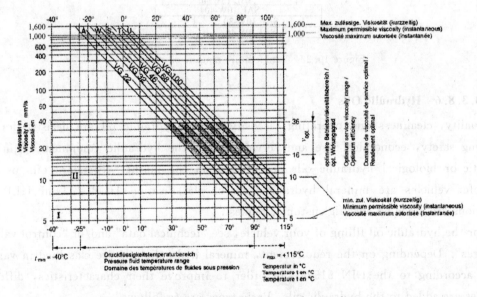

Figure 10.23　Viscosity-temperature-selection diagram

VG22 — in arctic climate conditions or for extremely long lines;

VG32 — in winter in Central Europe;

VG46 — in summer in Central Europe;

VG68 — in tropical climate conditions or for systems with high thermal load;

VG100 — DW Ⅲfor systems with excessive thermal load.

HV-oils offer the advantage of covering several viscosity classifications.

(1) Mixing of different hydraulic oils. Hydraulic oils cannot be mixed with each other or can only be mixed conditionally. When mixing hydraulic oils of different manufacturers or of

different types, clogging or formation of deposits can occur. In certain circumstances, they can cause hydraulic system problems and damage. That is why Goldhofer does not accept any responsibility for the use of mixed oils. Information whether hydraulic oils are mixable can be given by the oil supplier.

(2) Change to another hydraulic oil. When changing to another hydraulic oil, consultation of Goldhofer is absolutely necessary. Seals, filters and hoses of the hydraulic system are designed for mineral oils.

(3) Hydraulic oil recommendation. Hydraulic oils must be used which meet minimum requirement of hydraulic oil type HLP according to DIN 51524, part 2. By way of example, you can see some brands of hydraulic oils on the chart below(see Table 10.1). Equivalent hydraulic oils can also be used. The supplier must guarantee the use of the oil.

Table 10.1 Hydraulic oil recommendation

Hersteller Oil Company	Bezeichnung Designation	Hersteller Oik Company	Bezeichnung Designation
Mobil	Mobil DTE 22(VG22) Mobil DTE 24(VG32) Mobil DTE 25(VG46)	FUCHS	Renolin MR5 Renolin B5 Renolin MR10 Renolin B10 Renolin MR15 Renolin B15
ARAL	Vitam GF22 HLP22 Vitam GF32 HLP32 Vitam GF46 HLP46	BP	BP energol HLP 22 BP energol HLP 32 BP energol HLP 46 BP energol HLP 68
ESSO	Nuto H15 Nuto H32 Nuto H46	TEXACO	Rando oil A32 Rando oil A32 Rando oil B46
Castrol	Castrol hyspin AWS 22 Castrol hyspin AWS 32	SHELL	Shell tellus 22 Shell tellus 32 Shell tellus 46

10.3.8.7 Transmission Oil

Quality, cleanness and viscosity of the transmission oils are decisive for the operating safety, economical use and service life of the transmission/gear. Always use the transmission oils which are prescribed for the transmission/gear in question (see also quantities of filling and/or technical data in WA Part 2). Transmission oils of CLP grade are mainly used (see also documentation from sub-suppliers).

10.3.8.8 Lubricating Greases

Lubrication is essential for the adequate working of the bearings. During lubrication, a lubricating film is applied between antifriction bearing and bearing tracks in order to avoid wear by metal contact. A long service life and efficiency of the bearings directly depend on the efficiency of the lubricant and the regularity of lubrication.

By way of example, you can see some branded lubricating greases on the chart below (see Table 10.2). Equivalent greases can also be used. The supplier must guarantee the use of the grease.

Table 10. 2 Lubricating greases

Hersteller Company	Bezeichnung Designation (DIN 51828)	Hersteller Company	Bezeichnung Designation (DIN 51825)
Mobil	Mobilux 2	BP	BP energrease LS 2
ARAL	Aral Aralup HL 2	TEXACO	Multifak 2
ESSO	Beacon 2	SHELL	Shell alvania fett R 2
Castrol	Castrol Spheerol AP2		

10.3.9 Maintenance Intervals

The following maintenance plan lists in an overall view the maintenance intervals and the work to be carried out. Due to the special working conditions of the AST-3, the maintenance work is divided into singular or regular service, depending on the number of service hours, and into regular special services. The maintenance work of one maintenance interval also includes the maintenance work of the lower maintenance intervals, as far as these coincide with the higher interval.

Winter service or times with danger of frost:

Check suitable operating fluids and lubricants more often. Check fluid levels more often. Check cooler, fuel tank and system repeatedly with regard to condensation and drain it if necessary.

Clean vehicle more often, to prevent accumulation of thawing salts. Check vehicle constantly with regard to corrosion, observe hydraulic pipes and hoses, cable connections and plugs carefully. Pay attention to the battery to be fully charged. The tires should be in good condition and have sufficient tread. Treat door and window seals and door lock with special antifreeze. Check working order of the windshield washers more often. Check windshield wipers and change them if necessary.

Position and overall view of the construction parts, explications and possible additional information with regard to maintenance work, see the corresponding section of this maintenance manual. If necessary we also refer to the documentation of the producer (external documentation).

The indicated intervals refer to the number of service hours of the engine shown on the instrument panel.

10.3.9.1　Explication of the Symbols for Singular Maintenance Intervals
　　　　　　(See Table 10.3)

Table 10.3　Maintenance plan for singular maintenance intervals

●	■	(tools)	(magnifying glass)	(bin)	(brush)	(book)	Notice
General Information							
•	•	Visual inspection — condition, completeness — damages, leaks	•				
•	Nach 50 km	Wheels and tires — wheel nuts — pressures — condition and tightness	•			WA part 2 tightening torques and control values	
•	•	Fluid levels operating fluids	•			WA part 2 fluid quantities	
	•	All screw connections	•				
•		Controls cabin — functions	•			BA part 1, chap. 2	
•		Emergency release aircraft — functions	•			BA part 1, chap. 7	
Electric System							
•		Lighting /et of signals — condition, function	•				
	•	Protections — condition and fixing	•				

Continued

		Item			Remarks
•	•	Batteries — condition and fixing — condition of charge, acid	•		
		Hydraulic System			
•	•	Hydraulic oil level			WA part 2 fluid quantities
	•	Hydraulic pipes and hoses — condition and tightness — laying			
		Braking System			
•	•	Functioning and tightness of the complete system	•		
	•	Pipes and hoses — condition and tightness — laying	•		
•		Emergency release brake — function			
		Steering			
•	•	Functioning and tightness of the complete system	•		
•		Emergency steering — function			BA part 1, chap. 7
		Driving System			
•		Emergency driving — function	•		BA part 1, chap. 7
	•	Wheel gear — transmission oil		◆	
…	…	…	…	…	…

●—Before first operation；　■—After the first 50 operating hours or after the first week

New Words

1. designation [dezig'neiʃn] *n.* 标号
2. nitrogen ['naitrədʒən] *n.* 氮(元素)
3. stipulate ['stipjʊleit] *v.* 规定,保证
4. corrosion [kə'rəʊʒ(ə)n] *n.* 腐蚀,侵蚀
5. abrasion [ə'breiʒ(ə)n] *n.* 磨损
6. wax [wæks] *n.* 蜡,蜡状物
7. viscosity [vi'skɔːsəti] *n.* 黏性
8. impurities [im'pjuəritis] *n.* 杂质
9. condensation [kɒnden'seiʃ(ə)n] *n.* 浓缩物
10. thawing [θɔːiŋ] *n.* 熔化,融化
11. tread [tred] *n.* (轮胎面)花纹
12. swivel ['swivl] *vt.* 旋转,回转
13. triangular [trai'æŋgjələ] *adj.* 三角(形)的
14. slanting ['slæntiŋ] *adj.* 倾斜的,歪斜的

Phrases and Expressions

1. steam-jet 蒸汽喷射
2. torque wrench 转矩扳手,扭力扳手
3. maintenance-free 不需维护的
4. distributor box 配电箱
5. pressure limiting valve 限压阀
6. stop pad 限位垫,限位块
7. SPS＝Service Process Solution 服务处理结果
8. aircraft recognition 飞机识别
9. lock nut 锁紧螺母,防松螺母

Notes to the Text

1. The selection of the lube oil viscosity is made according to SAE classification (Society of Automotive Engineers). The ambient temperature is a determining factor for the selection of the correct SAE classification.

润滑油黏度的选择是根据美国汽车工程师学会分类法而定的。环境温度是正确选择润滑油黏度的确定因素。

2. Quality, cleanness and operating viscosity of the hydraulic oil are decisive for the operating safety, economical use and service life of the hydraulic equipment. Mineral, synthetic or biological hydraulic oils are offered on the market.

液压油的品质、清洁和工作黏度对液压装备的安全操作、经济使用和使用寿命起决定性的作用。矿物型、人工合成型和生物型液压油在市场上都可以购买到。

3. Lubrication is essential for the adequate working of the bearings. During lubrication, a lubricating film is applied between antifriction bearing and bearing tracks in order to avoid wear by metal contact. A long service life and efficiency of the bearings directly depend on the efficiency of the lubricant and the regularity of lubrication.

润滑对轴承的正常工作是非常重要的。润滑通过在轴承及轴承座之间形成润滑油膜,避免金属直接接触,从而减轻磨损。轴承的使用寿命及工作效能直接依赖于润滑剂的效能和规律的润滑。

4. The maintenance work of one maintenance interval also includes the maintenance work of the lower maintenance intervals, as far as these coincide with the higher interval.

一个维修间隔的维修工作包括低一级维修间隔的工作,直到与高一级维修间隔的工作一致。

5. Position and overall view of the construction parts, explications and possible additional information with regard to maintenance work, see the corresponding section of this maintenance manual.

零件与维修工作相关位置、总体构造、说明及其他信息可在维护手册的相应部分查阅。

6. Basically, the operator is required to keep to the maintenance periods and the corresponding maintenance services and note them down on the proof of maintenance.

此外,操作者起码要记住维修周期和相应的维修服务,并填写维修记录。

Chapter 11　Towbar Aircraft Tractor

Section 11.1　General Description

11.1.1　Unit Description

The model GT - 50DZ tow tractor is capable of generating approximately 40,000 - 48,000 pounds (18,160 - 21,792 kg) of tractive effort. The tow tractor is available in weights of 50,000 pounds (22,680 kg) and 60,000 pounds (27,216 kg). It is powered by a vertical, in-line six-cylinder diesel engine and a funk six-speed forward and three-speed reverse, semi-automatic transmission. The tow tractor can develop a top speed of about 15 mile/h (24 km/h) without a vehicle in tow.

The model GT - 50DZ tow tractor (see Figure 11.1) is comprised of the following major components and systems: engine, transmission, driveline assembly, fuel system, air intake system, lubrication system, cooling systems, steering system, brake system, and electrical system.

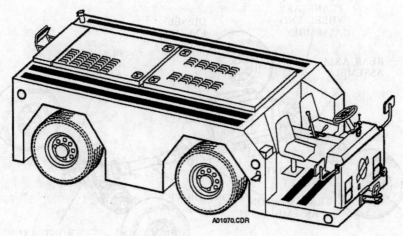

A01070.CDR

Figure 11.1　Model GT - 50DZ tow tractor

(shown without optional cab)

11.1.2　Engine

The tow tractor engine is a four-stroke cycle, six-cylinder turbocharged Deutz diesel engine. Engine displacement is 436 cubic inches (7 L). The engine is an in-line cast iron block. It contains a single camshaft actuating all valves and the fuel injector pump. The vertically aligned gear train, located at the front end of the engine block, contains drive gears for the lubricating oil pump, crankshaft, camshaft, and fuel pump.

11.1.3　Transmission

The tow tractor is equipped with a series of 2,000 funk transmission. It has six speeds forward and three speeds reverse. Other features are as listed below.

(1)Forward and reverse speeds are obtained through electrically controlled solenoids and hydraulically actuated multiple-disc clutches. The transmission can be shifted under full engine power.

(2) The clutches are hydraulically applied and spring-released. Each clutch uses a composition friction plate and a polished steel reaction plate.

(3) The power from the engine is transmitted to the transmission through a torque converter. The converter multiplies torque during heavy pull-down loads. When loads are light, the convener transmits the engine power directly.

11.1.4　Drive Line Assembly

The engine is coupled by the transmission, which drives the front drive shaft and the rear drive shaft as shown in Figure 11.2. The drive shafts transmit torque to the front and rear differentials. The axles from each differential transmit torque to the planetary wheel end assemblies.

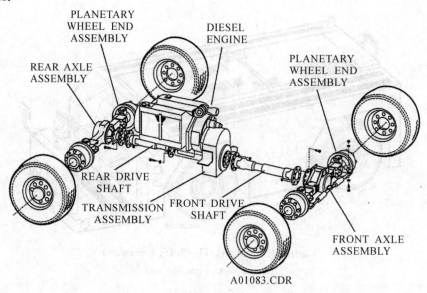

A01083.CDR

Figure 11.2　Driveline assembly

11.1.5   Fuel System

Fuel from the fuel tank is pulled by the fuel transfer pump through a water separator (standard equipment) and a filter. From the filter, fuel flows to the housing for the injector pumps. Fuel enters this housing at the top and through an inside passage to the fuel transfer pump. From fuel transfer pump, fuel under pressure, fills the housing for the fuel injection pumps. Pressure of the fuel in the housing is controlled with a bypass valve. Fuel pressure at full load is 30 psi (205 kPa). If fuel pressure exceeds this limit, the bypass will open allowing excess fuel to return to the inlet of the fuel transfer pump. A spin-on-type fuel filter is used on the engine.

11.1.6   Air Intake System

The air intake system consists of an air filter with a dry-type filter element, tubing, connectors, and clamps, air intake assembly routes filtered outside air to the engine air intake manifold. The air then enters when the intake valves are open.

11.1.7   Exhaust System

The exhaust system consists of two exhaust manifolds, ducting, clamps, and muffler assembly.

After ignition, the hot, burned exhaust gases, from the cylinders, are pushed out the exhaust valves to the exhaust manifold, through ducting to the muffler, and exited out the tailpipe.

11.1.8   Lubrication System

The engine lubrication system includes an oil pump, oil pump bypass valve, oil filter, oil alter bypass valve, oil cooler, oil cooler bypass valve, and an oil manifold.

A six-lobe rotary oil pump is located inside the front cover of the engine and is driven directly by the crankshaft. The oil bypass valve, located in the oil pump cover, controls pressure of oil coming from the pump. When pressure of oil flowing into the engine exceeds 55 – 80 psi (380 – 550 kPa), the bypass valve will open, controlling the output pressure of the pump.

The oil pump draws oil from the oil pan through the screened suction bell of the oil pump. The oil discharged from the pump flows to an oil passage in the engine front cover. It is then routed through the oil cooler and on to the oil filter. The oil filter will bypass oil if excessive restriction in the fitter element is present. Oil from the filters is then routed to the engine oil manifold. After engine lubrication, oil drains back to the oil pan.

11.1.9   Cooling System

A radiator and fan are used to effectively dissipate the heat generated by the engine. The radiator is mounted directly to the engine. A centrifugal-type water pump is used to circulate the engine coolant (see Figure 11. 3). A blocking-type thermostat is used in the water outlet passage to control the flow. This provides fast engine warm up and a regulated coolant temperature.

The tow tractor is equipped with a heat exchanger for the hydraulic steering system, and a heat exchanger for the transmission. These are mounted within the tractor's engine bay and provide a means of dissipating heat from the steering and hydraulic fluids.

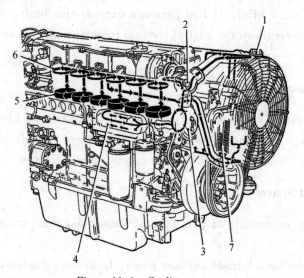

Figure 11. 3 Cooling system

1—coolant filler; 2—thermostat housing; 3—coolant pump; 4—lube oil cooler; 5—cylinder cooling;
6—cylinder head cooling; 7—heat exchanger with return line to thermostat housing

11. 1. 10 Steering System

The model GT – 50DZ tow tractor is equipped with a hydrostatic-type power steering system. Hydraulic power is supplied by a hydraulic pump mounted on the back of the transmission.

Note:

The model GT – 50DZ tow tractor has two hydraulic pumps. The pumps are mounted in tandem on the back of the transmission.

These pumps supply hydraulic power to the steering system and the brake booster. A third hydraulic pump is installed on the opposite side of the transmission for the lift cab option and supplies hydraulic pressure for the up and down directional control of the cab.

The steering system is composed of six major components. These components are described as follows (see Figure 11. 4).

(1)Hydraulic pump. This is a gear-type positive displacement pump directly driven off the transmission auxiliary drive. A suction hose from the pump inlet (suction side) is connected to the hydraulic reservoir.

(2) Flow control valve. This valve allows a constant volume of hydraulic fluid to flow to the steering system at high and low engine speeds while diverting excess flow directly back to the reservoir. The valve also contains an integral pressure relief valve which is factory preset at 2,250 psi (15,502 kPa).

(3) Steering valve. The steering valve consists of a manually operated directional control valve and meter in a single valve. Rotation of the steering wheel (which is attached to the valve) directs fluid to the steering cylinders mounted on the steering axle.

(4) Steering cylinders. The steering cylinders are double-acting hydraulic cylinders attached to the steering axle. When fluid, under pressure, is directed into the cylinders by the steering valve, the cylinders extend or retract to accomplish the required turning direction.

(5) Hydraulic reservoir. This reservoir contains the hydraulic fluid necessary to operate the steering system, hydraulic power to the brake booster, and provide a reserve. The reservoir contains an integral filler cap and level gauge, strainer, and vent.

(6) Hydraulic-filter. The moduflow filter assembly is designed for fluid to flow from the inside to the outside of the filter element. A diverter/bypass valve ensures full-flow filtration until bypass pressure is reached. A telis-all indicator, mounted on the side of the filter housing, allows a visual indication of the relative dirt accumulation.

11.1.11　Brake System

The master cylinder and power booster assembly supply hydraulic brake fluid pressure to the brake cylinder at each wheel (see Figure 11.5). Hydraulic power to the brake booster is supplied by a tandem-mounted, gear-type positive displacement pump driven off the hydraulic steering pump.

When the brake pedal is depressed, operating fluid pressure in the power booster forces a push rod in the power booster to actuate the primary and secondary pistons of the master cylinder. Movement of these pistons increases brake line pressure, causing the brakeshoes to actuate.

The primary and secondary chambers of the master cylinder function as two separate fluid systems. One chamber provides actuating fluid for the rear brakes, while the other chamber provides actuating fluid for the front brakes.

The hydraulic booster assembly includes a battery-powered backup booster pump. This backup pump supplies boost pressure in an emergency situation when the pressure in the brake booster cylinder is too low. This may happen if the engine is stalled while the tractor is in motion. The backup booster pump can also be used any time the service brakes are needed and the tractor engine is not running, providing the tractor's battery is charged and the battery cables are connected.

Parking brake. The parking brake is a mechanical caliper type, which operates on the transmission output flange. It is applied by actuating the parking brake lever in the operator' s compartment. The lever is connected to the parking brake caliper by a cable.

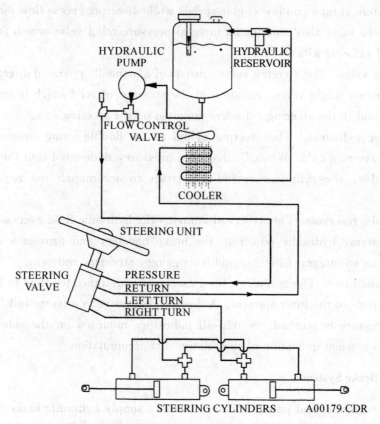

Figure 11. 4　Hydraulic steering system schematic

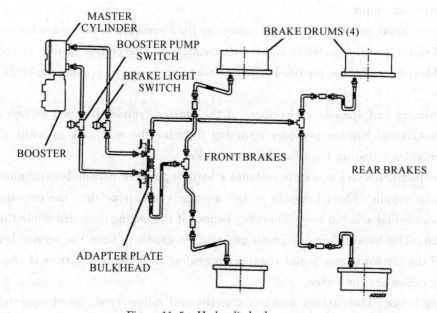

Figure 11. 5　Hydraulic brake system

11.1.12　Lift Cab (Optional)

The cab is a two-door, weathertight enclosure mounted on a steel-welded subframe. The controls are mounted inside the cab and are linked, either hydraulically, mechanically, or electrically, to the tractor body.

Pressurized hydraulic fluid is routed from the gear-type hydraulic pump, through the unloading valve, to the solenoid-operated directional control valve. This valve allows fluid flow for either lifting or lowering the cab, depending on its setting, controlled by the cab toggle switch located on the control panel.

(1) In cab lifting, the directional control valve routes fluid to the double pilot-operated check valve assembly, which contains two internal check valves. Pressure in the supply line opens one side of the internal check valves, admitting fluid to the top side of the cylinder. Simultaneously, pilot pressure, routed from the supply side of the assembly, opens the second internal check valve. This valve vents fluid from the low side of the cylinder, back to the reservoir.

(2) In lowering the cab, the directional valve routes the pressurized fluid to the lower side of the cylinder, reversing the flow through the double pilot-operated check valve assembly and cylinder.

11.1.13　Electrical System

The following paragraphs describe the theory of operation of the electrical system of the model GT – 50DZ tow tractor.

(1) Engine starting and battery-charging circuits. With the engine shift lever in neutral and engine STOP/RUN/START switch S1 in START position, relay K1 is energized, and the contacts of engine starting contactor K1 close. Starting solenoid LI is energized and applies 12 V battery voltage to engine starting motor Bl. After the engine starts and fuel pressure rises above 5 psi (34 kPa), the normally closed contacts of fuel pressure switch S3 open to prevent engagement of the starter while the engine is running. The battery-charging alternator G1 maintains battery BT1 in a charged condition and supplies power to the various 12 V PC engine accessory circuits.

(2) Brake booster motor circuit. This circuit supplies power to the brake booster backup pump motor in an emergency situation, when the brake booster cylinder has too low a hydraulic pressure to supply adequate boost to the master cylinder. This may happen when engine is stalled while the tractor is in motion, or any time the engine is not running. When this condition exists, the backup hydraulic brake booster pump flow switch S18 (normally open when the engine is running) is closed. Therefore, when the brake pedal is depressed,

causing switch S17 to close, it energizes backup brake booster pump motor control relay K4. The normally open K4 contacts then close, and apply 12 V DC battery voltage to backup brake booster pump motor B2 and brake booster pump motor on lamp DS13. When the brake pedal is released, relay K4 de-energizes. The K4 contacts then open, lamp DS13 extinguishes, and pump motor B2 stops running. It takes approximately 68 – 72 pounds of foot pressure on the brake pedal to activate pump motor B2.

(3)Engine run relay circuit. At the same time that the STOP/RUN/START switch (S1) is being actuated to crank the engine, battery voltage is applied through fuse F2 and the closed contacts of S1 (RUN) to the engine run relay K2. Engine run relay K2 is energized, closing the K2 contacts in the engine starting and battery-charging circuits. The contacts of K2 closing energize the engine run relay L2, allowing the engine to start and run.

(4) Horn circuit. When horn button S5 is depressed, horn relay K3 energizes. Therefore, the normally open K3 contacts close and energize the horn Hi with 12 V DC from battery BT1. The action of relay contacts K6 is covered in the warning light circuit description.

(5)Engine instrument circuit. When the engine is running and fuel pressure is 5 psi (34 kPa) or above, the engine fuel pressure switch S4 is closed and total (elapsed) time hourmeter M1 is energized. With engine switch S1 in the RUN position, the circuits for the following instruments are energized: fuel level gauge GA1, engine oil pressure gauge GA2, engine water (coolant) temperature gauge GA3, battery voltage meter M2, and tachometer gauge GA4.

(6)Light circuit. Parking lights switch S9, when placed in the ON position, applies 12 V DC from battery BT1 through fuse F1 to the following lamps: left tail lamp DS7, right tail lamp DS8, left front parking lamp DS9, and right from parking lamp DS10. Hydraulic brake pressure switch S10 energizes the stop element of taillight lamps DS7 and PS8 through hazard flasher switch S12, when the brake pedal is pressed. Head lights switch S11 energizes head lamps DS5 and DS6, tail lamps DS7 and DS8, and parking lamps DS9 and DS10. When placed in the ON position, hazard lights switch S12 energizes flasher FL1. Rasher EL1 then turns power on and off to tail lamps DS7 and DS8 and front hazard lamps DS9 and DS10.

(7) Warning light circuit. The following steps 1) thru 4) describe the warning light circuit:

1)If engine oil pressure drops below approximately 10 psi (69 kPa), engine low oil pressure switch S14 closes and engine low oil pressure warning lamp DS12 illuminates. Conversely, when engine oil pressure rises above approximately 10 psi (69 kPa), switch S14 opens and lamp DS12 extinguishes.

2)When coolant temperature is below approximately 205℉(96℃), switch S15 is open and lamp DS11 is extinguished. This is the normal operating condition on the other hand, if engine coolant temperature exceeds approximately 205°F (96℃), engine high coolant temperature switch S15 closes and the coolant temperature warning lamp DS11 illuminates.

3) When the parking brake is applied, parking brake switch S6 closes and hand brake on lamp DS1 illuminates.

4) When the engine is running and fuel pressure is 5psi (34 kPa) or above, fuel pressure switch S21 is dosed. When the transmission is not engaged, the closure of S21 will not affect anything, as the K5 contacts in series with S21 are open when the transmission shifter is in the neutral position. However, when the transmission is not in neutral (is engaged), the K5 contacts are closed. Relay K5 is in the engine starting and battery-charging circuits. Therefore, when the parking brake is applied and the engine is running, and the transmission is engaged, the hand brake on lamp DS1 and transmission engaged (parking brake warning) lamp DS17 illuminate to visually warn the driver. Moreover, at the same time DS17 illuminates, relay K6 is energized. The normally open K6 contacts then close and energize horn relay K3 in the horn circuit. In turn, the normally open K3 contacts close and energize the horn H1, giving the driver an additional warning.

5) When the parking brake is released, lamps DS1 and DS17 extinguish and horn H1 stops sounding. Lamp DS17 and horn H1 can also be de-energized (with the parking brake on) by operating the transmission selector to the neutral position or stopping the engine.

(8) Auxiliary light circuit. This circuit is explained in steps 1) and 2), as follows:

1) Before the tow tractor is reversed (backed up), the driver manually closes switch S19 and thereby applies battery voltage to backup lamps DS15 and DS16. The lamps then illuminate to provide a visual warning that the tow tractor is being reversed.

2) When panel lamp switch S20 is operated in the ON (closed) position, it applies battery voltage to the dash panel illumination lamp DS14. The lamp then illuminates, to provide the operator a better view of the controls and indicators during low-light conditions.

(9) Engine coolant heater (optional). In cold weather climates, under 40°F (4.4℃), an optional engine coolant heater is used to maintain the engine coolant at an operating temperature of 100 – 120°F (38 – 49℃). The coolant is heated and circulated through the engine. A 115 V AC receptacle is located near the engine control panel for external AC connection.

(10) Engine fuel/water separator and fuel heater (optional). The fuel/water separator and healer option are installed in place of the standard fuel/water separator. This separator filters and removes contaminants and water from the fuel before it enters the engine. The fuel/water separator may include a fuel healer, which is used below 40°F (4.4℃), to warm the fuel prior to it entering the engine, to ease starting. The fuel/water separator heating element uses engine briery voltage to operate and is controlled by a switch located on the engine control panel.

(11)Cab circuits (optional). The following paragraphs apply to units equipped with an optional cab：

1)Cab heater and ventilating fan circuits. When engine run relay K2 is energized，12 V DC is made available to the heater and ventilating fan motor circuits. The fan motors can then be controlled by the appropriate fan switch.

2)Windshield wiper and dome light circuits (optional). When engine run relay K2 is energized，12 V DC is made available to the optional windshield wipers and dome light circuits. The windshield wiper switch controls power allied to the windshield wiper motor. The dome light switch is used to turn the interior dome lamp on or off.

New Words

1. relay ['ri:lei] n. 继电器
2. energized ['enədʒaizd] adj. 通电的，导通的
3. torque [tɔːk] n. 扭矩
4. tandem ['tændəm] n. 串联

Phrases and Expressions

1. in-line six-cylinder diesel engine 直列六缸柴油发动机
2. turbocharged 涡轮增压
3. shift lever 换挡杆
4. power booster 助力器
5. brake shoes 刹车片,制动蹄
6. solenoid-operated 电磁控制
7. toggle switch 切换开关
8. directional valve 换向阀

Notes to the Text

1. Forward and reverse speeds are obtained through electrically controlled solenoids and hydraulically actuated multiple-disc clutches.

通过电控电磁阀和液压多片离合器实现前进和后退。

2. A third hydraulic pump is installed on the opposite side of the transmission for the lift cab option and supplies hydraulic pressure for the up and down directional control of the cab.

第三个液压泵安装在驾驶室举升选择传动装置的对边,为驾驶室升降方向控制提供液压。

3. The moduflow filter assembly is designed for fluid to flow from the inside to the outside of the filter element. A diverter/bypass valve ensures full-flow filtration until bypass pressure is reached.

ModuFlow 过滤器总成设计成使流体从滤芯内流向滤芯外。由一个分流器/旁通阀确保流体在压力达到旁通压力前全部过滤。

4. The parking brake is a mechanical caliper type, which operates on the transmission output flange. It is applied by actuating the parking brake lever in the operator's compartment. The lever is connected to the parking brake caliper by a cable.

驻车制动器是一种作用在变速器的输出法兰上的钳式机械制动器。它是通过在驾驶室操纵驻车制动杆来操作的。该制动杆通过钢索连接到驻车制动钳。

Section 11.2　Operation

11.2.1　Pre-operational Checks

Caution: Before using the equipment, the operator should carefully read the instructions in this section and become familiar with the tow tractor's controls and indicators, described in Figure 11.6 and Table 11.2 of this section.

Never run the engine with 115 V AC receptacle connected to the optional engine coolant heater. A flow check valve stops coolant flow through the heater while the engine is running. The heater will be damaged if 115 V AC is connected to the heater receptacle while the engine is running.

Note: In temperatures below 40°F (4.4℃), ensure that in units with an engine fuel heater option, heaters are operating 10 minutes prior to engine starting and turned off before ENGINE STOP/RUN switch is placed in the RUN position.

In temperatures below 40°F (4.4℃), ensure that units with an engine coolant heater option are connected to engine external 115 V AC power source 1 hour prior to engine starting and operation.

In order to ensure safe, reliable operation, perform the following checks prior to operating the tractor:

(1)Set parking brake (Pull handle up as far as it will go).

(2)Walk around and inspect tractor for damage or missing equipment. Check for the following:

1) Damaged or under-inflated tires. Cold-tire pressure should be 130 psi (895 kPa). Look for cuts, cracks in sidewalls, or foreign objects in treads.

2) Loose wheel nuts.

3)Leaking planetary wheel end housings.

4)Proper operation of hitches.

5)Damage to lights.

6)Fire extinguishers fully charged and in place.

(3)Look underneath tractor for leaks of coolant, hydraulic fluid, transmission fluid, or lubricating oil. If significant leaks in any of these systems are found, repair as soon as possible.

(4)Check engine oil.

Warning: Do not loosen or remove radiator cap while coolant is hot. Hot coolant under pressure can cause severe burns.

(5)Check engine coolant level. Coolant level should be approximately 1 – 1/2 in (4 cm) below top of radiator fill neck. Add coolant as necessary, but do not overfill. Make a visual check for leaks.

(6)Inspect engine compartment for loose hardware, loose wires, and leaking lines or fittings.

(7)Check hydraulic system reservoir by removing filler cap to ensure there is enough fluid for operation. A very low fluid level indicates a serious leak in the system. Repair as soon as possible.

(8)When mechanical checks are completed, move to the driver's seat in the cab. Check the operation of hazard flasher, taillights, headlights, backup lights, and dash panel lights.

(9) Check fuel level gauge to make sure tank is full.

Note: Fuel contamination owing to moisture condensing on the inner surfaces of the fuel tank can be minimized or eliminated by keeping the fuel tank as full as possible. The fuel tank should always be topped off at the end of the day's operation, as pan of the post-operation maintenance routine.

(10)Check operation of 12 V DC brake backup booster pump as follows:

Warning: If any malfunction of the brake system occurs, the tractor should be taken out of service until it is properly repaired.

Caution: If the brake backup booster pump is in operation longer than 1 min, possible overheating and damage to the motor may occur. Therefore, the driver should not apply heavy brake pressure any longer than necessary.

1) Make sure parking brake is still applied. Do not start engine. Turn engine stop-run-crank switch to RUN position.

2) Depress brake pedal firmly. Brake booster pump ON. The driver should be able to hear the DC motor running and ON lamp illuminated.

Note: It requires a minimum of approximately 65 – 72 lb (31 – 33 kg) of force, to activate the pressure switch to turn on the pump motor.

3) Turn STOP – RUN – CRANK switch to STOP position and repeat step 2).

(11) Start engine and allow engine to idle. When engine coolant temperature reaches 160°F (71.8°C), check transmission oil level. Add oil as required. Do not overfill.

11.2.2　Operating Procedures

Warning: Make sure all personnel are clear of the tow tractor before starting the engine.

(1) Check that parking brake is applied and transmission lever is in NEUTRAL (N) position.

Caution: To avoid overheating of starter motor, do not operate starting motor more than 30 seconds without allowing a minimum cooling period of 2 minutes.

(2) To start engine, turn engine STOP – RUN – CRANK switch to RUN position. Observe the glow plug lamp, when the lamp extinguishes, turn the STOP – RUN – CRANK switch to the CRANK position.

(3) Depress brake pedal and release parking brake.

(4) Select desired position on transmission shift lever.

Warning: Do not back up (reverse) tractor without first operating backup lights switch to on position. Operate switch to OFF position when tractor is no longer in reverse. Do not sheft transmission while handling aircraft.

(5) Release brake pedal and depress accelerator pedal to desired tractor speed.

Caution: The parking brake is not intended to be used for normal stopping.

(6) To stop tow tractor, release accelerator pedal and depress brake pedal.

Table 11.1, which follows, lists normal engine operating conditions.

Table 11.1　Normal operating conditions

Engine Oil Pressure (3,000 r/min)	40 – 60 psi (275 – 413 kPa))
Idle	12 – 15 psi (83 – 103 kPa))
Coolant Temperature Engine	160 – 185°F (71 – 85°C)
Governor Speed	3,000 r/min
Idle Speed	700 r/min

(1) Halt tow tractor at its parking location.

(2) Position transmission shift lever to NEUTRAL (N) position.

(3) Apply parking brake.

(4) Turn off all accessory equipment (lights, etc.).

(5) To stop tractor engine, turn engine STOP – RUN – CRANK switch to STOP position.

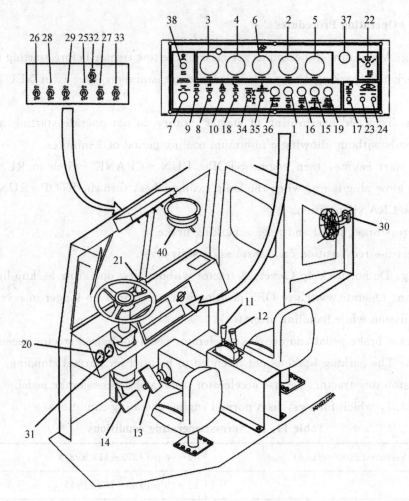

Figure 11.6   Operator's panel and controls

(see Table 11.2)(sheet 1 of 4)

Table 11.2   Controls and indicators

(see Figure 11.6)

Item	Name of Control or Indicator	Function
1	Lamp, hand brake on (parking brake)(DS1)	The hand brake on lamp illuminates when the hand (parking) brake is actuated
2	Gauge, fuel level(GA1)	The fuel level gauge indicates the level of fuel in the tractor fuel tank
3	Gauge, engine oil pressure(GA2)	The engine oil pressure gauge is used to monitor engine lube oil pressure

Continued

Item	Name of Control or Indicator	Function
4	Gauge, water temperature (engine coolant temperature) (GA3)	The engine coolant temperature gauge is used to monitor engine coolant temperature
5	Meter, time elapsed (hourmeter)(M1)	The time elapsed meter (hourmeter) indicates total engineoperating time
6	Meter, battery voltage (voltmeter)(M2)	The battery voltage meter (voltmeter) indicates battery voltage when the tractor is in the CRANK or RUN position
7	Switch, engine STOP – RUN – CRANK(S1)	The engine STOP – RUN – CRANK switch is a three-position Spring-return switch used to electrically start and stop the tractor engine
8	Switch, parking lights ON/OFF(S9)	The parking lights switch is used to turn on the front and rear parking lights
9	Switch, head lights ON/OFF(S11)	The head lights switch is used to turn on the headlights on the tractor (It also turns on the parking lights if they are not already on)
10	Switch, hazard lights ON/OFF (S12)	The hazard lights switch is used to mm on the front and rear hazard lights (If optional turn signal switch is famished; S12 is removed)
11	Transmission shift lever	The transmission shift lever is used to activate the six-speed forward and three speed reverse Funk automatic transmission
12	Parking brake lever	The parking brake lever is used to actuate the parking brake
13	Accelerator	The accelerator is used by the operator to regulate the speed of the tow tractor
14	Brake pedal	The brake pedal is used by the operator to manually activate the brake system on the tow tractor
15	Lamp, engine coolant temperature (DS11)	The engine coolant temperature warning lamp illuminates when engine block temperature exceeds 239°F (115°C)
16	Lamp, engine low oil press (DS 12)	The engine low oil pressure warning lamp illuminates when engine lubricating oil pressure drops to 5.8 psi (40 kPa)

Continued

Item	Name of Control or Indicator	Function
17	Switch, front pintle light ON/OFF	The front pintle light switch is used to illuminate the front pintle lights on the tractor
18	Switch, panel lights ON/OFF (S20)	The panel lights switch operates the panel lights in operator's compartment
19	Lamp, brake booster pump on (DS13)	The brake booster pump on lamp illuminates whenever the booster pump is supplying brake pressure to the brake system in an emergency situation
20	Gauge, tachometer (GA4)	The tachometer indicates engine rpm
21	Lamp. parking brake warning (DS17)	Warning lamp illuminates with the parking brake applied, in addition to lamp DS1
* 22	Switch, cab heater fan	Operates fan motor for cab heating
* 23	Switch, windshield defroster, front	Operates blower motor for front defroster tube
* 24	Switch, windshield defroster, rear	Operates blower motor for rear defroster tube
* 25	Switch, rear wiper motor	Operates rear windshield wiper motor
* 26	Switch, beacon	Operates rotating towing beacon
* 27	Switch, from wiper (left)	Operates left front windshield wiper motor
* 28	Switch, light, dome	Operates the cab's dome light
* 29	Switch, defogger fan	Operates defogger fan
* 30	Defogger fan	Circulates air in cab area to defog windows
* 31	Speedometer	Indicates vehicle speed
* 32	Switch, overhead wiper	Operates overhead wiper motor
* 33	Switch, front wiper (right)	Operates right from windshield wiper motor
* 34	Selector, steering mode COORD/TRACK	Selects vehicle steering mode
* 35	Switch, emergency steering pump ON/OFF	Operates emergency steering pump
36	Lamp, generator no charge	Warning lamp illuminates when alternator has failed
* 37	Gauge, transmission oil temperature	The transmission oil temperature gauge is used to monitor transmission fluid temperature

Continued

Item	Name of Control or Indicator	Function
38	Lamp, glow plugs	Lamp illuminates when glow plugs are hot and engine start can occur
* 39	Phone jack (not illustrated)	The headphone jack is mounted on the rear wall of the cab, between the operators seats
* 40	Light, dome	Illuminates cab area

* Optional Equipment

New Words

1. tachometer [tæ'kɒmitə(r)] *n.* 转速表
2. hourmeter ['aʊmiːə] *n.* 摩托小时计数器

Phrases and Expressions

1. check valve 止回阀、单向阀
2. Leaking planetary wheel end housings 行星齿轮端盖漏油
3. Proper operation of hitches 挂钩的正确操作
4. dash panel 仪表板
5. lube oil 润滑油
6. windshield defroster 挡风玻璃除霜器
7. receptacle 插接器，插座

Notes to the Text

1. Never run the engine with 115 V AC receptacle connected to the optional engine coolant heater. A flow check valve stops coolant flow through the heater while the engine is running. The heater will be damaged if 115 V AC is connected to the heater receptacle while the engine is running.

当115 V 交流电插接器连接到发动机冷却液加热器时，绝对不要启动发动机；当发动机运转时由一个单向阀阻止冷却液流经加热器；当发动机在运转时，如果115 V 交流电连接到加热器的插接器，那么加热器将损坏。

2. Note：Fuel contamination owing to moisture condensing on the inner surfaces of the fuel tank can be minimized or eliminated by keeping the fuel tank as full as possible. The fuel tank should always be topped off at the end of the day's operation, as pan of the post-operation maintenance routine.

注意：因水分在燃油箱内表面凝结所造成的燃油污染可以通过保持燃料箱尽可能满的油量来减小或消除。作为操作维护例行程序的一部分，在每天日常操作结束时，燃油箱应始终处于加满状态。

Section 11.3   Specifications

11.3.1   Introduction

The following specification tables provide important data as to the characteristics of the GT - 50DZ/50DZH tow tractor and its related components.

11.3.2   Specification Tables

Specifications for the GT - 50DZ/50DZH tow tractor and related components are listed in Table 11.1 - Table 11.3.

Table 11.3   Specifications

Equipment	Specifications
A. complete unit	Tow tractor GT - 50D2/50DZH
1. Manufacturer	Stewart & Stevenson Services, Inc. Houston, Texas
2. Weight GT - 50DZ/50DZH	50,000 - 60,000 lb (22,680 - 27,216 kg)
3. Tractive effort	40,000 - 48,000 lb (18,160 - 21,792 kg) (varies with surface conditions)
4. Speed (max.)	15 mile/h (24 km/h)
5. Length (front/rear hitches) Length (less hitches) Length (w/ground power unit)	224 in (568 cm) 188 in (478 cm) 266 in (675 cm)
6. Width	102 in (259 cm)
7. Height (less cab) (with cab model 416)	65 in (165 cm), includes fuel fill pipe. 86 in (215 cm)
8. Wheel base	90 in (229 cm)
9. Turning radius (outside)	21 ft 6 in (6.6 m)
10. Ground clearance	7 in (17.8 cm)
B. Engine	
1. Manufacturer	Deutz
2. Type	4-cycle, in-line 6-cylinder diesel. Model BF6M 1013 turbocharged
3. Number of cylinders	6

Continued

Equipment	Specifications
4. Bore	4. 25 in (10. 8 cm)
5. Stroke	5. 12 in (13 cm)
6. Effective displacement	436 in³ (7 L)
7. Speed (idle) (governed)	700 r/min 2,500 r/min
8. Compression ratio	17. 6 : 1
9. Firing order	1. 5. 3. 6. 2. 4
10. Number of main bearings	7
11. Governor	Mechanical/speed limiting
12. Starting motor	Deutz
13. Alternator — battery charger	Deutz
14. Regulator	Deutz
C. Transmission	
1. Manufacturer	Funk Mfg. Co. , Inc.
2. Model	2263E13NB, Spec YZ15176
3. Speeds	6 Fwd, 3 reverse
4. Controls	Electric, modulated
5. Low ratio	5. 2 : 1
D. Axle — front	
1. Manufacturer/model	Rockwell Int. , model 5MRDIA
2. Type	Planetary drive (limited slip)
3. Overall reduction	24. 50 : 1
E. Axle — rear	
1. Manufacturer/model	Rockwell Int. , model 5MRA
2. Type	Planetary drive (limited slip)
3. Overall reduction	24. 50 : 1
F. Wheels and tires	
1. Wheel type	Heavy-duty (10-stud hole)
2. Tire size	365/80R20PXZA
3. Inflation pressure	130 psi (895 kPa)
G. Brakes	

Continued

Equipment	Specifications
1. Service brakes, front and rear service brakes, front and rear	400×120 hydraulic drum 400×120 hydraulic drum
2. Paridng	Mechanically actuated caliper located on transmission output flange-mourned disc.
H. Steering hydraulic system	
1. Type	Hydrostatic
2. Pump	Direct-driven gear type
3. Steering cylinders	Twin, double-acting
4. Hydraulic pressure	2,250 psi (15.487 kPa)
I. Electrical system	
1. Staiting system	12 V DC
2. Charging system	12 V DC, 95 A
J. Capactty of lube systems (approximate values)	
1. Engine lube oil	22 quarts (21 L)
2. Transmission lube oil front	4 gallons (15 L)
3. Axle lube oil Differential carrier Planetary ends (2)	12 quarts (11.4 L) 2.5 quarts (2.4 L)
4. Engine cooling system	2.5 gallons (9.8 L)
5. Engine fuel tank	43 gallons (163 L)
6. Hydraulic steering system	12.9 gallons (48 L)
K. Engine protective system	

The abnormal operating conditions listed below will cause illumination of warning lights located in operator's instrument panel.

1. High water (coolant) temperature	239℉(115℃)
2. Low engine oil pressure	5.8 psi (40 kPa)
L. Engine 00. grades	

For the SAE viscosity grades of engine oil to use at various temperatures, refer to the Deutz operator's manual.

M. Transmission oil grades	

Note: Stewart & Stevenson recommends the use of ISO VG10 MOBIL 1310 in the transmission (see Table 11.4).

Table 11.4　Recommended transmission oils

Prevailing Ambient Temperatures	Oil Specifications
Above −10°F (−23°C)	Hydraulic transmission fluid type C - 3 (except grade 30)
Below −10°F (−23°C)	Hydraulic transmission fluid type C - 3 (except grade 30) Auxiliary preheat required to raise temperature in the sump to above −10°F (−23°C)
Above 32°F (0°C)	Hydraulic transmission fluid type C - 3 or C - 3 grade 30

11.3.3　Points of Service, Specification, and Capacities

Engine lube oil specifications are shown in Table 11.5. For location of unit service points, specification and capacities, see Figure 11.7 and Table 11.6..

Table 11.5　Engine lube oil specifications

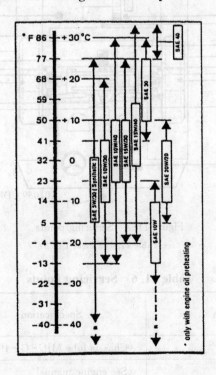

Note:

Approved API oils:

Naturally aspirated engines:　CC

Turbocharged engines： COCF
CECF4

Approved CCMC oils：
Naturally aspirated engines： D4
Turbocharged engines： D4D5

D5 (sulfate ash 11.8% by mass) corresponds to SHPD.

SHPD (Super High Performance Diesel) oils.

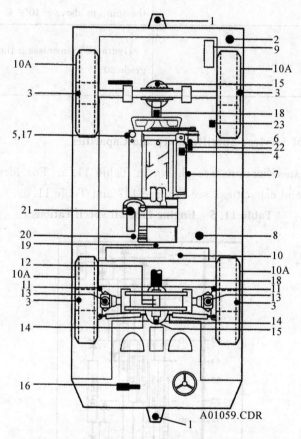

Figure 11.7　Servicing points
(see Table 11.6) (sheet 1 of 2)

Table 11.6　Servicing points

Item	Descripton	Specification	Capacity (Approximate)
1	Hitch pin	Chassis lube MIL‑G‑10924	2 pins
2	Fuel fill cap	See engine manual	43 gallons (163 L)
3	Wheel end assy. (fill plug in center of hub; drain plug at perimeter.)	API GL 5, 80W90	2.6 quarts (2.4 L)

Continued

Item	Descripton	Specification	Capacity (Approximate)
4	Engine oil filler cap		
5	Engine coolant filler cap		
6	Final fuel filler element	P/N 117 – 4423	
7	Engine oil dipstick	See engine manual	See engine manual
8	Hydraulic system filter element	P/N 932409	
9	Engine air cleaner element	P/N 70459A	
10	Hydraulic reservoir filler/ dip stick	Dexron II	18 gallons (68 L)
10A	Tire pressure (cold)	365/80R20PXZA	130 psi (895 kPa)
11	Tie-rod ends	Chassis lube MIL – G – 10924	
12	Bolster pivot	Chassis lube MIL – G – 10924	2 fittings
13	Trunnion pivot, front axle, (top and bottom)	NLGI No. 2, rust and corrosion inhibitors, able to withstand extreme pressure	4 fittings
14	Pivot ends, steering cylinders	Chassis lube MIL – G – 10924	4 fittings
15	Differential fill plug	API GL 5, 80W90	2. 4 gallons (9 L)
16	Brake master cylinder	Brake fluid, DOT – 3	Add fluid as required
17	Engine coolant	See engine manual in chapter 6	2. 38 gallons (9 L)
18	Drive shafts	Chassis lube MIL – G – 10924	6 fittings
19	Transmission fill plug		4 gallons (15 L)
20	Transmission dipstick	See transmission manual	
21	Transmission fluid filter	P/N 4003483	
22	Engine oil filter	P/N 117 – 4421	
23	Fuel/water separator	P/N 2020 SM – OR	

New Words

1. bore [bɔ:(r)] *n.* 缸径
2. stroke [strəuk] *n.* 活塞行程

Phrases and Expressions

1. length (front/rear hitches) 计前后挂钩长度
2. length(less hitches) 不计挂钩长度
3. wheel base 轴距
4. ground clearance 离地间隙
5. effective displacement 排量
6. hitch pin 连接销
7. fuel fill cap 油箱盖
8. hydraulic system filter element 液压系统滤清器滤芯
9. tie-rod ends 横拉杆球头
10. bolster pivot 支撑轴
11. trunnion pivot 主销
12. pivot ends，steering cylinders 转向油缸活塞杆头

Section 11.4　Preventive Maintenance Schedule

11.4.1　Introduction

This preventive maintenance schedule (see Table 11.7) defines preventive maintenance tasks and time intervals for each task. This schedule applies to all GT - 50DZ tow tractors. The following information will help you to use the preventive maintenance schedule.

11.4.2　How to Use the Preventive Maintenance Schedule

In order to use the preventive maintenance schedule，perform the following steps：

(1)Refer to Table 11.6. The scheduled maintenance tasks are listed in rows on the far left-hand side of the table. The frequency of maintenance is listed in columns at the top of the table.

(2)Immediately to the right of the maintenance tasks are references，which direct the technician to the required maintenance procedures. The references are defined as follows：MI 01001，MI 02001，etc. ：Refer to Maintenance Instructions located in Section 11.5.

(3)From the reference column，follow the row to the right until you see an "X" in one of the blocks.

(4)From the "X"，follow the column up until you reach the recommended frequency，or time interval between maintenance. Both calendar-day and hourly time intervals are included.

Warning：Ensure that personnel are well clear of the tractor wheels before beginning this procedure. Death or serious injury can result if someone is crushed between the wheels and the tractor.

Caution: To prevent damage to unit, do not exceed the recommended time interval between inspection, servicing or lubrication. Increase the frequency of inspection, servicing, and lubrication if the unit is operated in harsh environmental conditions.

Each "X" indicates a need for preventive maintenance at the specified time interval. Perform the maintenance item at or before the indicated time interval.

Table 11.7   Preventive maintenance schedule

Preventive Maintenance Tasks	Reference	Daily	Weekly 50 H	Monthly 200 H	6 Months 500 H	Yearly 1,000 H
Fluid levels, check	MI 99003	X				
Unit, visual inspection	MI 01004	X				
Fuel/water separator, drain water (raycor separator)	MI 04007	X				
Battery, service	MI 09012		X			
Tire pressures, check	MI 02007		X			
Hoses/clumps, inspect/replace			X			
Engine air cleaner, check restriction indicator	MI 08004		X			
Differential and wheel end assembly check oil levels*	MI 02009		*			X
Water pump & alternator v-drive belts, inspect/adjust	MI 99004			X		
Transmission oil and filter, change oil, replace filter	MI 02010			X		
Fuel filter, change element	MI 04010			X		
Engineoil and filter, change oil, replace filter	MI 06003			X		
valve lash**				**		X
Service brake reservoir, check level	MI 03007				X	
Steering assembly and driveshaft, lubricate	MI 0200S				X	
Fuel water separator, replace (raycor separator)	MI 04009				X	
Parking brake, inspect/adjust	MI 03006					X

Continued

Preventive Maintenance Tasks	Reference	Daily	Weekly 50 H	Monthly 200 H	6 Months 500 H	Yearly 1,000 H
Service brake shoes, inspect/replace	MI 03001					X
Service brake system, bleed wheel cylinders	MI 03009					
Differential and wheel end assemblies change oil	MI 02011					X
Radiator, inspect and clean core and fins (exterior)						X
Thermostats and seals, check						X
Cooling system, clean, flush, and replace coolant						X
Hydraulic oil system, change oil, replace filter	MI 12001					X
						X

* Check first 50 h, then according to schedule.

** Check first 200 h, then according to schedule.

Phrases and Expressions

1. visual inspection 目视检查
2. V-drive belts V 形传动带
3. fuel filter, change element 燃油滤清器，更换元件

Section 11.5   Troubleshooting

11.5.1   Introduction

The following troubleshooting tables provide information for diagnosing and correcting problems with the GT – 50DZ tow tractor. This introduction will help you to quickly identify the problem and to make the necessary repairs to bring the unit back into service.

11.5.2   How to Use the Troubleshooting Tables

Before beginning any trouble diagnosis or corrective action, perform the following steps：

(1) Determine which of the following categories most closely relates to the problem

your unit is experiencing:

1) Engine starting failures (see Tables 11.8, 11.9).

2) Hydraulic steering (see Table 11.10).

3) Hydraulic brakes (see Table 11.11).

4) Indicator lamp and lights (see Table 11.12).

5) Gauges (see Table 11.13).

(2) Turn to the applicable troubleshooting table and read the information carefully.

Note: The probable causes for problems are listed in descending order from "most likely" to "least likely" to occur.

(3)Carefully perform the troubleshooting steps and recommended corrective action.

(4) When needed, the tables refer to Maintenance Instructions (MI XXXXX) for corrective action. Refer to the appropriate Maintenance Instruction, being careful to select the correct procedure for the assembly part number used on your unit.

11.5.3 Obtaining Factory Assistance

Where any references require contacting Stewart & Stevenson Services, Inc., follow these steps:

(1) Ensure that all steps of the troubleshooting chart(s) have been checked.

(2) Ensure that any Maintenance Instructions used apply to the part number assembly installed on your unit and that the instructions have been strictly complied with.

Warning: Disconnect batteries before performing continuity or resistance checks on electrical components. Severe burns could result from coming in contact with a live circuit. Remove all rings, watches, and jewelry before performing voltage checks.

Table 11.8   No response when engine start switch is energized

Cause	Corrective Action
1. Transmission not in neutral (N)	Place transmission selector lever in neutral (N) position
2. Loose battery connectors	1. Clean and tighten battery cables 2. Repair or replace damaged battery cables as needed
3. Battery dead	1. Service and charge batteries. Refer to MI 09012 2. Replace Batteries with damaged terminal(s), case, or cell plates
4. 20 amp fuse (F2) open	Replace fuse (F2)
5. 20 amp fuse (F4) open	Replace fuse (F4)

Continued

Cause	Corrective Action
6. Loose starter connections	1. Check starter motor (B1 and solenoid (L1) connections 2. Clean and tighten loose connections
7. Defective starter	Remove and repair or replace starter motor
8. Electronic speed switch assy. (A1) contact close	Replace electronic speed switch assembly (Al)
9. Engine run/start switch (s1) malfunctioning	Test switch (S1). Replace switch (S1)
10. Engine starring contactor (K1) defective	1. Tighten connectors 2. Replace contactor (K1)
11. Engine run relay (K2) defective	Replace relay (K2)
12. Energized only when transmission shifter in neutral position relay (K5) defective	1. Check and lighten connectors 2. Replace relay (K5)
13. Engine running relay (K6) defective	1. Check and tighten connectors 2. Replace relay (K6)
14. Emergency engine shutdown switch (S7) malfunctioning (optional)	1. Test switch 2. Check and tighten connectors 3. Replace switch (S7)
15. Engine seized	Refer to Deui2 engine manual

Table 11.9 Engine cranks but will not start

Cause	Corrective Action
1. Fuel low	1. Fill fuel tank 2. Bleed fuel lines. Refer to MI 04010 for fuel system bleeding procedure
2. High lube oil viscosity	1. Drain crankcase and fill with correct viscosity oil for current ambient temperatures 2. If optional engine heater is installed, operate heater to thin oil
3. Low battery output	1. Service and charge batteries. Refer to MI 09012 2. Clean and tighten dirty or loose battery cable terminals 3. Repair or replace damaged battery cables 4. If battery healer is provided, operate heater to warm battery 5. Replace damaged batteries
4. Loose starter connections	Clean and tighten starter connections
5. Starter defective	Replace starter

Continued

Cause	Corrective Action
6. Air in fuel lines	1. Check for air leak 2. Bleed air from system. Refer to MI 04010 for fuel system bleeding procedure
7. Contaminated fuel	1. Inspect fuel tank and filler for din or other contaminants and replace or clean as necessary 2. Refer to MI 04008 and MI 04010. If optional equipment is installed, refer to MI 04007 and MI 04009
8. Faulty fuel pump	Cali Stewart& Stevenson for Deutz service information or refer to Deutz System Operation. Testing and Adjustment Manual
9. Fuel injection timing incorrect	Adjust injection timing. Refer to Deutz System Operation, Testing and Adjustment Manual
10. Intake air flow restricted	1. Inspect air intake filter for restrictions. Refer to MI 08004 2. Remove debris or replace filter. Refer to MI 08005
11. Engine run solenoid (L2) defective	1. Check and tighten connectors 2. Check wiring circuit to ensure solenoid is energized 3. Replace solenoid (L2)
12. Electronic control unit assembly (ECU) defective	1. Check circuit and wiring integrity 2. Reset circuit breaker (CB1) if nipped 3. Replace electronic control unit (ECU)
13. Glow plugs (GPI – 6) defective	1. Check circuit and lighten connectors 2. Replace glow plugs (GPI – 6)
14. Low engine compression	Call Stewart& Stevenson for Deutz service information

Warning: Exercise extreme caution when performing maintenance on the hydraulic system. Fluid escaping under high pressure can cause serious personal injury.

Table 11.10 Hydraulic steering troubleshooting

No Response When Steering Wheel is Turned Slowly

Cause	Correcttve Action
Low fluid level	1. Check fluid level in reservoir. Add fluid as necessary 2. Check for leaks in system 3. Repair or replace defective lines or components

Continued

Slow or Hard Response

Cause	Corrective Action
1. Filter bypassed	1. Check hydraulic filter indicator 2. If filter is in bypass position, change filter element refer to MI 12001
2. Line restriction	1. Check lines for damage 2. Repair or replace lines as required
3. Damaged component	Repair or replace component as required

Hydraulic Pressure Low

Cause	Corrective Action
1. Hydraulic pump	Note: Make the following pressure checks with the engine running at proper idle speed. Wheels must be turned hard over, to bottom out cylinder Check hydraulic pressure: a. Install pressure gauge in hydraulic system b. Check that hydraulic pressure is approximately 2,250 psi (15,503 kPa) c. If pressure is significantly less than 2,250 psi (15,503 kPa), break connections and check pump discharge pressure for 2,250 psi (15,403 kPa)
2. Flow control valve malfunctioning	1. If pump discharge pressure is 2,250 psi (15,503 kPa), check relief valve by turning clockwise to increase pressure, counterclockwise to reduce pressure 2. If no change is noted, tap flow control valve to unstick or replace flow control valve
3. Steering valve malfunctioning	1. With steering wheel (steering valve) held stationary, check that pressure does not exist at steering cylinders 2. If pressure exists, steering valve is faulty
4. Steering cylinders malfunctioning	If no pressure exists, check steering cylinders: a. With gauges attached to the tee connection of each cylinder, rotate steering wheel and check pressure at both cylinders. One cylinder will register approximately 2,250 psi (15,503 kPa) and the other should register little or no pressure while the steering wheel is being turned b. If significant pressure exists at both points, one cylinder is faulty. Steeling cylinders must be removed and checked to determine their condition

Warning: Ensure that personnel are well clear of the tractor wheels before beginning this procedure. Death or serious injury can result if someone is crushed between the wheels and the tractor.

Table 11.11　Hydraulic brake troubleshooting

Brake Pedal Soft or Spongy

Cause	Corrective Action
Fluid level low	1. Cheek fluid level in master cylinder reservoir.　Refer to MI 03007 2. Check hydraulic lines for leaks 3. Repair or replace defective lines or components 4. Bleed air from brake system.　Refer to MI 03009

Excessive Pedal Travel, Abnormal Stopping Distance

Cause	Corrective Action
1.　Brake lining worn	1. Check brake pressure of hydraulic pump.　Relief valve is set at 800 psi 2. Repair or replace defective pump 3. Check brake drums for scoring 4. Resurface brake drums if necessary.　Do not exceed minimum dimension
2.　Wheel cylinder defective	1. Check for brake fluid leakage at wheel cylinder 2. Replace defective wheel cylinder
3.　Master cylinder defective	1. Check master cylinder for internal leakage 2. Replace defective master cylinder

Hard Brake Pedal

Cause	Corrective Action
Hydraulic pump defective	1. Check brake pressure of hydraulic pump.　Relief valve is set at 800 psi 2. Repair or replace defective pump.　Relief

Noisy Operation (Grinding Sound)

Cause	Corrective Action
Brake lining	1. Check brake lining for imbedded debris.　Refer to MI 03008 2. Replace brake linings if necessary

Noisy Operation (Squealing Sound)

Cause	Corrective Action
Brake lining	1. Check brake lining for glazing 2. Sand or replace brake linings if necessary

Long Stopping Distance (Normal Brake Pedal)

Cause	Corrective Action
Brake lining	1. Check brake lining for water, grease, or oil on lining 2. If linings are wet with water, dry linings by applying brakes while unit is in motion 3. If linings are oily or greasy, replace brake linings

Table 11.12   Indicator lamp and light troubleshooting

Indicator Light or Lamp not Functioning	
Cause	Corrective Action
1. Defective bulb	1. Place switch to ON position 2. Remove lens cover and check bulb 3. If bulb is defective, replace bulb 4. If bulb is good, check voltage at bulb socket
2. No voltage to bulb socket	1. Check voltage from switch 2. If voltage from switch is correct, check wiring to socket 3. Repair or replace defective wiling
3. No voltage from switch	1. Check voltage to switch 2. If voltage to switch is correct, replace switch
4. No voltage to switch	1. Check wiring to power source 2. Repair or replace defective wiring 3. Check power source for 12 V DC 4. If power source voltage is incorrect, repair or replace power source

Table 11.13   Gauge troubleshooting

Gauge Inoperative	
Cause	Corrective Action
1. Gauge defective	1. Check voltage to gauge 2. If voltage to gauge is 12 V DC, check ground from gauge 3. If ground is good, replace gauge
2. No ground from gauge	1. Check wiring from gauge and gauge connections 2. Repair or replace any defective wiring or connections
3. No voltage to gauge	1. Check wiring to gauge and gauge connections 2. Repair or replace any defective wiring or connections

New Words

1. fuse [fjuːz] *n.* 保险丝
2. viscosity [viˈskɒsəti] *n.* 黏度

Phrases and Expressions

1. engine seized 发动机卡死
2. glow plugs 电热塞
3. tighten connector 收紧连接器
4. engine run solenoid (L2) defective 发动机电磁线圈运行故障

5. filter bypassed 油滤旁通

6. filter indicator 油滤堵塞指示器

7. steering valve malfunctioning 转向阀失灵

8. brake lining worn 制动衬片磨损

9. master cylinder defective 主油缸故障

Notes to the Text

1. Engine cranks but will not start.

发动机曲轴转动但是不能启动。

2. Warning：Disconnect batteries before performing continuity or resistance checks on electrical components. Severe burns could result from coming in contact with a live circuit. Remove all rings，watches and jewelry before performing voltage checks.

警告：检查电气元件的通断或测量电阻值前断开蓄电池。与带电线路接触可能导致严重烧伤。测量电压前摘下所有的戒指、手表、首饰。

Chapter 12　Large Modular Aircraft Deicer

Setction 12.1　General Description

12.1.1　General

The modular deicer provides mobile deicing and anti-icing for aircraft of all sizes. It is designed to be easily adaptable to each customer's specific needs.

The unit can be adapted for use on various chassis used worldwide. This adaptability includes a variety of tank capacities and deice and anti-ice configurations. The various configurations are capable of handling a wide variety of fluids currently in use for deicing and anti-icing. The unit can also be adapted to use many different types of power modules using various types of fuels.

The controls for operating the modular deicer systems, including the power module, are located on control panels in the truck cab and in the basket.

An intercom is included in the modular deicer for communication between the cab and the basket.

The modular deicer has built-in safety features to prevent damage to the unit as well as the operator. All safety precautions and warnings should be followed as prescribed in this manual as well as on the truck itself.

12.1.2　Hydraulic System

12.1.2.1　Pump (See Figure 12.1)

The hydraulic pump is axial piston pressure compensated, load sensing and is used for all systems. The pressure compensator feature is designed to protect the system against the failure of other components. In normal operation, the pressure system will never be high enough to cause the pump to compensate. This pump receives oil from the reservoir and delivers it to a closed valve at the inlet of each system where it maintains pressure while pumping only enough oil to replace leakage.

12. 1. 2. 2　Load Sensing System

Each individual hydraulic system is fitted with an opening, either fixed or variable, which is sized to provide the desired flow at a specific pressure to drop across the opening. When a system is actuated and the valve is opened, the pump maintains a specific pressure above the operating pressure and provides the correct flow through the opening. The integrated circuit manifold assembly is located on the backside of the pedestal (see Figure 12. 2). Located in the right side of the integrated circuit manifold assembly is a two position, two way solenoid valve (30SOL) that blocks pump flow to the boom valve while de-energized. This immediately stops the pump flow when the power is turned off or the emergency stop switch is pushed.

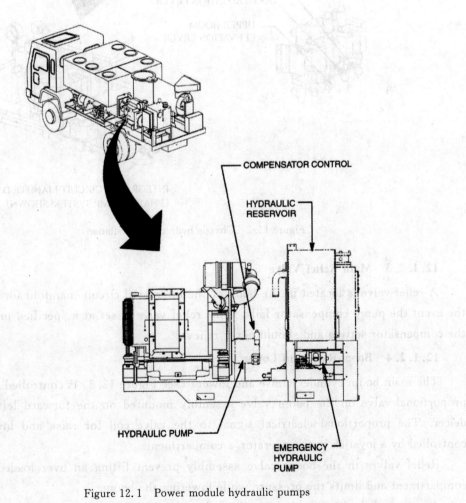

Figure 12. 1　Power module hydraulic pumps

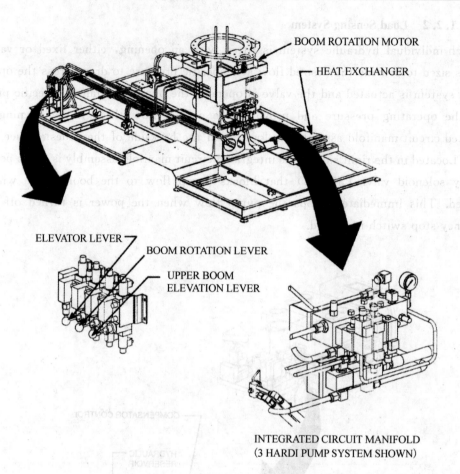

BOOM ROTATION MOTOR

HEAT EXCHANGER

ELEVATOR LEVER

BOOM ROTATION LEVER

UPPER BOOM
ELEVATION LEVER

INTEGRATED CIRCUIT MANIFOLD
(3 HARDI PUMP SYSTEM SHOWN)

Figure 12. 2　Chassis hydraulics installation

12. 1. 2. 3　Main Relief Valve

A relief valve is located in the body of the integrated circuit manifold for protection in the event the pump compensator fails. This relief valve is set at a specified pressure above the compensator setting and should never relieve.

12. 1. 2. 4　Boom Raise and Lower

The main boom cylinder (raise and lower) (see Figure 12. 3) is controlled by oil from a proportional valve on the boom valve assembly mounted on the forward left side of the deicer. The proportional electrical signal to the valve coil for raise and lower speed is controlled by a joystick in the operator's compartment.

Relief valves in the boom valve assembly prevent lifting an over loaded operator's compartment and limits the pressure while lowering the boom.

A counter-balance valve is located on the cap end of the boom cylinder to prevent the boom from lowering in the event of a hose failure. When the cylinder is being retracted hydraulically the counter-balance valve is piloted open. The boom raise and lower valve has a manually operated lever that can be used to override the valve and lower the operator in case of a coil or joystick failure.

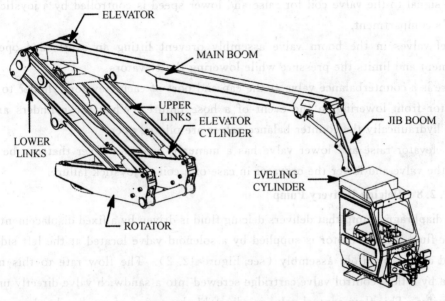

Figure 12. 3   Boom components

12. 1. 2. 5   Basket/Cab Leveling Raise and Lower (Standard Boom)

The basket/cab leveling (slave) cylinder is controlled by oil from the master cylinder that is attached to the upper boom and lower elevator. The master cylinder volume is matched to the leveling slave cylinder so that as the boom raises or lowers oil is ported from the master cylinder to the slave cylinder in the correct proportions. The slave cylinder may be overriden with the basket/cab leveling switch located in tie operator's basket/cab. This switch is active only with the boom in the support cradle.

A relief/check valve assembly is located in the rotator to protect the master/slave system.

A counter-balance valve is located on the cap end of the slave cylinder to prevent the basket/cab from lowering in the event of a hose failure.

12. 1. 2. 6   Boom Rotation

The boom is rotated by a geroller type hydraulic motor through a gear reduction. Flow for this motor is supplied from a valve on the boom valve assembly. The rotation speed (flow rate) is controlled by the electrical signal supplied to the valve coils from the joystick in the operator's compartment.

A dual cross over relief assembly is mounted in the left end of the boom valve assembly. This relief assembly is for the protection of the boom rotation motor. This valve is equipped with a manually operated rotation lever for overriding the valve in the event of a coil failure.

12. 1. 2. 7   Elevator Raise and Lower

The elevator raise and lower cylinders are controlled by oil from a proportional valve on the boom valve assembly mounted on the forward left side of the deicer. The proportional electrical signal to the valve coil for raise and lower speed is controlled by a joystick in the operator's compartment.

Relief valves in the boom valve assembly prevent lifting an overloaded operator's compartment and limits the pressure while lowering the elevators.

There is a counterbalance valve at the cap end port of each elevator cylinder to prevent the elevator from lowering in the event of a hose failure. When the cylinders are being retracted hydraulically the counter balance valves are piloted open.

The elevator raise and lower valve has a manually operated lever that can be used to override the valve and lower the operator in case of a coil or joystick failure.

12. 1. 2. 8　Deicing Delivery Pump

The diaphragm pump that delivers deicing fluid is driven by a fixed displacement motor. Hydraulic fluid for this motor is supplied by a solenoid valve located at the left side of the integrated circuit manifold assembly (see Figure 12. 2). The flow rate to this motor is controlled by a flow control valve cartridge screwed into a sandwich valve directly under the solenoid valve. This flow control valve is adjustable by means of an alien head screw.

12. 1. 2. 9　Anti-ice/Refiller

The diaphragm pump that delivers the anti-ice fluid is driven by a fixed displacement motor. Hydraulic fluid for this motor is supplied from a solenoid valve on the left end of the integrated circuit manifold. The flow rate to this motor is controlled by a flow control valve cartridge screwed into a sandwich valve located on top of the manifold. This flow control valve is adjustable by means of an alien head screw. Pressure to the drive motor is regulated by a pressure reducing valve cartridge located directly under the solenoid valve. This valve cartridge provides constant pressure anti-ice fluid to the nozzle independent of the flow rate. When the nozzle is closed, the hydraulic system maintains a constant static pressure at the nozzle.

12. 1. 2. 10　Blower Drive

The blower fan for the fluid heater is driven by a fixed displacement motor. Hydraulic fluid for low speed start up operation of the blower fan is supplied from a solenoid valve cartridge at the right-hand side of the integrated circuit manifold, near the pressure gauge. The speed of the motor is controlled by the flow control valve cartridge located at the lower right-hand corner of the front face of this same manifold. Hydraulic fluid for high speed operation of the blower fan is supplied from a solenoid valve cartridge at the top right-hand side of the integrated circuit manifold. High speed operation of the motor is controlled by the flow control valve cartridge located at the left side of the low speed flow control valve cartridge.

12. 1. 2. 11　Fluid Heater Fuel Pump Drive (Optional)

The fuel pump for the diesel fired fluid heater is driven at a constant speed by a hydraulic motor. Hydraulic oil to drive the motor is supplied by a solenoid valve through a pressure compensated flow control valve. These valves are located at the right side end of the integrated circuit manifold. Constant fuel pressure is maintained by a relief valve between the fuel pump and the burner nozzle.

12.1.2.12　Emergency Pump (See Figure 12.1)

The modular deicer has an optional electrically powered hydraulic pump for use in an emergency such as the failure of the engine or main pump while the operator is elevated in the operator's compartment. This pump can be used to rotate and lower the boom. A check valve prevents the emergency pump flow from flowing through the main pump in the event of a failure.

12.1.2.13　Heat Exchanger

Two oil to air heat exchangers are installed in the return to tank line of the hydraulic system to prevent the hydraulic oil from overheating. One heat exchanger is located under the hydraulic reservoir and the other is located under the heater module. The exchangers have electric motor driving fans to move air through the fins of the oil coolers. A temperature switch located inside the hydraulic reservoir controls the fan motors.

12.1.2.14　Filtration

Fluid returning to the hydraulic reservoir passes through a 10 micron filter with a built-in bypass relief valve. Air entering the hydraulic reservoir is filtered through a breather mounted on top of the reservoir. Fluid from the case drain port of the main hydraulic pump passes through a 10 micron filter with a built-in bypass relief valve.

12.1.3　Electrical System

12.1.3.1　Cab Electrical (See Figure 12.4)

The cab electrical consists of a cab control box in which most of the modular deicer controls/indicators are located. The top window wiper motor harness and chassis power harnesses branch off from the cab control box harness. The control box is mounted on a pedestal, and is connected by a rectangular connector coming from the main junction box in the left rear of the modular deicer.

12.1.3.2　Power Module Electrical (See Figure 12.5)

The power module contains the main junction box that ties all other areas together. A battery equalizer, two 12 V batteries and an oil cooler fan are also located on the power module. The batteries are connected in series for 24 V and some 12 V loads are used off the center tap. The battery equalizer maintains an equal charge on each battery.

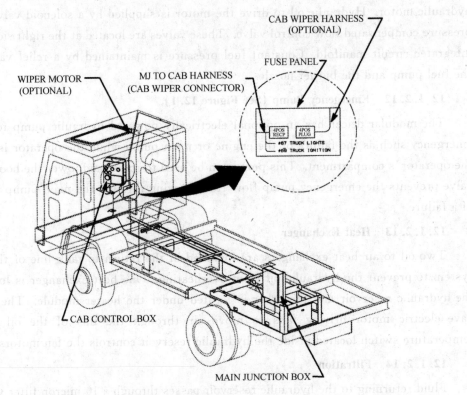

Figure 12.4　Cab electrical

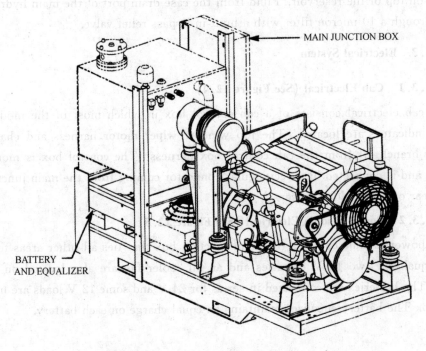

Figure 12.5　Power module primary electrical

12. 1. 3. 3　Heater Module Electrical (See Figure 12. 6)

The healer module electrical consists of the controls (timers, relays and maintenance lights) in the heater control box. A harness from the main junction feeds power to the heater control box. A harness also feeds the heater and associated heater controls in the inlet and outlet plumbing.

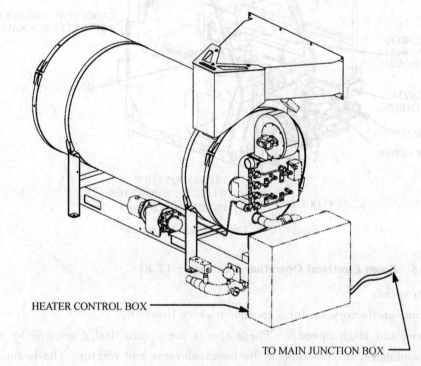

HEATER CONTROL BOX

TO MAIN JUNCTION BOX

Figure 12. 6　Heater module electrical

12. 1. 3. 4　Enclosed Basket Electrical (See Figure 12. 7)

The enclosed basket primary electrical components consist of the following:

(1) Upper and lower control panels;

(2) Video monitor (optional);

(3) Boom control joystick;

(4) Nozzle control joystick (electric or optional hydronic);

(5) Spray motor;

(6) Nozzle spotlight;

(7) Cabin heater;

(8) Cabin heat control box;

(9) Flood lights.

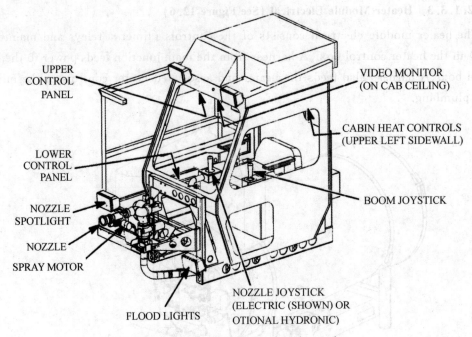

UPPER
CONTROL
PANEL

VIDEO MONITOR
(ON CAB CEILING)

CABIN HEAT CONTROLS
(UPPER LEFT SIDEWALL)

LOWER
CONTROL
PANEL

NOZZLE
SPOTLIGHT

NOZZLE

SPRAY MOTOR

BOOM JOYSTICK

NOZZLE JOYSTICK
(ELECTRIC (SHOWN) OR
OTIONAL HYDRONIC)

FLOOD LIGHTS

Figure 12. 7　Enclosed basket electrical

12. 1. 3. 5　Boom Electrical Operation (See Figure 12. 8)

1. Boom speeds

The booms are factory set for 2 speeds on every function:

Mid Speed and High Speed — These speeds are automatically selected by proximity switches depending on the position of the boom, elevator and rotator. The boom, elevator and rotator are designed to be in "Mid Speed" at the beginning and end of travel. The boom is set to 45 s, $+/-$ 5 s. The elevator and rotator speeds are set to 35 s (end to end) $+/-$ 5 s.

2. Basket/cab leveling

The basket/cab is automatically leveled when the boom is out of the rest position.

The basket/cab may be leveled when the boom is in the rest position by using the basket/cab level switch.

3. Beacon light

An amber (optional red) beacon light is mounted on the aft end of the boom. This beacon light flashes as long as the chassis ignition is turned on.

An optional, green beacon light is mounted near the mid-point of the main boom. This beacon light flashes when Anti-icing fluid is applied.

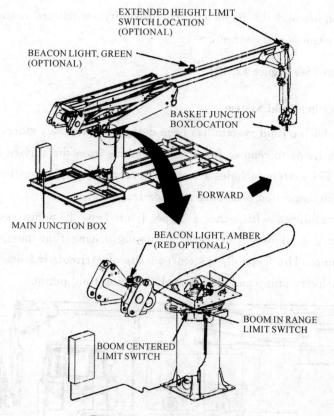

Figure 12. 8   Boom electrical

12. 1. 3. 6   Body Electrical Installation (See Figure 12. 9)

The truck body contains the outside marker lights and their lamps (12 V or 24 V depending on the chassis voltage). A boom stowed proximity switch and optional front and rear video cameras are also located on the truck body.

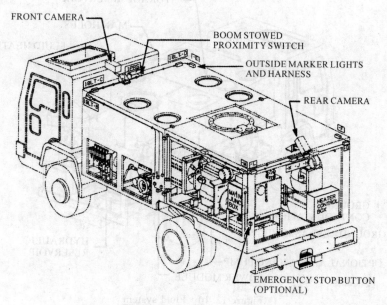

Figure 12. 9   Body electrical

The marker lights and the boom stowed proximity switch are connected to the main junction box by a harness and connectors

12. 1. 4　Fluids (See Figure 12. 10)

12. 1. 4. 1　Deicing Fluid System

The modular deicing fluid system has been designed to receive, store, heat and dispense hot aircraft deicing fluids to remove frost, snow and ice from the surface of commercial and military aircraft. The system includes a storage reservoir, pumping system, fluid heater and aerial platform with associated plumbing and nozzles.

The deicer is equipped with either a Type Ⅰ or Type Ⅱ pump system. The Type Ⅰ system consists of a combination delivery/circulating deicing fluid pump with an optional anti-icing fluid pump. The Type Ⅱ system consists of a circulating deicing fluid pump, a separate deicing delivery pump and an optional anti-icing fluid pump.

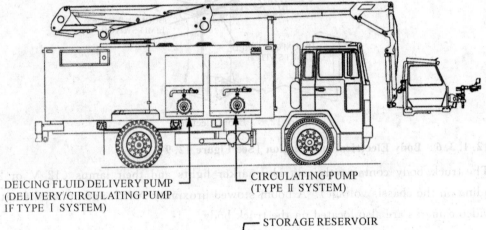

DEICING FLUID DELIVERY PUMP (DELIVERY/CIRCULATING PUMP IF TYPE Ⅰ SYSTEM)

CIRCULATING PUMP (TYPE Ⅱ SYSTEM)

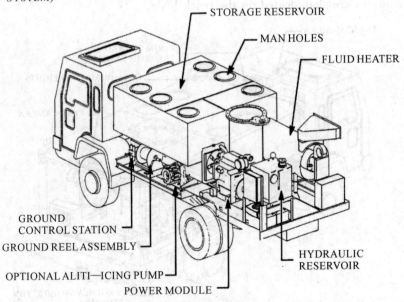

STORAGE RESERVOIR

MAN HOLES

FLUID HEATER

GROUND CONTROL STATION

GROUND REEL ASSEMBLY

OPTIONAL ALITI—ICING PUMP

POWER MODULE

HYDRAULIC RESERVOIR

Figure 12. 10　Fluid system

1. Storage reservoir

The storage reservoir comes in a variety of sizes for deicing fluid only or a combination of deicing and optional anti-icing fluid. The reservoir is constructed of stainless steel to reduce maintenance and increase the life of the tank as well as protect sensitive anti-icing fluids. Exterior man holes have been supplied for cleaning and maintenance.

2. Pumping system

The deicing delivery pump is a two stage centrifugal or diaphragm pump used to circulate deicing fluid through the heater. A second diaphragm pump then delivers the deicing fluid to the basket and ground reel hose. Plumbing from the reservoir to these pumps includes service valves, stainless plumbing, and high temperature hose.

3. Fluid heater

The heater is self controlled and regulating during operation and will automatically turn off when the reservoir has reached deicing temperatures. The internal heating flame is fully enclosed allowing heating operations around aircraft. The heater will raise the fluid temperature very quickly. The fluid is returned to the reservoir during heater operation, rapidly heating the stored deicing fluid.

4. Aerial platform

Boom movement is controlled by the basket joystick, or the control levers located at the ground control station. Aircraft deicing can be accomplished by using the basket nozzle or the ground level hose reel nozzle.

12.1.4.2   Anti-icing Fluid System (Optional)

An anti-icing system may be installed for coating the aircraft skin immediately after deicing operations to increase hold-over times.

1. Storage reservoir

When this system is installed, the main reservoir is partitioned off to contain the anti-icing fluid.

2. Pumping system

An additional high volume anti-icing pump, associated plumbing, and special nozzle are used to dispense the anti-icing fluid. This fluid is not heated and can only be dispensed from the basket. The anti-icing pump transfers anti-icing fluid from the reservoir, increases the pressure to working levels, and transfers the fluid to the basket.

Note: There is no On-Off switch for the anti-icing pump. The anti-icing pump is active whenever the auxiliary engine is running and sufficient fluid is available.

12.1.5   Power Module (See Figure 12.11)

12.1.5.1   Function

The power module supplies all electrical and hydraulic power for the modular deicer Unit. The engine and all major electrical and hydraulic components are mounted on a skid-like frame. This module can be built up outside the machine and then picked up as a unit and installed in the modular deicer.

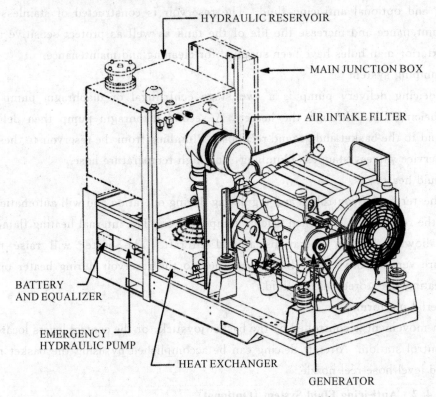

Figure 12.11　Deutz BF4M1012 power module

12.1.5.2　Components

(1) Auxiliary engine assembly;

(2) Electric generator;

(3) Battery(s);

(4) Battery equalizer;

(5) Main electrical junction box;

(6) Hydraulic pump (mounted directly on rear of engine);

(7) Hydraulic reservoir with return flow filter;

(8) Heat exchanger (hydraulic oil cooler and fan);

(9) Emergency electric driven hydraulic pump.

12.1.5.3　Espar Auxiliary Engine Coolant Heater (Optional) (See Figure 12.12)

An optional coolant heater, manufactured by Espar, can be used to preheat the auxiliary engine coolant to facilitate starting in very cold weather conditions. The heater burns fuel, drawn from the auxiliary engine fuel tank, to heat the engine coolant. Electrical power for the coolant heater is supplied by the auxiliary engine batteries. When activated, the heater will run for 60 min, then cycle off automatically.

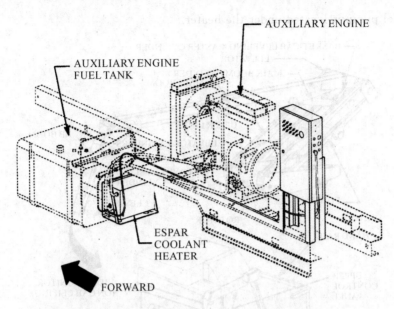

Figure 12. 12   Espar coolant heater (optional)

12. 1. 6   Aerial Device

12. 1. 6. 1   Enclosed Basket Assembly (See Figure 12. 13)

The enclosed basket is comprised of a fully enclosed cab, safety handrail, attached flood lights, and an integrated nozzle. The cab contains upper and lower control panels, boom control joystick, nozzle joystick, cabin heat controls, and an optional video monitor.

12. 1. 6. 2   Boom Assembly

The boom assembly consists of an elevating, rotating boom mounted on top of the deicer in a rotator assembly, with a personnel basket attached to the end of the boom. The boom is comprised of an elevator, main boom, and jib boom. Separate cylinders power each of these sections. UMASTERH and KSLAVEM leveling cylinders are also provided to level the basket or optional enclosed basket (cab).

12. 1. 7   OPTIMAX HEAT SYSTEM™ (See Figure 12. 14)

The modular deicer uses the OPTIMAX HEAT SYSTEM™ for the heater module assembly. This system will raise the fluid temperature from normal storage up to 185 – 190°F (85 – 88°C). This will require the operator to use gloves when handling the deicing guns. The fluid is returned to the fluid storage reservoir during heater operation, rapidly heating the stored deicing fluid. The OPTIMAX HEAT SYSTEM™ is self controlled and regulating during operation and will automatically turn off when the reservoir deice fluid has reached deicing temperatures.

The heater system is located behind the fluid tank on the right hand side of the unit.

The heater fuel pump is located under the heater.

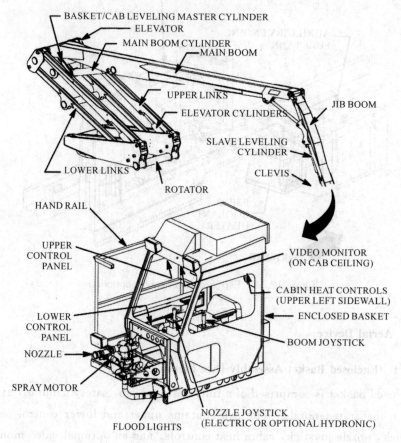

BASKET/CAB LEVELING MASTER CYLINDER
ELEVATOR
MAIN BOOM CYLINDER
MAIN BOOM
UPPER LINKS
ELEVATOR CYLINDERS
JIB BOOM
SLAVE LEVELING CYLINDER
LOWER LINKS
CLEVIS
ROTATOR
HAND RAIL
UPPER CONTROL PANEL
VIDEO MONITOR (ON CAB CEILING)
CABIN HEAT CONTROLS (UPPER LEFT SIDEWALL)
ENCLOSED BASKET
LOWER CONTROL PANEL
BOOM JOYSTICK
NOZZLE
SPRAY MOTOR
NOZZLE JOYSTICK (ELECTRIC OR OPTIONAL HYDRONIC)
FLOOD LIGHTS

Figure 12. 13　Aerial device

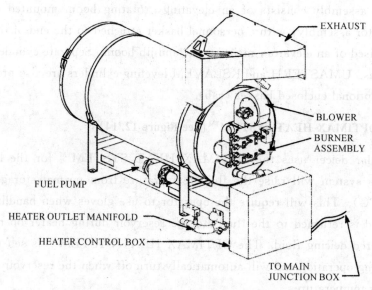

EXHAUST
BLOWER
BURNER ASSEMBLY
FUEL PUMP
HEATER OUTLET MANIFOLD
HEATER CONTROL BOX
TO MAIN JUNCTION BOX

Figure 12. 14　OPTIMAX HEAT SYSTEM™

12.1.8   Fuel System (See Figure 12.15, 12.16)

12.1.8.1   System Function

The modular deicer fuel system supplies fuel to the power module auxiliary engine and the fluid heater. It is separate and independent of the truck fuel system and engine. The system can be adapted to either gasoline or diesel fuel.

12.1.8.2   Fuel Supply

A 125 gallon (475 L) fuel tank can be mounted on either the left or right side to accommodate various truck configurations. Four separate fuel lines are connected to the tank. One line supplies fuel to the power module auxiliary engine and another supplies fuel to the fluid heater. The remaining two lines are fuel return lines for the auxiliary engine and the fluid heater.

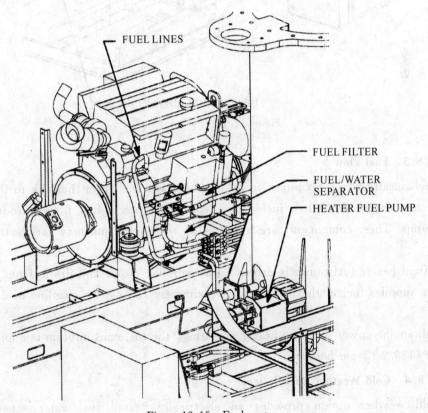

FUEL LINES

FUEL FILTER

FUEL/WATER
SEPARATOR

HEATER FUEL PUMP

Figure 12.15   Fuel system

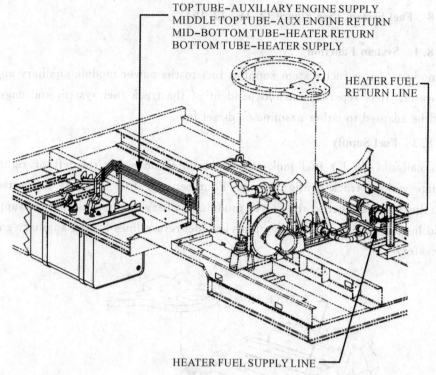

TOP TUBE—AUXILIARY ENGINE SUPPLY
MIDDLE TOP TUBE—AUX ENGINE RETURN
MID—BOTTOM TUBE—HEATER RETURN
BOTTOM TUBE—HEATER SUPPLY

HEATER FUEL
RETURN LINE

HEATER FUEL SUPPLY LINE

Figure 12. 16　Fuel system

12. 1. 8. 3　Fuel Flow

A mechanically driven lift pump on the engine drafts fuel from the tank to the engine. The fuel then passes through a fuel/water separator, then a fuel filter enroutes to the injector pump. These components are mounted on the engine and may vary with different engines.

The fluid heater fuel pump is driven by a hydraulic motor and drafts from the tank. This pump supplies fuel to the heater at approximately 125 psi for gasoline or 300 psi for diesel.

At full engine power with the heater operating, the fuel consumption is approximately 36 gallons (136. 2 L) per hour.

12. 1. 8. 4　Cold Weather Fuel System

A cold weather option provides an electrically heated fuel/water separator and electrically heated fuel supply hoses for the engine and heater. This system is battery powered with 24 V DC.

12. 1. 9　Exhaust System (See Figure 12. 17)

12. 1. 9. 1　Heater Exhaust

The heater exhaust is ducted through a rectangular port directly to the exhaust screen on aft end of the heater module.

12. 1. 9. 2   Auxiliary Engine Exhaust

The auxiliary engine exhaust is routed from the engine to the muffler, then around the pedestal to the forward end of the heater exhaust screen. Guards are mounted on the muffler and exhaust to shield deicer components and protect personnel.

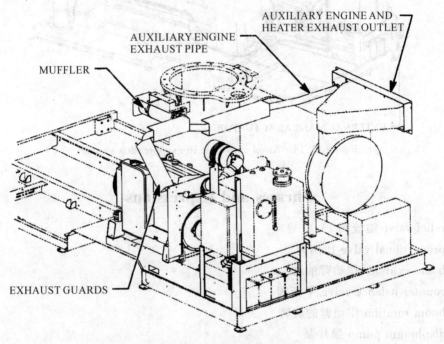

Figure 12. 17   Exhaust system

12. 1. 10   Automatic Fire Suppression System (See Figure 12. 18)

Should a fire occur in the auxiliary engine compartment, an optional Ansul Automatic Fire Suppression System is available that also automatically shuts off the auxiliary engine and heater. A strike button, located on the left side of the hydraulic fluid tank, is provided for manual activation of the fire suppression system. An optional external strike button is provided at the left rear corner of the deicer (see section 1 – 2, item 10, "EMERGENCY PROCEDURES").

Note: Preliminary testing and final hook-up must be accomplished on the Ansul Automatic Fire Suppression System prior to placing the modular deicer in service. Refer to the Ansul Installation, Inspection and Maintenance Manual are provided in Chapter 6 of this volume for preliminary testing and hook-up procedures.

Note: The preliminary testing and final hook-up of the Ansul Automatic Fire Suppression System must be performed by authorized personnel in accordance with applicable local, state and federal regulations. If additional assistance is required, contact your nearest Ansul representative.

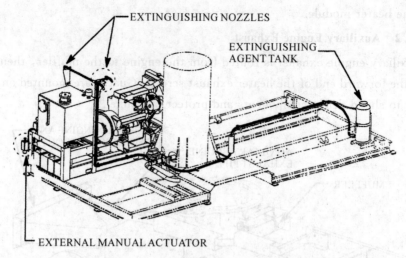

Figure 12. 18　Ansul automatic fire suppression system

Phrases and Expressions

1. relief valve 溢流阀
2. proportional valve 比例阀
3. boom cylinder 作动臂油缸
4. counter-balance valve 平衡阀
5. boom rotation 作动臂的旋转
6. diaphragm pump 膜片泵
7. fixed displacement motor 固定排量液压马达
8. solenoid valve 电磁阀
9. pressure compensated flow control valve 压力补偿流量控制阀
10. boom stowed proximity switch 作动臂收回开关
11. battery equalizer 电池均衡器
12. manual activation 手动激活,人工启动

Notes to the Text

1. The proportional electrical signal to the valve coil for raise and lower speed is controlled by a joystick in the operator's compartment.

升降速度控制阀线圈的比例电信号由驾驶室/操作台的操纵杆控制。

2. The boom is rotated by a geroller type hydraulic motor through a gear reduction.

作动臂是由 geroller 牌液压马达通过齿轮减速器驱动旋转的。

3. The basket/cab leveling (slave) cylinder is controlled by oil from the master cylinder that is attached to the upper boom and lower elevator. The master cylinder volume is matched to the leveling slave cylinder so that as the boom raises or lowers oil is ported from the master cylinder to the slave cylinder in the correct proportions.

吊篮/操控室的调平(伺服)油缸由来自主油缸的液压油控制,该油缸的上面连接作动臂而下面连接升降机。主油缸的容积与调平伺服油缸相匹配以便当作动臂上升或下降时从主油缸排至伺服油缸的液压油保持正确的比例。

Section 12. 2   Operation

Warning：Ensure deicer body access doors are firmly secured in the open position when accessing controls and equipment.

12.2.1   Safety Features

(1) The Emergency Stop switches will immediately shut down the power module auxiliary engine, heater, fluid pumps, and hydraulics for boom movements.

Optional：Some units will also apply chassis brakes if the boom is out of rest.

(2) The power module auxiliary engine will automatically shut down in the event of low oil pressure, high coolant temperature, or high hydraulic oil temperature.

(3) The power module auxiliary engine starter can not be engaged while the engine is running.

(4) The heater will shut down if any of the following conditions occur：

1) Low fluid flow；

2) Auxiliary engine not at high throttle；

3) Low blower speed；

4) Heater outlet temperature exceeds 230°F (110℃)；

5) Fluid tank exceeds 180°F (82℃)；

6) Exhaust stack sensor switch (7TAS) exceeds 1,400°F (760℃).

(5) The boom will not lift more than the rated basket load (300 lb (136 kg)) from the rest position.

Optional：Some units may be equipped with a warning light in the basket and at the ground control station to indicate a boom overload.

(6) The basket is equipped with safety belts and lanyards.

(7) Rubber bumpers are provided on the front corners of the basket.

(8) Basket door is spring closed and self latching.

(9) Chassis mounted torsion bar provides automatic boom stability.

(10) All systems, except for heater ignition and chassis lights, use 24 V DC operating voltage.

(11) Boom enable switch is incorporated on boom joystick control.

(12) All cylinders used on the aerial device incorporate a holding valve to prevent rapid lowering of boom in the event of hydraulic line failure.

Optional：Some units may be equipped with a battery powered hydraulic pump for emergency boom lowering.

(13) Fluid tank is compartmentalized to prevent large fluid movements.

(14) Boom out of rest light is provided on cab control panel.

(15) Manual boom control levers are provided at the ground control station on the forward left side of the modular deicer. They allow ground personnel to operate the boom for servicing or if the basket operator is disabled.

(16) An optional automatic fire extinguishing system is provided which also automatically shuts down the power module auxiliary engine.

(17) Optional low fluid level warning lights are provided on the cab control panel to help to prevent running the fluid tanks empty.

(18) An optional chassis speed limiter limits the modular deicer to 3. 5 mile/h (6 km/h) with the boom out of rest.

(19) A boom control lockout provides an interlock that disables the basket controls unless the door to the manual boom levers is closed.

(20) An optional battery disconnect switch provides isolators of both positive and negative from the 12 V chassis batteries. The switch is designed so that a lever must be installed and turned in order for the machine to operate. Once disabled, the lever may be removed so that no one else can turn the system on.

12. 2. 2　Deicer Controls and Indicators

12. 2. 2. 1　The Primary Controls and Indicators Consist of a Cab Control Box, Enclosed Basket Upper and Lower Control Panels, Boom Joystick, and Nozzle Control Joystick (See Figure 12. 19)

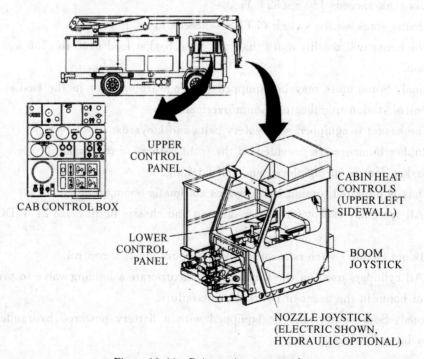

Figure 12. 19　Deicer primary controls

12.2.2.2   Cab Control Box (See Figure 12.20)

In Figure 12.20:

1 — Panel light — illuminates when chassis lights are turned on.

2 — Auto shutdown light (green) (optional) — illuminates when auto shutdown system (optional) is activated by turning off the auxiliary engine ignition switch. If the fluid heater is on, this light remains illuminated until the heater purge cycle is completed (1 – 6 min). If the fluid heater is not on, this light remains illuminated until the engine cool down cycle is completed (51 s).

3 — Fluid level indicators (optional) — indicate tank fluid levels for deicing fluid, anti-icing fluid, or water.

4 — Low fluid level warming lights (red) (optional) — illuminate when fluid level is low to prevent running the fluid tanks empty.

5 — Deicing fluid pump light (green) — illuminates when fluid pump is "ON".

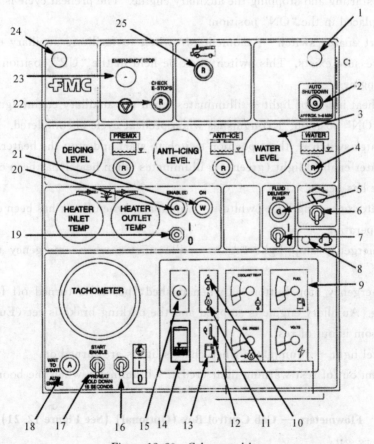

Figure 12.20   Cab control box

6 — Overhead wiper switch (optional) I — turns overhead wiper "ON" and "OFF".

7 — Deicing fluid delivery pump switch — three position, momentary switch, center "OFF", momentary up "ON", momentary down "OFF" (shutdown).

8 — Intercom plug-in — receptacle for headset plugs.

9 — Auxiliary engine instruments — separate gauges for indicating fuel quantity, coolant temperature, oil pressure, and system voltage from generator or battery output.

10 — High engine temperature light — illuminates when auxiliary engine temperature is high.

11 — Low oil pressure light — illuminates when auxiliary engine oil pressure is low.

12 — High hydraulic oil temperature light — illuminates when hydraulic oil temperature is high.

13 — Low fuel light — illuminates when auxiliary engine fuel is low.

14 — Low coolant light (optional) — illuminates when auxiliary engine coolant level is low.

15 — Generator light — illuminates when generator is inoperative.

16 — Ignition switch — three position switch, "OFF", "ON", and momentary "START" for starling and stopping the auxiliary engine. The preheat cycle is activated when this switch is placed in the "ON" position.

17 — Start enable switch — momentary switch to facilitate auxiliary engine starting while no oil pressure exists. This switch must be held in the "UP" position while starting the auxiliary engine.

18 — Preheat indicator light — illuminates when the auxiliary engine Ignition switch is placed in the "ON" position. Extinguishes when preheat cycle is completed.

19 — Heater switch — three position switch to start and stop the heater.

20 — Heater enabled light (green) — illuminates when heater has been enabled and the firing sequence has begun.

21 — Heater operating light (white) — illuminates when flame has been established and the heater is operating.

22 — Emergency stop light (red) — illuminates when emergency stop has been engaged.

23 — Emergency stop button — when pushed, power is turned off for all systems except lighting. Auxiliary engine is stopped and the parking brake is set (European models only) if the boom is out of rest.

24 — Panel light — illuminates when chassis lights are turned on.

25 — Boom out of rest warning light (red) — illuminates when the boom is not in the rest position.

12. 2. 2. 3 Flowmeters — Cab Control Box (Optional) (See Figure 12. 21)

In Figure 12. 21:

1 — Deicing fluid counter — indicates, in liters or gallons, amount of deicing fluid dispensed. A red reset button is provided adjacent to the counter to reset the counter to zero.

2 — Deicing fluid totalizer — indicates, in liters or gallons, amount of deicing fluid dispensed since unit was commissioned. The reset button for this counter is disabled.

3 — Anti-icing fluid counter — indicates, in liters or gallons, amount of anti-icing fluid dispensed. A red reset button is provided adjacent to the counter to reset the counter to zero.

4 — Anti-icing fluid totalizer — indicates, in liters or gallons, amount of anti-icing fluid dispensed since unit was commissioned. The reset button for this counter is disabled.

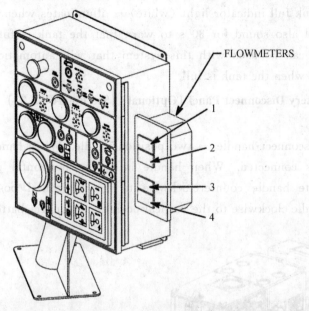

Figure 12.21　Flowmeters (optional)

12.2.2.4　Tank Full Indicator and Anti-Icing Refiller Switch panel (Optional) (See Figure 12.22)

In Figure 12.22:

1 — Anti-icing refiller switch — two position, momentary switch. Used to activate anti-icing pump if pump is not running or to increase pump speed up to 60 gallon/min (227.1 L/min) for suction refill. Switch must be held in the "ON" position for pump operation.

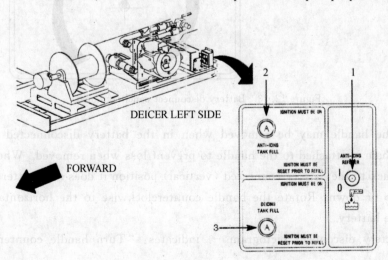

Figure 12.22　Tank full indicator and anti-icing refiller switch panel (optional)

2 — Anti-icing tank full indicator light (white) — illuminates when Anti-icing tank is full. An audio alarm will also sound for 30 s to warn that the tank is full. Optional Facility control connections can be used with this system that will automatically shut down the external fluid pump when the tank is full.

3 — Deicing tank full indicator light (white) — illuminates when deicing tank is full. An audio alarm will also sound for 30 s to warn that the tank is full. Optional Facility control connections can be used with this system that will automatically shut down the external fluid pump when the tank is full.

12.2.2.5   Battery Disconnect Panel (Optional) (See Figure 12.23)

In Figure 12.23:

1 — Battery disconnect handle — two position handle. When handle is in the vertical position, battery is connected. When handle is in the horizontal position, battery is disconnected. Rotate handle counterclockwise to the horizontal position to disconnect battery. Rotate handle clockwise to the vertical position to connect battery.

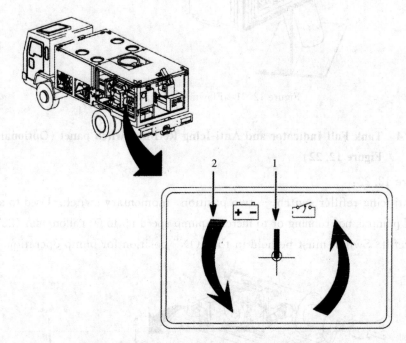

Figure 12.23   Battery disconnect panel (optional)

Note: The handle may be removed when in the battery disconnected (horizontal) position. A chain is attached to the handle to prevent loss when removed. When re-inserted and rotated back to the battery connected (vertical) position it does not matter if the handle is pointing up or down. Rotate the handle counterclockwise to the horizontal position to disconnect the battery.

2 — Battery disconnect pictogram — indicates: "Turn handle counterclockwise to

disconnect battery. "

12.2.2.6   Enclosed Basket Upper Panel (See Figure 12.24)

In Figure 12.24:

1 — Emergency stop button — when pushed, power is turned off for all systems except lighting. Auxiliary engine is stopped and the parking brake is set (European models only) if the boom is out of rest.

2 — Left window washer switch — momentary switch for dispensing window washing fluid on the left side window.

3 — Center window washer switch — momentary switch for dispensing window washing fluid on the center window.

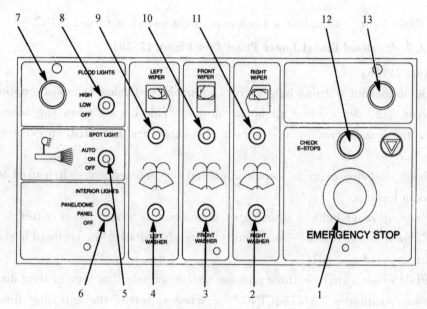

Figure 12.24   Enclosed basket upper panel

4— Right window washer switch — momentary switch for dispensing window washing fluid on the right side window.

5 — Nozzle spot light switch — three position switch.

Lower position is "OFF".

Center position is "ON" — in this position the nozzle spot light remains "ON" when fluid is dispensed.

Upper position is "ON-AUTO" — in this position the nozzle spot light will turn off automatically when the red trigger is squeezed on the nozzle control joystick to prevent light reflection when fluid is dispensed.

6 — Interior light switch — three position switch for panel and dome lights.

Lower position is "OFF".

Center position is "PANEL".

Upper position is "DOME" (Panel lights are also "ON" in this position).

7— Panel light — illuminates when panel light switch is turned "ON".

8 — Flood light swich — three position switch "OFF, LOW, HIGH".

9 — Right window wiper switch — three position wiper switch — "OFF" "SLOW" "FAST".

10 — Center window wiper switch — three position wiper switch — "OFF" "SLOW" "FAST".

11 — Left window wiper switch — three position wiper switch — "OFF" "SLOW" "FAST".

12 — Emergency stop light (red) — illuminates when emergency stop has been engaged.

13 — Panel light — illuminates when panel light switch is turned "ON".

12.2.2.7 Enclosed Basket Lower Panel (See Figure 12.25)

In Figure 12.25：

1 — Boom lockout warning light (red) (optional) — flashes to indicate optional boom lockout is engaged, deactivating all boom motion. This occurs when the boom collision sensors on the nozzle arm or underneath the cabin are activated, preventing further movement.

2 — Boom collision override switch (optional) — momentary switch used to reset or override boom lockout.

3 — Boom in range light — illuminates when boom is within 15° of center.

4 — Cabin rotation switch — momentary switch for rotating the enclosed basket (cabin) left or right.

5 — Fluid select switch — three position switch for selecting type of fluid dispensed.

1) Lower position is "ANTI-ICING" — when selected, the anti-icing fluid valve is opened (fluid pump indicator light does not illuminate).

2) Center position is "OFF".

3) Upper position is "DEICING" — when selected, the deicing pump is activated and the fluid pump indicator light illuminates.

6 — Cabin level switch — momentary switch for leveling the enclosed basket (cabin).

7 — Emergency hydraulic pump switch — momentary switch used to turn on emergency hydraulic pump. Switch must be held in "ON" position for emergency hydraulic pump operation.

8 — Panel light — illuminates when panel lights switch on upper panel is turned on.

9 — Boom overload light (optional) — illuminates when basket load exceeds 300 lb(136 kg). Boom controls remain functional.

10 — Fluid delivery pump indicator light — illuminates when deicing pump is activated.

11 — Boom enabled light — illuminates when boom joystick is enabled (red trigger depressed on joystick).

12 — Boom centered light — illuminates when boom is centered.

13 — Panel light — illuminates when panel lights switch on upper panel is turned on.

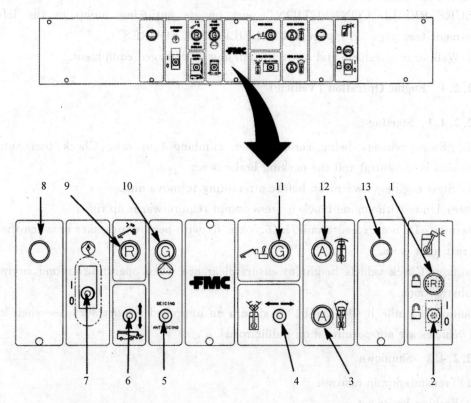

Figure 12. 25   Enclosed basket lower panel

12. 2. 3   Preparing the Unit for Deicing

(1) Always perform daily maintenance checks. Proper maintenance is necessary to ensure the safe and optimum performance of the modular deicer.

Pre-start checks.

1) Check the oil level and coolant levels on the power module auxiliary engine and vehicle engine.

2) Check condition of wiring and hoses.

3) Check battery connections and water level.

4) Look underneath vehicle and check for leaks. If significant leaks are found maintenance must be performed to eliminate problem.

5) Check tire pressures.

6) Check fuel gauges located on cab control station for fuel supply.

7) Test all emergency step switches.

(2) Fill the tank with proper amounts of deicing fluid using either the "DEICING SUCTION / PRESSURE REFILL CONNECTION" port in the right side compartment or

the "DEICING PRESSURE REFILL" port in the left hand compartment aft of the hose reel. Fill the aft portion of the tank with anti-icing fluid through the "ANTI-ICING SUCTION / PRESSURE REFILL CONNECTION" port on tie anti-icing pump in the left side compartment (see page 317 – page 321 "REFILL PROCEDURES").

(3) Walk around vehicle and inspector missing or damaged equipment.

12. 2. 4　Engine Operation (Vehicle)

12. 2. 4. 1　Starting

(1) Enter vehicle, being careful when climbing into cab. Check that automatic transmission is in neutral and the parking brake is set.

(2) Start engine — warm up before attempting to move unit.

Note: Units with engine block heaters do not require warm up time.

Warning: Do not exceed 4 mile/h (6.5 km/h) with person in basket or when the boom out of rest light is on.

Caution: Check vehicle height to ensure clearance when operating around aircraft and other obstructions.

Caution: 25 mile/h (40 km/h) maximum on improved airport surfaces when loaded, loaded vehicles are not permitted on public roads.

12. 2. 4. 2　Shutdown

(1) Transmission in neutral.

(2) Parking brake set.

(3) Ignition off.

(4) Lights off, if used.

12. 2. 5　Power Module Auxiliary Engine Operation (See Figure 12. 26)

12. 2. 5. 1　Cab Control Box

All controls for starting and stopping the auxiliary engine are located on the cab control box. The cab control box is mounted on a pedestal to the right of the drivers' seat.

The cab control box controls/indicators are grouped according to function from bottom to top of the control box in the order of the start sequence. For example, the auxiliary engine must be started prior to running the heater, therefore, auxiliary engine controls are on the bottom followed by heater controls on top.

12. 2. 5. 2　Auxiliary Engine Controls

Auxiliary engine controls consist of an "ignition switch" and a "start enable switch".

1. Start enable switch

This switch momentarily bypasses the engine shutdown system to facilitate starling while no oil pressure exists. This switch can be held during cold starting to allow the engine to warm up and develop sufficient oil pressure. Some options that are available modify this two position switch. "COLD START" is available on diesel engines. This option changes the start enable switch to a three position switch.

(1)Center position is "OFF".

(2)Upper position is still the "START ENABLE" position and is momentary.

(3) Lower position is also momentary but, when hold down, the glow plug heaters on the engine are energizing. This switch is generally held down for about 15 seconds prior to starting.

For Ford gas engines, this lower position is for priming the carburetor prior to starting. Selecting this position injects fuel into the carburetor.

Regardless of the version, this "START ENABLE" switch must be held in the "UP" position prior to starting the engine.

2. Ignition switch

This switch is always a 3 position switch. The lowest position is "OFF".

The center position is the "ON" position. The emergency pump system and the gauge lights can be operated in this position. On a Ford gas engine, the ignition switch should remain in this position 3 seconds prior to starting.

The upper position on the ignition switch is the "START" position and is a momentary switch.

Note: The ignition switch should always be placed in the "OFF" position whenever the deicer is not in use to prevent draining the batteries.

3. Auxiliary engine gauges

(1)Tachometer. The tachometer is located directly above the ignition and start enable switches. Engine speed is automatically controlled, with idle speed set for 1,000 r/min and normal operation set for a "HI" throttle position of 2,000 r/min.

(2) Engine condition instruments. These instruments are located to the right of the tachometer and consist of 4 gauges:

1)Coolant temperature.

2)Fuel level.

3)Oil pressure.

4)Voltmeter.

(3) Warning lights.

There are 5 red warning lights on the cab control box associated with the auxiliary engine:

1)High engine temperature.

2)Low oil pressure.

3)High hydraulic oil temperature.

4)Low fuel.

5)Generator (no out put).

12.2.5.3 Auxiliary Engine Start Sequence

(1) Place ignition switch in the "ON" position.

The preheat indicator light will illuminate. wait until preheat indicator light extinguishes before starting auxiliary engine.

(2) Ensure that the preheat indicator light has extinguished and hold the "START ENABLE" switch in the "UP" position.

(3) While holding "START ENABLE" switch "UP", push ignition switch to "START".

(4) The engine is running when the low oil pressure light goes out and the tachometer indicates a steady rotate speed.

(5) Once the engine is running, release the ignition switch to its "ON" position.

(6) The "START ENABLE" switch can remain held in the "UP" positron until indicators show engine is up to speed (engine will idle at 1,000 r/min).

The best indicators for a running auxiliary engine are:

1— Tachometer — steady rotate speed (1,000 r/min for idle);

2— Generator light — out (not illuminated);

3— Low oil pressure light — out (not illuminated).

The oil pressure and voltmeter gauges can also be indicators that the engine is running. The oil pressure gauge should indicate in the green band and the voltmeter should indicate in the green band at approximately 24 V.

Note: Once engine is running, the starter will be electrically locked out from re-engaging.

Note: The tachometer will have a floating indicator when the ignition is turned off. Once the ignition switch is pushed to "ON", the needle will go to zero.

12.2.5.4 Auxiliary Engine Manual Shutdown

(1) Turn the heater and deicing fluid delivery pump off if they are operating. The engine will continue to run at 2,000 r/min until purge cycle is completed (approximately three minutes).

(2) Allow the engine to idle (1,000 r/min) for approximately two minutes to properly cool down.

(3) Turn the ignition switch off.

12.2.5.5 Auxiliary Engine Automatic Shutdown (Optional) (See Figure 12.26)

(1) Turn auxiliary engine ignition switch off (Auto shutdown light on cab control box will illuminate).

(2) If the fluid heater is running, the automatic shutdown system will turn off the heater. The heater will then run through the purge cycle (1 – 6 min) prior to shutdown.

(3) The auxiliary engine will then cycle to idle rpm to cool down for 51 s prior to shutdown. The auto shutdown light extinguishes at the completion of this cycle.

Note: The ignition switch should always be placed in the "OFF" position whenever the deicer is not in use to prevent draining the batteries.

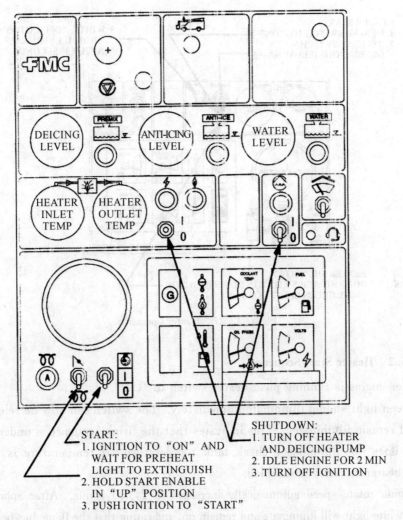

START:
1. IGNITION TO "ON" AND WAIT FOR PREHEAT LIGHT TO EXTINGUISH
2. HOLD START ENABLE IN "UP" POSITION
3. PUSH IGNITION TO "START"

SHUTDOWN:
1. TURN OFF HEATER AND DEICING PUMP
2. IDLE ENGINE FOR 2 MIN
3. TURN OFF IGNITION

Figure 12.26　Auxiliary engine start and shutdown

12.2.6　Heater Operation (See Figure 12.27)

12.2.6.1　Control Functions

(1) The heater control consists of a 3 position, center off, momentary switch for starting and stopping the heater.

(2) Two indicator lights accompany the switch.

1) The green light directly above the switch indicates that the heater has been enabled and the firing sequence has begun.

2) The white light above and to the right of the switch indicates that flame has been established and the heater is operating.

（3）The two gauges to the left of the switch indicate fluid temperature at the heater inlet and heater outlet.

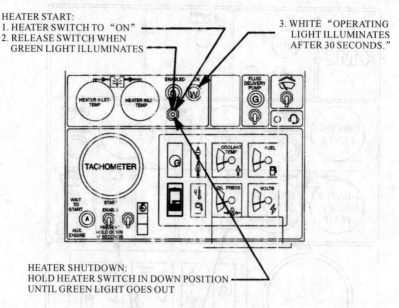

Figure 12.27　Heater operation

12.2.6.2　Heater Start Sequence

（1）After engine is running push heater switch to "ON".

The green light should illuminate immediately. The switch can now be released and the light should remain illuminated. This indicates that the firing sequence is under way. If the green light does not illuminate, check tank temperature. If temperature is above 180℉ (82℃) the heater circuitry is disabled.

（2）Engine rotate speed automatically increases to 2,000 r/min. After approximately 30 seconds the white light will illuminate and remain on, indicating that the flame has been established and the heater is operating. If white light does not illuminate, call for service.

（3）Monitor the temperature gauges to determine fluid temperature.

Note：The enclosed basket（cabin）fluid heater start controls include a heater misfire warning light（red）. if this light illuminates, attempt a restart. If this is unsuccessful, call for service.

12.2.6.3　Heater Cycle

（1）The heater will shut down automatically when the fluid temperature reaches 180℉ (829℃)（normally 20 - 30 min）and both the white light and green light will extinguish.

（2）After shut down, the heater fluid pump and air blower will continue to run for 3 min to reduce the heater temperature and purge any fumes and unburned fuel that may remain after shut down. After this purge cycle is complete the engine will return to idle speed.

(3) Monitor the temperature gauges on the cab control panel or, if equipped, on the basket control panel to determine fluid temperature. The heater will not restart automatically if the fluid temperature falls below acceptable levels, it will be necessary for the operator to restart the heater if the fluid temperature is too low.

Note: If the heater outlet temperature reaches 230°F (110°C) the heater will turn off and the heater "OVER TEMP" light, located on the front of the heater control box, will illuminate (see Figure 12. 28). If this occurs, it will be necessary to manually reset the heater switch by pressing the red button located on the heater outlet marrow. If an over temperature condition continues to shut down the heater, remove the vehicle from service and refer to the maintenance section for further action.

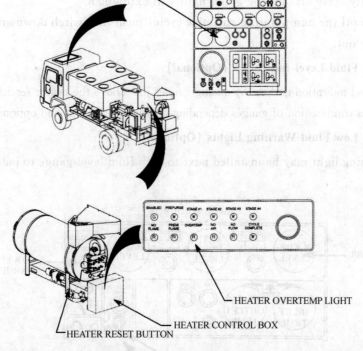

Figure 12. 28　Heater overtemp

12. 2. 6. 4　Heater Shut Down (Manual)

To turn off the heater hold the heater switch down until the green light goes out (white light will also go out). The engine will continue to run at 2,000 r/min until the purge cycle is complete (approximately 3 min).

Caution: After turning off the heater, do not shut off the engine until the purge cycle is completed.

12.2.7 Deicing Fluid Delivery Pump (See Figure 12.29)

12.2.7.1 Fluid Pump Controls

The deicing fluid delivery section of the cab control box and basket control panel consists of a three position toggle switch and a green "ON" light. The switch is maintained in its center position and pushed momentarily up or down for pump operation.

12.2.7.2 Deicing Fluid Pump Operation

(1) To turn on the pump, push the switch up to "ON" until the green light illuminates, then release the switch (Engine rotate speed will increase to 2,000 r/min). If fluid is not delivered from the basket or ground deicing nozzle for a period of two minutes, the pump will automatically cycle off and the green light will extinguish.

(2) To turn off the pump during an active cycle, push the switch down and hold until the green light goes out.

12.2.7.3 Fluid Level Indication (Optional)

All fluid level indication is done by analog gauges just above the heater section. The unit may have a gauge or a combination of gauges depending on the type of unit and options purchased.

12.2.7.4 Low Fluid Warning Lights (Optional)

A red warning light may be installed next to each fluid level gauge to indicate a low fluid level condition.

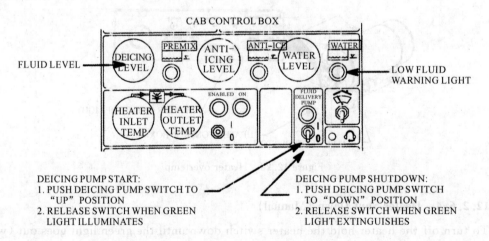

Figure 12.29　Deicing fluid delivery pump operation

12.2.8 Deicer Operating Mode (See Figure 12.30)

12.2.8.1 Operating Modes

1.1 Man (optional)

This option may be provided on units with an enclosed basket. Controls are provided within the enclosed basket for all deicer functions, including vehicle movement. This allows one man to control the deicer from the enclosed basket.

2. 2 Men

This is normal operation, with the cab driver controlling the vehicle. A switch is provided on units equipped with the 1 man option to select 1 man or 2 men drive mode.

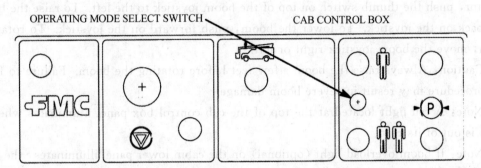

Figure 12. 30　Operating mode select switch

12. 2. 8. 2　Operating Mode Controls

An "OPERATING MODE SELECT SWITCH" (1 man option only) is provided on the cab control box to select either 1 man or 2 men drive mode. The switch is placed in the position for 1 man drive and in the "down" position for 2 men drive. A light is provided for each switch position to indicate which mode is selected.

12. 2. 9　Normal Operating Procedures

12. 2. 9. 1　Start Auxiliary engine (Cab Control Box)

(1) Start heater if needed (cab control box).

(2) Start deicing fluid delivery pump (cab control box).

(3) Enter enclosed basket (cabin), close cabin door, and fasten seat belt.

Note: Insure cabin door is closed. Enclosed basket controls will not function unless cabin door is fully closed.

(4) For night operation, turn on panel lights nozzle spot light, and flood lights (cabin upper panel).

(5) Turn on cabin heater and adjust thermostat as required. The cabin heat control box is located on the left upper cabin wall.

(6) Check intercom operation and adjust volume.

The intercom control box is located to the right of the seat on the aft cabin wall.

(7) Turn on overhead monitor (optional) using power switch on face of monitor.

(8) Select "NIGHT" or "DAY" mode, as appropriate, on monitory and adjust brightness and contrast as required.

(9) Select monitor "INPUT A" (rear camera) and "INPUT B" (forward camera) and

check video presentation.

(10)Insure aerial device is clear of personnel and obstructions.

(11)Operate the aerial device using the boom joystick control. Depress and hold red trigger to activate boom joystick (Joystick enabled light will illuminate). To raise the elevator, push the thumb switch on top of the boom joystick to the right. To lower the elevator, push the thumb switch on top of the boom joystick to the left. To raise the boom, pull back on the joystick. To lower the boom, push forward on the joystick. To rotate the boom, move the boom joystick right or left.

Caution: Always raise the boom a few feet before rotating the boom. Failure to follow this procedure may result in severe boom damage.

Note: A red light located at the top of the cab control box panel illuminates when the boom is out of its rest.

Note: If boom overload light (optional) on the cabin lower panel illuminates, the boom controls will remain functional but the cabin load should be reduced before further operation.

(12)If the boom collision sensor (optional) mounted underneath the cabin is activated by contacting an object, all boom motion will be deactivated and the boom lockout indicator light will illuminate. When this occurs, perform the following:
1)Hold the override switch in the override position.
2)Reposition the aerial device break contact with the object.
3)Release the override switch.

The boom lockout indicator light will extinguish. Boom motion functions will be resumed if the boom motion components have not been damaged.

Note:The boom collision sensors (optional) may be activated without contacting an object if the boom is shaken abruptly. If this occurs, reset the boom lockout by momentarily placing the override switch in the override position.

Note:If the override switch is inoperative or cannot be held in the "OVERRIDE" position, the boom must be repositioned using the "emergency boom operation-hydraulic pump failure" procedures described in the emergency procedures section of this chapter.

(13)Select anti-icing or deicing fluid with the fluid select switch (cabin lower panel).

(14)Rotate cabin, if necessary, with the cabin rotation switch. Move switch left to rotate cabin to the left. Move switch right to rotate cabin to the right (cabin lower panel).

(15)Position nozzle and control nozzle movement with the nozzle joystick. To raise the nozzle, pull back on the joystick. To lower the nozzle, push forward on the joystick. To rotate the nozzle, move the joystick right or left.

(16)To dispense fluid, pull the red trigger on the nozzle joystick. Adjust spray pattern with the thumb switch on top of the stick. Push thumb switch to the right for "steady stream". Push thumb switch to the left for "fan" pattern.

(17)The cabin is automatically leveled when the boom is out of the rest position, if the cabin is out of level when the boom is in the rest position, using the cabin leveling switch to level the basket. Push switch up to raise cabin nose or down to lower cabin nose (cabin lower control panel).

(18) To stow the aerial device, lower the boom first, then rotate the aerial device to bring the cabin to the front. The boom in range light will illuminate when the boom is within 15° of center and the boom centered light will illuminate when the boom is at the center position.

12. 2. 9. 2　Aerial Device (See Figure 12. 31 – Figure 12. 33)

Warning: Do not operate boom/aerial device when wind speed is above 46 mile/h (74 km/h).

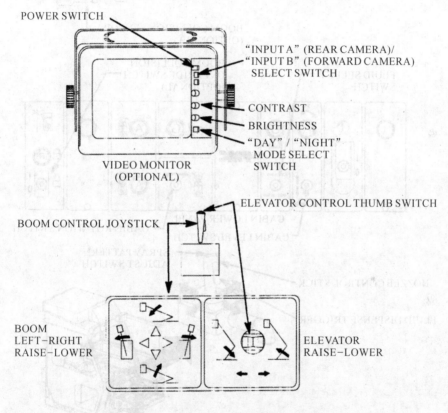

Figure 12. 31　Aerial device operation

Warning: Do not exceed 4 mile/h (6. 5 km/h) with person in basket/cabin.

Note: Always keep the hydraulic enable valve override in "NORMAL" position.

Note: The manual lever cover must be in "DOWN" position for basket controls to operate.

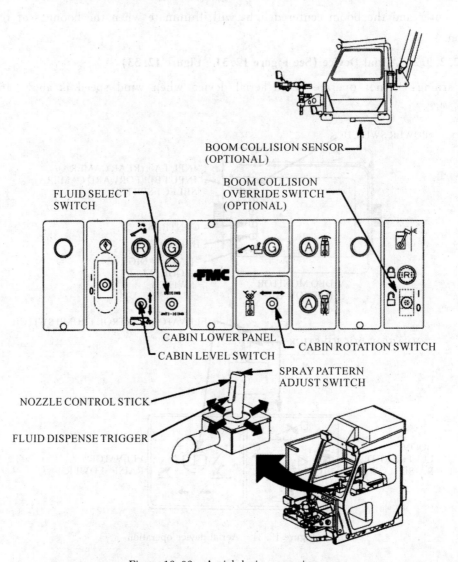

BOOM COLLISION SENSOR
(OPTIONAL)

BOOM COLLISION
OVERRIDE SWITCH
(OPTIONAL)

FLUID SELECT
SWITCH

CABIN LOWER PANEL

CABIN LEVEL SWITCH

CABIN ROTATION SWITCH

SPRAY PATTERN
ADJUST SWITCH

NOZZLE CONTROL STICK

FLUID DISPENSE TRIGGER

Figure 12. 32   Aerial device operation

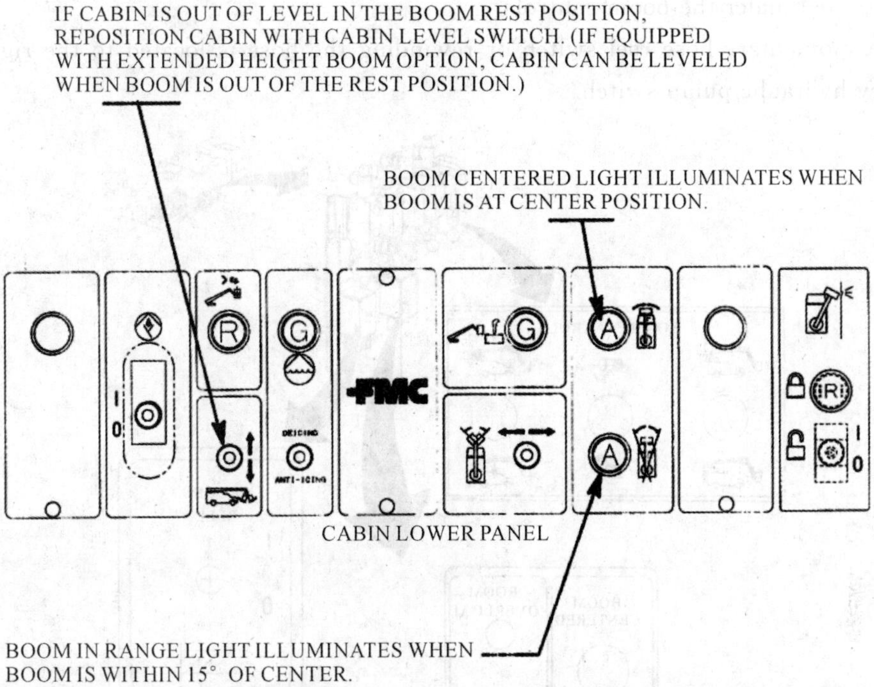

IF CABIN IS OUT OF LEVEL IN THE BOOM REST POSITION,
REPOSITION CABIN WITH CABIN LEVEL SWITCH. (IF EQUIPPED
WITH EXTENDED HEIGHT BOOM OPTION, CABIN CAN BE LEVELED
WHEN BOOM IS OUT OF THE REST POSITION.)

BOOM CENTERED LIGHT ILLUMINATES WHEN
BOOM IS AT CENTER POSITION.

CABIN LOWER PANEL

BOOM IN RANGE LIGHT ILLUMINATES WHEN
BOOM IS WITHIN 15° OF CENTER.

Figure 12.33　Aerial device operation

12.2.9.3　Aerial Device Operation — Ground Control Station (See Figure 12.34)

Warning: Do not operate boom/aerial device when wind speed is above 46 mile/h (74 km/h).

Caution: Always raise the boom a few feet before rotating the boom. Failure to follow this procedure may result in severe boom damage.

If necessary, the boom may be controlled by using the boom control levers at the ground control station.

(1) Ensure that the boom and basket are clear of all personnel and obstructions throughout the boom travel range.

(2) Pull up on the boom "ROTATION" lever (center lever) to rotate the boom in a clockwise direction and pull down on the lever to rotate the boom in a counter-clockwise direction.

(3) Pull up on boom "RAISE-LOWER" lever (right lever) to raise the boom and pull down on the lever to lower the boom.

(4) Pull up on the elevator "RAISE-LOWER" lever (left lever) to raise the elevator and pull down on the lever to lower the elevator.

(5) A boom in range light and a boom centered light are located under the boom controls to aid in centering and stowing the boom. The boom in range light illuminates when the boom is within 15°. The boom centered light illuminates when the boom is centered.

(6)An emergency hydraulic pump switch and an optional boom overload warning light are also located under the boom controls.

(7)A momentary hose reel switch or rewinding the hose is located to the right of the emergency hydraulic pump switch.

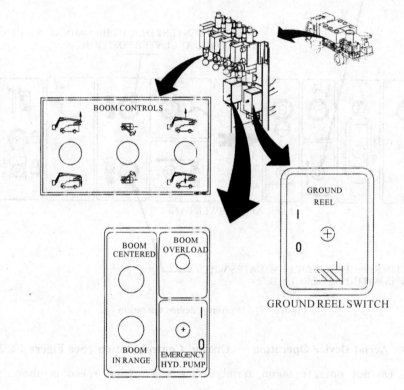

Figure 12. 34   Ground control station

12. 2. 9. 4   Ground Reel Hose and Nozzle Operation (See Figure 12. 35)

(1)The spray pattern and flow rate of the ground reel hose nozzle are preset.

(2)Aim the nozzle with the arrow pointing away from you and squeeze the lever located on the aft side of the grip.

Note:Your unit may be equipped with an optional Akron or task force nozzle. The flow rate and spray pattern are not preset on these nozzles.

(3)If equipped, the hose may be rewound automatically by using the optional ground reel rewind switch.

Note:A ground reel shut-off valve is located to the left (forward end) of the ground reel. This valve may be used to shut off fluid to the ground reel hose in the event of a nozzle or hose malfunction.

Warning:Because of the outward appearance, the spray head on the ground reel gun could easily be pointed toward the operator. Be sure to hold the ground reel gun with arrow pointing away from the operator.

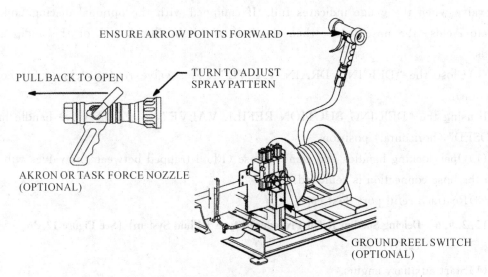

ENSURE ARROW POINTS FORWARD

TURN TO ADJUST
SPRAY PATTERN

PULL BACK TO OPEN

AKRON OR TASK FORCE NOZZLE
(OPTIONAL)

GROUND REEL SWITCH
(OPTIONAL)

Figure 12. 35   Ground reel hose and nozzle

12. 2. 9. 5   Deicing Pressure Refill Procedure (Type Ⅰ fluid System) (See Figure 12. 36, 12. 37)

Note: Deicing fluid is pressure refilled through either the "DEICING SUCTION / PRESSURE REFILL CONNECTION" port forward of the deicing delivery pump in the right side compartment, or the "DEICING PRESSURE REFILL CONNECTION" port in the left hand compartment aft of the hose red.

Note: The "DEICING PRESSURE REFILL CONNECTION" port in the left hand compartment is also used for draining the deicing fluid tank. This port does not contain a check valve and fluid remaining in the deicing tank will drain when the port drain valve is opened. During refilling, fluid in the tank will back flow to the filling source if the supply source is not pressurized.

(1) Remove dust cap (if equipped) on the refill port.

(2) With the locking levers fully extended, push the hose coupling onto the refill port and, when seated, force the levers inward to lock in place.

(3) Open the valve on the refill port.

If using the "DEICING SUCTION/PRESSURE REFILL CONNECTION" port, open the "DEICING SUCTION REFILL VALVE" located above the refill port. The valve handle points down when in the "OPEN" position (This valve remains in the "CLOSED" position for normal operation) (see Figure 12. 19).

If using the "DEICING PRESSURE REFILL CONNECTION" port, open the "DEICING DRAIN VALVE" (The valve handle points toward the operator when in the open position) (see Figure 12. 18).

(4) Open hose valve to begin refilling.

(5) Observe the sight level gauges on the forward left side of the deicer and close the

hose valve when the gauge indicates full. If equipped with the optional deicing tank full indicator, close the nose valve when the indicator light illuminates or the audio alarm sounds.

(6) Close the "DEICING DRAIN VALVE" (This valve remains closed for normal operation).

If using the "DEICING SUCTION REFILL VALVE", place the valve handle in the "CLOSED" (horizontal) position.

(7) Open locking handles and remove hose (Fluid trapped between the valves will spill when the hose connection is removed).

(8) Reattach refill port dust cap.

12.2.9.6　Deicing Suction Refill Procedure (Type Ⅰ Fluid System) (See Figure 12.36, 12.37)

(1) Start auxiliary engine.

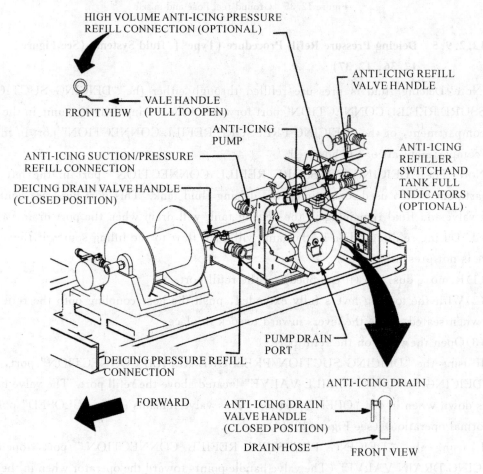

Figure 12.36　Modular deicer-left side

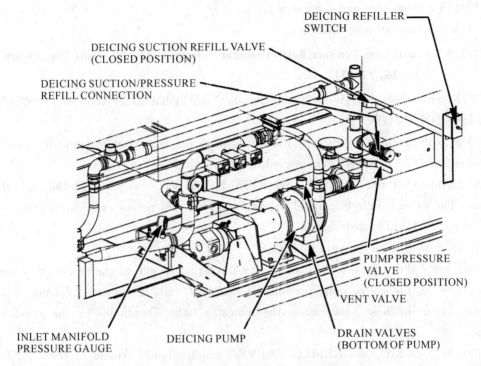

DEICING REFILLER SWITCH

DEICING SUCTION REFILL VALVE (CLOSED POSITION)

DEICING SUCTION/PRESSURE REFILL CONNECTION

INLET MANIFOLD PRESSURE GAUGE

DEICING PUMP

PUMP PRESSURE VALVE (CLOSED POSITION)

VENT VALVE

DRAIN VALVES (BOTTOM OF PUMP)

Figure 12.37   Modular deicer-right side

(2) Remove dust cap (if equipped) on the "HDEICING SUCTION/PRESSURE REFILL CONNECTION" port.

(3) With the leaking levers fully extended, push the hose coupling onto the "DEICING SUCTION/PRESSURE REFILL CONNECTION" port and, when seated, force the levers inward to lock in place.

(4) Open the "DEICING SUCTION REFILL VALVE" (The valve handle points down when in the "OPEN" position).

(5) Push and hold the "DEICING REFILLER" switch in the "ON" position (Pump will start and run).

(6) Open the "PUMP PRESSURE VALVE" located under the "DEICING SUCTION / PRESSURE REFILL CONNECTION" (Valve handle points down when in the "OPEN" position).

(7) Observe the sight level gauges on the forward left side of the deicer and close the hose valve when the gauge indicates full. If equipped with the optional deicing tank full indicator, close the hose valve when the indicator light illuminates or the audio alarm sounds.

(8) Close the "PUMP PRESSURE VALVE"(The valve handle is horizontal when in the "CLOSED" position).

(9) Close the "DEICING SUCTION REFILL VALVE" (The valve handle is horizontal when in the "CLOSED" position).

(10) Close hose valve and remove hose.

(11) Reattach refill port dust cap.

12.2.9.7　Anti-icing Pressure Refill Procedure (Type Ⅰ Fluid System) (See Figure 12.36, 12.37)

(1) Remove dust cap (if equipped) on the "ANTI-ICING SUCTION / PRESSURE REFILL CONNECTION" port.

(2) With the locking levers fully extended, push the hose coupling onto the refill port and, when seated, force the levers inward to lock in place.

(3) Ensure that the "ANTI-ICING REFILL VALVE" handle is in the "CLOSED" position (The valve handle is parallel to the fluid tank and points away from the operator when in the "CLOSED" position).

(4) Open hose valve to begin refilling.

(5) Observe the sight level gauges on the forward left side of the deicer and close the hose valve when the gauge indicates full. If equipped with the optional deicing tank full indicator, close the hose valve when the indicator light illuminates or the audio alarm sounds.

(6) The "ANTWCING REFILL VALVE" handle should remain in the "CLOSED" position.

(7) Open locking handles and remove hose (Fluid trapped between the valves will spill when the hose connection is removed).

(8) Reattach refill port dust cap.

12.2.9.8　Anti-icing Suction Refill Procedure (Type Ⅰ Fluid System) (See Figure 12.36, 12.37)

(1) Start auxiliary engine.

(2) Remove dust cap (if equipped) on the "ANTI-ICING SUCTION/ PRESSURE REFILL CONNECTION" port.

(3) With the locking levers fully extended, push the hose coupling onto the refill port and, when seated, force the levers inward to lock in place.

(4) Open the "ANTI-ICING REFILL VALVE" by pulling out on the hanti-icing refill valvew handle (Handle points toward the operator when in the open position). This will activate the anti-icing pump and it will pump at 20 gallon/min.

(5) If the Anti-Icing pump does not run or a faster refill rate is desired, an optional "ANTI-ICING REFILLER" switch is located to the right of the pump. It may be used to operate the pump as follows:

1) Push and hold the "ANTI-ICING REFILLER" switch in the "ON" position (Pump will start or increase in speed up to a maximum of 60 gallon/min).

2) Release "ANTI-ICING REFILLER" switch to the "OFF" position when refilling is complete.

(6) Observe the sight level gauges on the forward left side of the deicer and close the hose valve when the gauge indicates full. If equipped with the optional deicing tank full indicator, close the hose valve when the indicator light illuminates or the audio alarm sounds.

(7) Return the "ANTI-ICING REFILL VALVE" handle to the "CLOSED" position.

(8) Ensure that anti-icing pump has stopped.

(9) Open locking handles and remove hose (Fluid trapped between the valves will spill when the hose connection is removed).

(10) Reattach refill port dust cap.

12.2.9.9　Top Refill (See Figure 12.38)

Warning: During top filling, fluids may be spilled on the walkway creating a hazard. Extreme caution should be exercised during top filling operations.

(1) Open latch on top tank fill cover and lift cover.

(2) Swing drop pipe or hose into position and fill as required.

(3) After filling is complete close cover and insure spring loaded latch is secure.

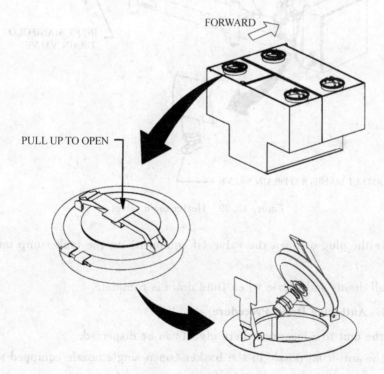

FORWARD

PULL UP TO OPEN

Figure 12.38　Top refill

12.2.9.10　Deicing Drain Procedure (Type I Fluid System)

The deicing fluid pump and the heater must be turned off prior to draining or purging.

(1) Move the unit to a location where glycol can be dispersed.

(2) Open the deicing nozzle in the basket (open single nozzle if equipped with enclosed

basket).

(3)Raise the boom 5 to 10 degrees above the horizontal.

(4)Open the deicing drain valve and drain as much fluid as possible (drain valve handle points toward the operator when in the open position) (see Figure 12. 36).

(5)Open both drain valves and the vent valve on the deicing pump (see Figure 12. 19).

(6)Open the drain valve on the heater inlet and outlet manifolds and drain as much fluid as possible (see Figure 12. 39).

Note：The heater, pump and ground reel hose will not drain completely. Some fluid will back drain from the boom nozzle and plumbing.

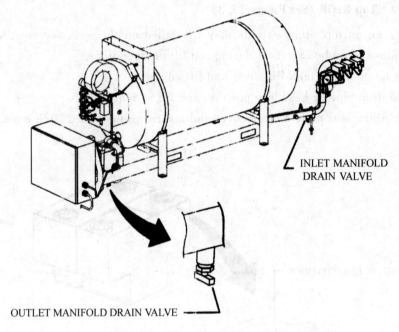

INLET MANIFOLD
DRAIN VALVE

OUTLET MANIFOLD DRAIN VALVE

Figure 12. 39　Heater drain valves

(7)Remove the plug or open the valve (if equipped) on the tank sump underneath the deicing tank.

(8)Close all drains and nozzle when fluid drain is complete.

12. 2. 9. 11　Anti-icing Drain Procedure

(1)Move the unit to a location where glycol can be dispersed.

(2)Open the anti-icing nozzle in the basket (open single nozzle equipped with enclosed basket).

(3)Raise the boom 5 to 10 degrees above the horizontal.

(4)Open the anti-icing drain valve (located to the right rear of the anti-icing pump) and drain as much fluid as possible (drain valve handle points toward the operator when in the open position) (see Figure 12. 40).

(5)Remove the anti-icing pump drain port plug.

(6)Pull the anti-icing refill handle to the half open position.

Note:Pump will not drain completely.

(7)Close all drains and nozzle and reinstall anti-icing pump drain port plug when fluid drain is complete.

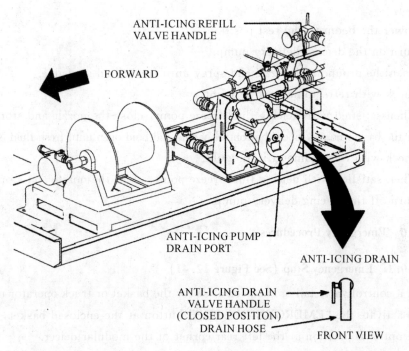

Figure 12.40   Anti-icing drain procedure

12.2.9.12   Purge Procedure

When the unit has been used for hot water deicing or aircraft washing and is to be stored in cold temperatures, it will be necessary to purge and winterize the deicing plumbing, pumps, heater and fluid tank.

(1)Drain the deicing system as described under item "deicing drain procedure", steps (1) thru (7).

(2)Add 100 gallons of glycol to the deicing tank.

(3)Start the power module auxiliary engine and run at slow speed.

(4)Open the vent valve on top of the deicing pump.

(5)Close vent valve when a steady flow of glycol flows from the vent.

(6) Turn on the deicing pump by pressing and holding the "DEICING REFILLER" switch or by using the deicing pump switch on the cab control panel.

If the inlet manifold pressure gauge indicates more than 75 psi, proceed to step (7).

If the pressure does not rise, turn the pump off and open the vent valve until a steady stream of glycol is visible. Close valve and repeat step (6).

(7)Run the delivery pump five to ten minutes to purge the heater.

(8) Open the heater "INLET MANIFOLD DRAIN VALVE" slightly and take a sample.

(9) Using a refractometer, check to be sure the freeze point is sufficiently low for the expected conditions.

(10) Add glycol if necessary and repeat steps (7)-(10) until freeze point is sufficiently low.

(11) Lower the boom to the rest position.

(12) Turn on the deicing delivery pump.

(13) Open the ground reel nozzle and spray until fresh fluid is visible.

(14) Check with refractometer if in doubt.

(15) When satisfied with deicing fluid freeze point, close the nozzle and store.

(16) At the basket nozzle, open the control handle and hold open until fresh fluid is visible.

(17) Check with refractometer if in doubt.

(18) When satisfied with deicing fluid freeze point, close the nozzle and store.

(19) Turn off the deicing delivery pump.

12. 2. 10　Emergency Procedures

12. 2. 10. 1　Emergency Stop (See Figure 12. 41)

Should a control valve fail or the need arise for the basket or truck operator to halt boom movement, activate the "EMERGENCY STOP" button at the enclosed basket, cab control box, or the optional position at the left rear corner of the modular deicer.

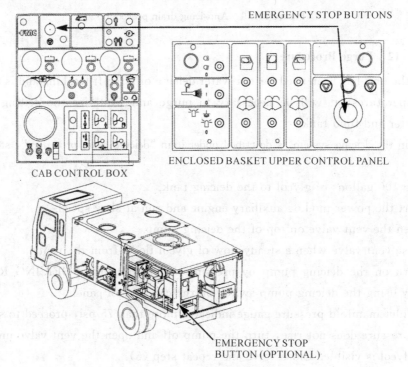

Figure 12. 41　Emergency stop

When the "EMERGENCY STOP" button is pushed, power is turned off for all systems except lighting. The auxiliary engine is stopped and the parking brake is set (European models only) if the boom is out of rest. The red "EMERGENCY STOP" light on the cab control box and the enclosed basket upper control panel illuminates as long as the ignition switch is on. To reactivate the system, rotate the "EMERGENCY STOP" button counter clockwise (approximately 1/4 turn) until the button pops out.

Note: Engine starting will be prevented until all emergency stop buttons are pulled out.

12. 2. 10. 2   Fire Suppression (See Figure 12. 42)

Should a fire occur in the auxiliary engine compartment an optional automatic fire extinguishing system is available that also automatically shuts off the auxiliary engine and heater. If the automatic system fails to activate, the system can be manually activated by pulling the safety pin and striking the external manual actuator button located on the rear of the vehicle (see Figure 12. 42).

Caution: The emergency hydraulic pump is not good for continuous use. Do not use more than 60 s at a time and allow 5 min for cooling time between use periods.

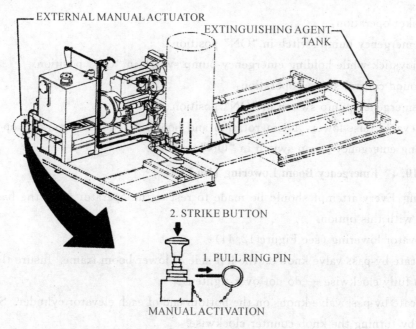

Figure 12. 42   Fire suppression

12. 2. 10. 3   Emergency Boom Operation — Hydraulic Pump Failure (See Figure 12. 43)

An emergency, electrically driven hydraulic pump is provided to operate the boom in the event of failure of the engine driven hydraulic pump. An optional emergency pump switch is provided on the cab control box, the basket control panel and at the ground control station. This switch should only be used when the engine is not running.

Caution: The emergency hydraulic pump is not good for continuous use. Do not use more than 60 s at a time and allow 5 min for cooling time between use periods.

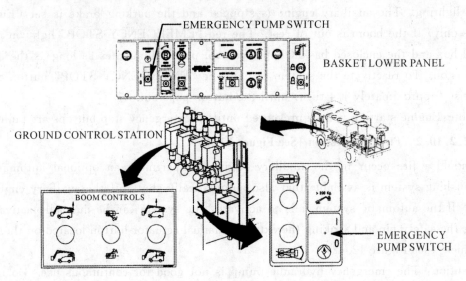

Figure 12.43   Emergency boom operation

(1)Basket operation:

Hold emergency pump switch in "ON" position.

Move joystick while holding emergency pump switch in "ON" position.

(2)Ground control station operation:

Hold emergency pump switch in "ON" position.

Use elevator raise-lower, boom rotation and boom raise-lower levers to operate boom while holding emergency pump switch in "ON" position.

12.2.10.4   Emergency Boom Lowering (No Power)

Warning:Every attempt should be made to rescue the operator from the basket before proceeding with this option.

1. Elevator lowering (see Figure 12.44)

(1)Locate by-pass valve knob on the left side of lower boom frame. Insure this is closed (turn knob fully clockwise — do not over-tighten).

(2)Locate by-pass valve knobs on the bottom end of each elevator cylinder. Slowly open each valve by turning the knob counter-clockwise.

(3)Slowly open by-pass valve knob on left side of boom frame until elevators lower.

(4)Close all by-pass valves when finished lowering by turning knobs clockwise (do not over-tighten).

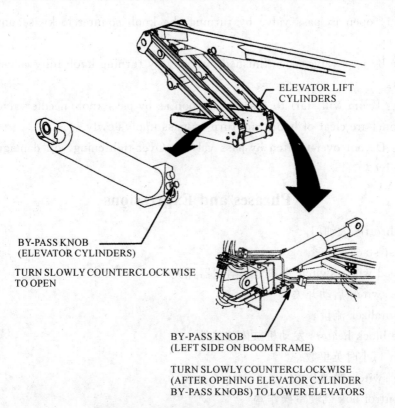

ELEVATOR LIFT
CYLINDERS

BY-PASS KNOB
(ELEVATOR CYLINDERS)

TURN SLOWLY COUNTERCLOCKWISE
TO OPEN

BY-PASS KNOB
(LEFT SIDE ON BOOM FRAME)

TURN SLOWLY COUNTERCLOCKWISE
(AFTER OPENING ELEVATOR CYLINDER
BY-PASS KNOBS) TO LOWER ELEVATORS

Figure 12. 44　Elevator lowering-no power

2. Upper boom lowering (see Figure 12. 45)

(1) Locate by-pass valve knob on lower end of boom lift cylinder.

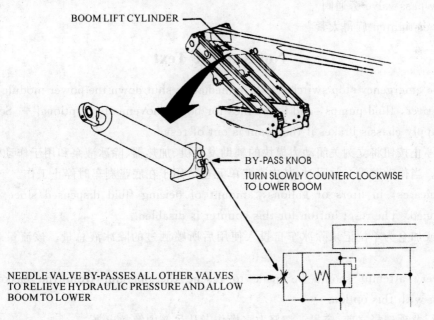

BOOM LIFT CYLINDER

BY-PASS KNOB

TURN SLOWLY COUNTERCLOCKWISE
TO LOWER BOOM

NEEDLE VALVE BY-PASSES ALL OTHER VALVES
TO RELIEVE HYDRAULIC PRESSURE AND ALLOW
BOOM TO LOWER

Figure 12. 45　Elevator lowering-no power

(2) Slowly open by-pass valve by turning the knob counterclockwise until the boom lowers.

(3) Close by-pass valve when finished lowering by turning knob fully clockwise (do not over-tighten).

Warning：Boom will start to descend as soon as by-pass knob(needle valve) is turned. Insure personnel are clear of boom and turn by-pass knob slowly.

Caution：Do not over-tighten by-pass valves. Over-tightening will damage the seal on the bypass valve.

Phrases and Expressions

1. high throttle 大油门

2. exhaust stack 排气管

3. overhead wiper switch 头顶上的雨刮器开关

4. enable switch 启动开关

5. fluid totalizer 流量表

6. engine block heaters 发动机缸体加热器

7. mile/h 每小时英里数

8. toggle switch 拨动开关

9. cab control box 驾驶室操作盒

10. override switch 超越控制开关,超控电门

11. aerial device 登高装置,举升装置

12. momentary switch 瞬时开关

13. by-pass valve 旁通阀

14. over-tighten 旋得太紧

Notes to the Text

1. The emergency stop switches will immediately shut down the power module auxiliary engine, heater, fluid pumps, and hydraulics for boom movements. Optional — Some units will also apply chassis brakes if the boom is out of rest.

紧急停止按钮将立刻关闭动力模块的辅助发动机、加热器、除冰液泵和用于作动臂作动的液压系统。当作动臂没有收回时,其中有些单元也适用于在底盘刹车时停止工作。

2. Indicates, in liters or gallons, amount of deicing fluid dispensed since unit was commissioned. The reset button for this counter is disabled.

以升或加仑为单位记录除冰车自投入使用后所喷洒过的除冰液总量。该流量表不能清零。

3. Every attempt should be made to rescue the operator from the basket before proceeding with this option.

在进行此项选择之前,应尽一切努力将操作者从吊篮中解救出来。

Chapter 13　The History of Ground Support Equipment

Just 50 years after the Wright Brothers' first flight, the British Overseas Airways Corp. began operating the first passenger jet service.

And consider how much the industry changed in little more than the following two generations:

In 1958, for example, more people crossed the Atlantic by ship than plane. In the United States, the 10 largest transportation companies were all railroads. Less than one in 10 Americans had ever even been on an airplane. The notion of "how far" it was from New York to Los Angeles was measured in days. Hawaii drew just 171,000 visitors.

With that in mind, here are some milestones in GSE development:

1705 — The Goldhofer family starts a forge in Amendingen, Germany.

1811 — Friedrich Krupp starts casting steel.

1860 — What eventually will become TLD begins with a silk-weaving business in Lyon, France.

1883 — John Bean invents a continuous spray pump to battle bugs on his 10 - acre almond orchard and the Bean Spray Pump Co. is born(see Figure 13.1).

Figure 13.1　Promotional material for the Bean Spray Pump Co.

1891 — Brothers August and Joseph Thyssen start out with a coal mine and steelworks.

At the start of the 1900s, Earl Estes starts the Dixie Manufacturing Co. The company's original product line includes horse collars and saddles. You no doubt recognize many of these other surnames, but what about Earl Estes? We'll pick up his story in another 70 - some years.

1903 — Wilbur Wright becomes the world's first ramp agent. This year (2013) marks 110 years since Orville Wright took off from the sand at Kill Devil Hills aboard the Wright Flyer into a freezing headwind of 27 miles per hour and flew about 35 yards(see Figure 13.2).

Take a look at one of history's most famous photos above, and you can see much more of Wilbur than Orville as the Wright Flyer makes its maiden voyage.

A closer look, however, reveals the picture also shows a work bench and a jumble of something to the right of the bench.

Figure 13.2 Orville Wright gets all the credit, but Wilbur was the world's first ramp agent. Take a closer look at that box under the right wing of the plane, and read more about its contents in the feature

That turns out to be world's first "GPU" and "chocks". We found out more about the details at an EAA Web site on vintage aircraft:

"On the right end of the airplane's foot-printed outline is a small footstool or bench, with a large C-clamp lying across the center support of the bench", H. G. Frautschy writes. "Ken Hyde of the Wright Experience believes they used the clamp to gently secure the wingtip of the machine to the bench, to prevent the Flyer from rocking too much from side to side in the breezy conditions as they prepared it for flight.

"To the bench's right, there is the starting battery, with its kinky, stiff wire sticking out of the wooden box. It was used to start the engine of the Flyer, which was also equipped with a Splitdorf dynamo. A battery was needed to supply enough electricity to generate a spark within the primitive make-or-break ignition system used for the engine. There's also a shovel and a small can, which, according to "The Papers of Wilbur and Orville Wright" edited by Marvin W. McFarland, contained "nails, tacks and a hammer in it, for emergency repairs".

1914 — The "aeroplane" is just starting to see military service at the start of World War
I. Over the next four years, the plane graduates from reconnaissance missions to bombing
runs. The Hucks Starter, an auxiliary power unit that provides the initial start-up power to
piston engines, proves that machines replacing workers isn't anything new.

1917 — C. C. Hobart, along with his wife, Lou Ella, and their three sons, Edward,
Charles and William, starts what will become Hobart Brothers. The company makes
generators, metal office furniture and air compressors.

1918 — Pilots buzz rural America as "barnstorming" becomes popular entertainment.

All a daredevil pilot needs is an open field to land on, in other words, the first farm he
spots(see Figure 13. 3). After cutting a deal with the farmer, the pilot performs various
stunts for the crowd.

Figure 13. 3   Aerial shot taken in the late-1920s of what would someday be ATL. Photo credit:
Photo provided courtesy of Hartsfield-Jackson Atlanta International Airport

Meanwhile, the farmer could lend an extra hand with his tractor. Which, of course, is
why even to this day, we refer to an aviation towing vehicle as a "tractor".

Farm tractors prove a reliable source for aircraft towing throughout most the early days
of aviation.

Meanwhile, James L. Entwistle, electrical engineer and MIT prof, starts the Entwistle
Co.

1923 — Clark Tructractor Co., better known today as forklift manufacturer, Clark
Material Handling Co., builds the Duat tow tractor to pull trailer loads of lumber, freight
and industrial material(see Figure 13. 4).

Figure 13. 4　Long before there were baggage carts to tow, small tractors like this one, a 1926 Duat from Clark Tructractor Co. , were pulling trailer loads of lumber, freight and industry material. This workhouse looks like very much like today's tugs

1926 — The American Brattice Cloth Corp. opens and sells flame-proof cloth cut in order for the mining industry. Some 60 years later, ABC Industries is known for ducting materials and other GSE products.

Godtfred Vestergaard starts his business at his home outside Copenhagen, Denmark. The company's original product is an aluminum mold for constructing mattresses. Later, the company starts making lifts for a university.

1928 — Two years after founding Kato Engineering, owners Elmer Jensen and Louis Wilkinson hire Cecil Jones who develops a rotary converter that lets rural families operate AC appliances with DC storage batteries.

1929 — Regent Manufacturing sets up shop.

1933 — J. C. Gorman and H. E. Rupp, two engineers out of work during the Great Depression, begin making pumps in a barn outside of Mansfield, OH. Their competitors ridicule their first line of "non-clogging" pumps. The company goes on to report $359 million in sales for 2011.

Clifford Hannay starts out with a few lathes and establishes what will become Hannay Reels. Company remains owned and managed by the 4th generation of the Hannay family.

1935 — E. P. "Ed" Grime starts the Malabar Machine Co. making items from customer drawings. In just a few years, Lockheed asks Grime to build the first tripod jacks specifically for aircraft.

1939 – 1945 — Aviation has an enormous impact on the course of World War II and the war has just as significant an impact on aviation. The United States has 3,600 military aircraft when Hitler marches into Poland in 1939. By the end of the war, U. S. military aircraft production reaches nearly 300,000 — turning out more than 96,000 aircraft at peak production in 1944.

Of course, the whole world is arming itself. As a result of all this aircraft, we finally begin to see a real market for GSE or, as the U. S. military refers to it to this day, the "forgotten enabler" (see Figure 13. 5).

Figure 13. 5　World War Ⅱ had a huge effect on aviation and GSE and vice versa

We start recognizing more names of well-known manufacturers:

The Northwestern Motor Company — the "NMC" of NMC-Wollard-introduces a tow tractor.

During the war, Stewart & Stevenson builds hundreds of tractors and self-propelled bomb ordnance loaders for the U. S. Air Force.

Hobart Brothers produces 100,000 welders and 45,000 generators to support the war receiving the Army/Navy E Award for its efforts.

Columbus Jack of Columbus, OH, gets its start selling most of its production to the military fighting World War Ⅱ (and later the Korean War).

Ford, as just an example of other automotive makers, bulks up its 9N tractor with cast-iron. Approximately 10,000 of these tractors dubbed Moto-Tugs see duty (see Figure 13. 6).

Figure 13. 6　This restored Ford Moto-Tug was a militarized farm tractor that Ford modified during
World War Ⅱ. The model 9N tractor was also a popular choice in the civilian world. In
a photo from Roosevelt Field, Long Island, the same model pulls a Beech 18

By this time John Bean's company was known as FMC ("Food Machinery Corp."). In

1941, FMC designed and built amphibious landing craft. While not GSE, the contract helped the company to gain a foothold in the military GSE market.

The David Clark Co. provides one of the more interesting stories in the aviation industry at this point. The company's first products are griddles and bras, but by the war years it specializes in pressurized suits for the Air Force and, later, space suits for all NASA missions (The company also made the suit Felix Baumgartner used last October to jump from 24 miles above the earth).

1945 – 1959 — Commercial aviation starts to take off. By this time, a host of international GSE manufacturers begin building specialized equipment:

Hobart Brothers sets up Hobart Ground Power after American Airlines asks the company to design a generator to start larger aircraft engines.

Air-A-Plane begins manufacturing PCA units.

Douglas Equipment opens.

Davis Taylor builds an electric cart for his own use in his poultry supply business. After numerous requests for the vehicle, he starts the Taylor Shop. Fred Dunn joins Davis Taylor's business in 1951, and several years later the company changes its name to Taylor-Dunn Manufacturing Co.

Garsite, LLC starts manufacturing aircraft refuelers, hydrant dispensers, fuel delivery trucks, above-ground fuel storage tanks, aviation fueling systems and vacuum pumper trucks(see Figure 13. 7).

Figure 13. 7   Triangle Aviation owned and operated Stanton Airfield in Minnesota after World War Ⅱ. The airfield was used for primary flight training for military cadets during the war

Tracma begins making tractors designed for aircraft towing to replace commonly used farm tractors. The name of the company becomes synonymous with "tractor" in French-speaking countries in the same way "tug" will be used in the United States(see Figure 13. 8).

ACE starts business making test equipment for the aviation industry, but also expands into GSE.

An airport manager and a friend of Axel Ackerman, who started out fixing automotive electrical systems in 1924, asks Axel to make a small 28 V DC rectifier for starting small aircraft. The AXA Power unit works trouble-free for another three decades.

Joe Cochrane creates Cochrane Airport Systems to build the first belt loaders specifically for the aviation industry — a natural since he was already making similar lettuce-packing loaders. The company eventually expands into cargo loaders and, after an ownership change, becomes known as Lantis Corp.

L. W. (Lu) Taylor and Harold Higbee start Enfab Inc. Innovative engineering leads to the creation of a proprietary fiberglass filter coalescer. The company is eventually renamed Velcon Filters.

Albret gets its start making maintenance platforms and aircraft docking systems. Later, the company adds passenger stairs.

Jim Kaplan starts Harlan Corp. to rent and rebuild lift trucks. Kaplan realizes that the parts most common to fail are not readily available so he redesigns the parts and develops sources for new designs. Ten years later, one of his customers in Venezuela asks Kaplan to make him a tow tractor. Harlan buys a Model E Clark lift and re-engineers it. Eventually, the company grows from building 10 tractors a month to 90.

Stewart & Stevenson enters the GSE business with GM Detroit Diesel.

Glen Cummins Sr. goes to work as the general manager of Berglund Motor Co. 's new division, Engine Distributors Inc. , to distribute Ford Motor Co. ' s industrial gasoline engines. Glen rises through the ranks and eventually becomes vice president as EDI reps more engine lines. In 1983, he buys the company. Today, his son and two grandsons own and operate EDI.

Figure 13. 8   While "tug" is American shorthand for a GSE towing vehicle, the France had us beat in
the 1950s when TracMa introduced a new type of tractor for the aviation market to
replace the agricultural models then in use. To this day, French-speaking countries call
a "tug" a "tracma"

Undoubtedly, the 1950s close on a high note for GSE. The first passenger boarding bridges in the United States are installed at San Francisco International Airport and LaGuardia Airport in 1959(see Figure 13. 9 – Figure 13. 12).

Figure 13. 9   In 1959, this passenger bridge was installed at San Francisco Airport. While we're not
          sure who gets "first place", we do know that passenger bridges were also installed at
          LaGuardia Airport that same year

Figure 13. 10   A collection of early passenger boarding bridges

Figure 13. 11   Passenger stairs? More like passenger stair. In the early days of commercial aviation,
          getting on and off a low-slung plane wasn't that difficult

Figure 13. 12　"A wing and a prayer" takes on all new meaning in this shot well before modern deicing technology

1960 — SCHOPF introduces its first aircraft tow tractor.

Engineers at FMC Corporation start building some of the first deicer vehicles that used aerial devices to spray aircraft. John Bean's spray pump serves as the foundation. Its early deicers could deice a plane in 15 minutes. FMC also develops a cargo handling system for the new containerized generation of jet aircraft. The self-propelled Flite-Line Loader allows one person to unload a plane's full cargo of containerized baggage in just 15 minutes.

S. L. Parker opens a metal fabricating business called Parker Industries making garbage containers under the trade name "Par-Kan".

Clyde W. Olson starts Clyde Machines Inc. and begins making hydraulic motors for tampers used by utility companies.

Unitron starts supplying the defense-aerospace, aviation and industrial markets with GPUs, PCAs and other power systems.

1962 — Richard Stern and Yves Helleboid form Devtec to distribute and service GSE outside of the United States. Much later, Devtec becomes TLD Asia and TLD America.

1963 — U. S. Airmotive GSE begins providing a full line of GSE parts and supplies for the industry.

1966 — Bud Bushnell buys the manufacturing rights to a material lift operated with compressed air. Customers are impressed with the "magic in the bottle" and Genie Industries gets its name.

SAS asks the Vestergaard Co. to modify some existing aircraft deicers. As a result, Vestergaard wins an order for new aircraft deicers. The "Beanstalk", as it's informally called, consists of a vertical, telescopic tower with a platform on top from which the operator applied fluid with a spray gun(see Figure 13. 13,13. 14).

Harold G. Hall opens Hall Industries as a contract screw machine shop.

Figure 13. 13   One of Vestergaard's first deicers was dubbed "The Beanstalk".

Figure 13. 14   Vestergaard introduced the Elephant Alpha equipped with the now familiar telescopic boom. It's a big change since its first generation of deicers. The "Bean Stalk", which featured vertical, telescopic tower with a platform on top from which the operator applied fluid with a spray gun

1967 — Lektro, which pioneered the electric golf cart, produces a small electric aircraft tug for a Oregon FBO using a chassis originally built for an eclectic cart for area mink ranchers.

1968 — Robert Watkins starts General Transervice Inc., an airport refueler maintenance company at PHL. GTI later develops the Rampmaster, a modular design that simplifies maintenance by separating the truck from the fuel tank.

1969 — Eagle Tugs introduces its bobtail cargo tractor, a model still in production.

John L. Grove forms a partnership with two friends and buys a small metal fabrication business in McConnellsburg, PA. The company sells its first JLG lift.

1970 — Remember Earl Estes and Dixie Manufacturing, which started back in the 1900s catering to the horse and buggy market. Robert Smith buys the company, now known as Estex, from the founder's widow. Smith grew up near ATL and figures the company's textile products could expand into the aviation industry. Products include baggage cart side curtains and covers. Delta Air Lines becomes its first customer and remains a major account.

The Dana Corp. 's flight department starts Danair. It first products were towbars for corporate jets. Danair is sold in 1980 and becomes Tronair.

1972 — ITW Military GSE begins specializing in military GSE.

Paul MacCready, an avid aviator who set soaring records in his glider in the 1940s, starts AeroVironment Inc. The company becomes a leader in unmanned aircraft and eventually well-known for electric GSE charging stations.

MacCready also makes the history books again in 1977 when the Gossamer Condor, becomes the first aircraft powered solely by the pilot's muscles. Later, the Gossamer Albatross flies across the English Channel.

SAGE Parts opens to distribute parts and service throughout the world for the GSE industry.

Trilectron begins manufacturing GSE.

Beta Fluid Systems starts producing military refueling equipment and then expands in the commercial market. Liquip International, which has 40 years of international refueling expertise, acquires Beta in 2006.

1973 — TUG Manufacturing Corp. starts making its eponymous "tugs", namely, the Model MA, which is still produced today(see Figure 13. 15).

Figure 13. 15　This is the first tug to come off Tug Technologies' line in 1973 has serial No. 001 to prove it. Two ground handling companies use the tug for more than 35 years. TUG restored the vehicle and donated it to the Museum of Commercial Aviation

1975 — David Clark Co. introduces the first headset specifically designed to provide hearing protection for pilots while providing clear, isolated reception and transmission at normal voice levels inside noisy aircraft. Ground support models follow.

1976 — Nicky Ghaemmaghami establishes Hydraulics International, which goes on to specialize in military GSE.

1979 — Jim Watkins starts WASP, Inc. (Watkins Aircraft Support Products) in Alexandria, MN. In a news article published in 1981, Watkins says, "Our first year of business we had just one customer, now we have 10. We feel we can provide a lot of jobs and bring other investment money into our community." In 1980, sales were $280,000. One year later, sales topped $1 million. Jim turns out to be right about those jobs — currently, six employees have each worked at WASP for 30 years. Watkins ends up building another plant in Nebraska. By 1996, sales grow to $33 million, and Jim sells the company to his employees in 1997(see Figure 13.16).

Figure 13.16   WASP delivers its first shipment of cargo dollies in this shot from 1982

1980s — Tracma and Air France introduce the first towbarless tractor.

Charlatte SA, which started 20 years before machining metal parts, creates Charlatte Menutention and becomes a leader in electric GSE throughout Europe.

1981 — Vestergaard builds a new type of deicer, the Elephant Alpha, equipped with the now familiar telescopic spray boom.

1983 — Fortbrand Services starts serving the GSE industry, but also expands into selling airfield equipment.

Hugh I. Hunt opens Ground Support Products, specializing in tires, rollers and casters.

1987 — FCX Systems, Inc. starts to design and manufacture solid-state frequency converters.

ERMA get its start selling GSE to Airbus.

After decades of manufacturing heavy equipment for the construction and transportation industries, Goldhofer introduces a towbarless tractor.

1987 — Matt Sheehan starts AERO Specialties, a manufacturer and distributor of new and used GSE throughout the world(see Figure 13.17).

Figure 13. 17   The National Musuem of Commercial Aviation is currently renovating this airstair
truck that serviced the Concorde. The museum is also in the process of creating an
exhibit on the history of GSE

1989 — Jamie Kaplan joins Harlan Corp. as president. Jamie develops the company's
low profile tractor that remains the company's highest-volume product.

1990s — George Prill publishes the first issue of GSE Today in February 1992.

Jim and Jamie Kaplan hire George Revere to help improving Harlan Corp. 's business
operations and market strategies. Since the late-1990s the company has expanded its product
line into the electric GSE market.

After working for various refueling companies for some 20 years, Terry Bosserman
starts selling refuelers from his house. A year later, Bosserman Aviation Equipment gets its
own address.

TLD creates its GSE division and acquires Tracma, Albret Industrie, Erma, Devtec (in
the United States and in Asia) and Lantis. By the end of the decade, TLD decides to
specialize in GSE and sells its aeronautical equipment division.

Charlatte expands to the U. S. market and opens Charlatte of America.

Elite Line Services begins providing GSE and airport equipment maintenance.

Phoenix Metal Products, Inc. begins designing and manufacturing GSE.

Ground Support Specialist LLC, starts manufacturing and remanufacturing GSE.

A. T. Juniper's commercial engine wash system originated from the military wash rigs
Juniper was designed in the 1980s. The rigs were first trailed commercially in the early 1990s
at Gatwick Airport with Virgin Atlantic using shepherd's hook type washing probes
directing the washing solution into the booster from positions behind the fan.

Stephen Parker, expands Par-Kan into GSE and other equipment for the aircraft
industry.

Patrick G. O'Brien starts MCM Engineering Inc. O'Brien was the chief engineer for
well-known GSE companies such as Hobart Brothers, Devtec (now TLD) and McCormick-
Morgan before starting MCM.

Premier Engineering & Manufacturing Inc. enters the deicing arena initially servicing a line of deicers that Premier's founder Jerry Derusha had helped build. Shortly after starting, Premier builds its own line of deicers and receives a contract for 64 units from United Airlines.

Alan J. Janis and Bruce K. Wayne open J&B Aviation Services Inc. The company initially capitalizes on its design for 400 Hz cable assembly, but expands extensively into other GSE, including PCA, baggage chutes and air-starts.

The Northwestern Motor Co. buys Wollard Airport Equipment Co.'s broad line of GSE. In 2000, the company becomes known as NMC-Wollard.

FMC buys Jetway Systems, the original creator of apron drive passenger boarding bridges, and the world's leading manufacturer of boarding bridges, solid state 400Hz inverters and fixed PCA. FMC also bolsters its airport equipment division with the acquisition of Trump deicers.

Metroplex Conveyor & Services begins fabricating and installing safety/maintenance and production platforms, and modifying and servicing machinery for the bakery and food industry as well as luggage conveyor systems inside airports. As a result of this last relationship, Metroplex develops a PCA hose trolley system that continues to be a large part of the company's business.

Lektro introduces two new electric vehicles, one for the military and the other for commercial aviation(see Figure 13.18).

Figure 13.18　Lektro first developed an electric vehicle to feed mink. Later, the company modified the same chassis and built its first electric vehicles for an FBO in Oregon

Air T buys the Simon Deicer division from Terex and subsequently renames the company Global Ground Support LLC.

The first hybrid-electric tow tractor is tested at the North Island Naval Air Station in Coronado, CA. The ISE Research Thunder Volt hybrid tractor used an Entwistle Co. MB-4 tow tractor chassis, and was the first of three such tractors placed into service with the United States Air Force and United States Navy.

1999 — Thyssen AG and Krupp merge to form ThyssenKrupp.

2000 — Illinois Tool Works creates the ITW GSE Ground Services division，which brings together Hobart Ground Power，AXA Power，Trilectron Industries and Air-a-Plane and J&B Aviation.

2001 — Cygnus Business Media buys GSE Today and renames it Ground Supportand later Ground Support Worldwide.

FMC Airport Systems re-enters the military market with the design and development of the Halvorsen loader，selected by the USAF to replace all its existing 25 K cargo loaders.

2006 — Columbus JACK acquires Regent Manufacturing.

2008 — John Bean Technologies Corporation（JBT Corporation）is formed，and becomes a publicly listed company on the New York Stock Exchange. FMC Airport Services becomes JBT AeroTech.

2009 — Velcon founder Lu Taylor's son，Dave，and grandson，Chase，sell company and launch Petroleum Equipment Aviation Refueling.

2012 — Lektro delivers its 3,700 towbarless tractor to DTW.

New Words

1. daredevil ['deədevl] *n.* 冒失鬼
2. stunt [stʌnt] *n.* 特技
3. shorthand ['ʃɔːthænd] *n.* 速记
4. tampers ['tæmpə(r)s] *n.* 打夯机
5. dubbed ['dʌbd] *v.* 给……起绰号
6. bolster ['bəʊlstə(r)] *vt.* 支持

Phrases and Expressions

1. tripod jacks 三脚架插孔
2. fiberglass filter coalescer 玻璃纤维凝聚过滤器

Notes to the Text

1. Wilbur Wright becomes the world's first ramp agent. This year（2013）marks 110 years since Orville Wright took off from the sand at Kill Devil Hills aboard the Wright Flyer into a freezing headwind of 27 miles per hour and flew about 35 yards.

威尔伯·莱特成为世界上第一个活动梯代理商。自从奥维尔·莱特从降魔山的沙地起飞驾驶莱特飞机在 27 英里每小时（1 英里每小时＝1.609 34 千米每小时）的寒风中逆风飞行了大约 35 码（1 码＝0.914 4 米），迄今已过去了 110 年。

2. Lektro，which pioneered the electric golf cart，produces a small electric aircraft tug for a Oregon FBO using a chassis originally built for an eclectic cart for area mink ranchers.

开创了电动高尔夫球车的 Lektro 为一家俄勒冈的航空地面固定基地运营商生产了一种小型电动飞机牵引车，它使用了原本为大型水貂农场主设计的电动车的底盘。

Chapter 14　Airport Ground Support Equipment:
Emission Reduction Strategies

Section 14.1　Background

Airports are vital national resources. They serve a key role in transportation of people and goods and in regional, national, and international commerce.

Increased levels of demand at airports in the United States may result in a growth in airport GSE activity and an associated increase in airport surface emissions. Local air quality and global climate change concerns, regulatory pressures, and the desire to be environmentally responsible have resulted in a growing number of airport programs around the United States looking to assess and reduce airport emissions.

Although much is known about aircraft fleets, operations, and emissions, comparatively little is known about GSE. The available GSE data are outdated, unreliable, and limited. Accurate GSE data are needed by the FAA and airport sponsors to plan adequately and to balance the growing demands of air travel with air quality concerns.

Proactive strategies that reduce surface emissions may help airports address air quality concerns. As such, research is needed to obtain additional information on GSE equipment and to identify programs and best practices that could reduce GSE emissions for GSE owners, operators, and airports.

Section 14.2　Air Emission Mitigation Strategies Applicable to GSE

This section identifies and discusses various approaches that have been implemented at airports to reduce air emissions from GSE. In addition, the available incentives, the benefits gained, and the potential barriers to attaining emission reductions associated with GSE are also discussed. For ease of comprehension, the prevailing approaches are described first followed by specific airport examples of GSE emission reduction measures.

14.2.1　Equipment-Related Approaches

Equipment-related approaches to reducing emissions from GSE characteristically comprise (a) the use of infrastructure or hardware systems as an alternative to GSE; (b) the use of add-on control devices on conventional-fuel GSE; and (c) the use of the advanced fuel combustion technologies for conventional-fuel GSE.

14.2.1.1　Infrastructure and Hardware Systems

In some cases, the primary functions of select types of GSE can be replaced by incorporating fixed point-of-use support equipment into airport terminal gate design. One common example involves terminal gate electrification through the use of (a) fixed preconditioned air (PCA) systems replacing diesel-powered air conditioning units (ACUs); and (b) 400 Hz electrical systems to replace diesel-powered ground power units (GPUs) and aircraft air start units (ASUs). Although many aircrafts use on-board jet fuel-powered auxiliary power units (APUs) to perform these necessary functions, the PCA and 400 Hz systems eliminate the need for such GSE and minimize APU use. Notably, APU usage at the gate cannot be eliminated completely as it is required during preflight checks and aircraft main engine startup.

Eligible airports can obtain funding under the FAA VALE (Voluntary Airport Low Emission) Program for these qualified infrastructure projects that reduce air emissions. For example, the Seattle-Tacoma International Airport (SEA) recently obtained VALE funding for the installation of PCA at 82 gates and the Gerald R. Ford International Airport (GRR) obtained VALE funding for PCA and 400 Hz power at five gates.

Another infrastructure GSE emission reduction example is the use of in-ground hydrant fueling systems in place of mobile refuelers thereby decreasing engine emissions associated with these trucks. Most fuel hydrant systems still require an interface between the in-ground system and the aircraft, commonly provided by an engineless fuel cart or a fuel pumping truck powered by a conventional-fuel engine.

Importantly, such infrastructure projects are less costly to install when designed as part of new facilities rather than as retrofits to existing facilities. For example, a gate electrification project may require an upgrade to the power supply to the terminal building, electrical improvements at the terminal gate, and power improvements within the gate area.

Installing a fuel hydrant system at existing airport facilities can also be relatively expensive as well as disruptive of operations in the terminal gate area because it requires belowground installation. Installing more advanced systems to replace GSE (e. g., a centralized conveyer belt-driven baggage distribution and delivery system to replace baggage tugs and belt loaders) is also possible. However, the costs and cost effectiveness for these types of infrastructure improvements are difficult to generalize and would need to be evaluated on a case-by-case basis.

14.2.1.2　Add-on Emission Control Devices

Engine exhaust after-treatment systems have been successfully used in on-road vehicles for more than 35 years to reduce emissions. In general, these control devices serve to collect and convert the exhaust emissions to more environmentally friendly compounds before they are discharged into the atmosphere. The following examples of exhaust after-treatments are applicable to GSE:

(1) Oxidation catalysts: At the most basic level, oxidation catalysts use a material such as platinum to more efficiently oxidize unburned hydrocarbons and CO in the engine exhaust to carbon dioxide (CO_2) and water.

(2) Three-way catalytic converters: These devices oxidize unburned hydrocarbons and CO to CO_2 and water, but also reduce NO_x to molecular nitrogen and oxygen. These devices have been particularly successfully in on-road vehicles in the form of catalytic converters but are currently only compatible with spark-ignition engines. The removal of lead and the lowering of the sulfur content in gasoline have further improved the effectiveness of these devices.

(3) Particulate traps: Particulate traps collect soluble and carbonaceous particulate matter in the diesel exhaust and during regeneration convert it to CO_2 and water. Because sulfur in fuel can interfere with the operation of the device, the technology requires the use of ultra-low sulfur diesel.

GSE equipped with non-road engines are characteristically "open-loop" systems that have no combustion control feedback system to adjust the air/fuel mixture. For this reason, only the oxidation catalyst technology is used on these engines (both compression ignition and spark ignition).

In those applications where the non-road engines have been retrofitted with a "closed-loop" combustion control system, the three-way catalytic converter can be used effectively to reduce emissions but limit the maximum power available for the GSE. Furthermore, because some types of GSE engines are tuned to run rich, adjusting the air/fuel ratio to run lean would limit the engine power available to the equipment.

Because these types of add-on control devices need to reach a critical temperature to allow the conversion of pollutants to take place, GSE with short duty cycles (i. e. , low load factors) may not achieve the temperatures needed for maximum conversion efficiency. For particulate traps, backpressure increases as particulate matter collects on the trap. If the operating cycle of the GSE does not include sufficient periods of high load (which promotes the necessary regeneration temperature), it can affect the performance of the equipment. The ideal condition is a high-load activity level to regenerate the trap regularly and maintain low backpressure on the trap.

One other potential constraint of note is the space requirement for the add-on devices. Since such equipment has to be retrofitted onto GSE not originally designed to accommodate it, one must consider how the placement of the add-on device can be accomplished without interfering with the intended operation and maintenance of the GSE.

14.2.1.3　Evolving Engine Technology for Conventional Fuels

In the mid-1990s, the U. S. EPA began to issue non-road engine emission standards that are being phased in over a number of years; prior to these standards, non-road engines were essentially non-regulated. In particular, initial standards for non-road compression-ignition (e. g. , diesel) engines were promulgated in 1996 and then more advanced standards were set in 2008. For non-road spark-ignition (e. g. , gasoline) engines, the U. S. EPA promulgated standards to take effect over the period from 2004 through 2008. These standards will result in significant emission reductions (in some cases, greater than 90 percent) from the non-regulated baseline as the cleaner engines meeting the emission standards to displace the uncontrolled equipment.

14.2.2　Alternative-Fuel GSE

Fuel-related solutions to reduce emissions from GSE include the use of alternative-fuel and electric-power GSE in place of conventional-fuel GSE, either through acquisition of new purpose-built equipment or retrofitting of existing equipment. Today, a variety of alternative combustion fuels are available for use in internal combustion engines that power GSE.

The primary alternative fuels known to be used in GSE include compressed natural gas (CNG), liquefied petroleum gas (LPG, also known as propane), ethanol, and biodiesel. These fuels typically generate lower air emissions than the conventional fuels; however, the relative energy content and on-airport infrastructure requirements to provide alternative fuel may reduce the overall air quality benefits associated with the use of this equipment. In addition, accounting for off-airport electric power generation impacts associated with charging electric GSE also reduces the air quality benefits of this equipment.

14.2.3　Operations/Maintenance-Related Approaches

Operators of GSE have developed specific operations and maintenance (O&M) programs for the GSE that they own. These procedures have been developed to reduce the overall cost of running GSE as well as to avoid operating delays associated with equipment breakdowns. However, there are potential air quality benefits to these O&M measures too, for example, (a) idling time restrictions; and (b) maintenance activities.

14.2.3.1　Idling Time Restrictions above and beyond Regulatory Requirements

Most GSE duty cycles consist of short periods of high-load operation followed by extended periods of idle or engine off. Over a long period of operation, the engine load factor (ratio of actual work performed to the maximum work that the engine is designed to do) can account for differing operating conditions and must be taken into account when estimating emissions. Although load factors for GSE have been developed, they may be highly uncertain on a generalized basis when attempting to account for differences across multiple units of the same type, across airlines, and across airports. Equipment idle time can vary considerably and, in extreme cases where idle periods represent the major portion of the duty cycle, the load factor approaches zero while the actual emission rate per unit work performed approaches infinity.

Idling of GSE is a common practice, particularly for diesel equipment, primarily as a convenience to the operator to maintain the equipment in a ready mode and avoid lengthy warm-up periods in cold climates. However, an engine at idle continues to emit pollutants, although at a different rate from that under higher load conditions. Imposing idling restrictions on GSE (e. g. , no idling longer than 5 min) could result in substantial emission reductions in the long term. Implementing such restrictions may be as simple as training operators to turn off the engine after use. Alternatively, an anti-idling device could be installed that automatically shuts off the engine after a pre-set period of time. For engines that need to be maintained in a "warm standby" condition for ready access or in the case where an equipment cab needs its interior temperature maintained for operator comfort, a small auxiliary unit can be integrated with the anti-idling device to keep the equipment in a ready condition while reducing overall emissions. Including the small auxiliary unit into such a system design would clearly limit the degree of emission reductions achieved in practice.

14. 2. 3. 2　Maintenance Activities

In general, maintenance activities have been developed to cost effectively limit equipment downtime in maintenance while avoiding inconvenient equipment breakdowns during operations. As part of the field surveys, maintenance activities that have potential air emissions were noted along with any available mitigation options.

14. 2. 4　Other Approaches

At least two other approaches to managing GSE emissions also exist: emission-related fees and tenant lease agreements.

14. 2. 4. 1　Emissions Fees

While no data were obtained indicating that GSE are being assessed emission fees at airports in the United States, several European airports assess fees on aircraft emissions.

12. 2. 4. 2　Tenant Lease Agreements

Several airports have begun attempting to incorporate emission reduction goals into tenant lease agreements. However, the use of such goals in lease agreements can be problematic and may not be a viable option because the goals may not be legally binding on an airline.

There are numerous constraints on U. S. airport proprietors that would limit their ability to reduce emissions associated with airport operations. These constraints include federal laws preempting certain actions by airport proprietors to regulate air carriers, the ban on passenger head taxes, the prohibition against diverting airport revenue for non-airport purposes, and the requirement to impose only reasonable and not unjustly discriminatory terms and conditions on aeronautical users (Reimer and Putnam, 2007).

14.2.5　Airport-Specific GSE Emission Reduction Measures

Presently, there are a great number and variety of GSE emission reduction measures in place (or planned) at airports of nearly every size and function located across the United States and internationally. The sponsors of these initiatives also range widely and include airlines, airport operators, and GSE providers.

Table 14.1 provides a partial sampling of these GSE emission reduction measures implemented over the past few years. For the purposes of this research project, this listing is not intended to be inclusive but rather to provide some examples of these programs. The GSE tutorial includes a more comprehensive and up-to-date compilation of these measures and programs.

Table 14.1　Sampling of GSE emission reduction measures implemented by airports

Implementer	GSE Program Details
	Example Airport Programs
Atlanta Hartsfield-Jackson International Airport (ATL)	At ATL, a new baggage system, extensive use of fueling carts in lieu of fueling trucks, and more than 200 new electric GSE units are expected to result in reductions in conventional fuel use and emissions associated with GSE. Virtually all of ATL's gates are equipped with preconditioned air and 400 Hz power, which greatly reduces the emissions that result from APU usage at the airport's gates (see also Delta Air Lines)
Boston Logan International Airport (BOS)	Delta Air Lines received a $3 million loan from Massport in 2009 for the purchase of 50 electric baggage cart tugs, 25 electric baggage conveyor belt vehicles, and charging stations as part of the replacement of terminal A. Massport has a number of other GSE emission reduction programs under way at BOS
Charlotte Douglas International Airport (CLT)	CLT introduced 10 battery-powered tugs on the Express ramps to replace their old diesel-engine counterparts, reducing N_2O emissions by as much as 70 tons
Dallas-Fort Worth International Airport (DFW)	DFW selected clean energy fuels in 2010 to construct and operate a new CNG refueling station at this airport. Most of the airport's fleet of more than 500 maintenance vehicles operates using CNG. The fleet is also fueled at another on-site CNG refueling station constructed in 2000

Continued

Implementer	GSE Program Details
Denver International Airport (DIA)	The Alternative-Fuel Vehicles (AFV) program was implemented with the construction of DIA. The GSE fleet at the airport includes 40 CNG bag tugs, nine electric bag loaders, and four electric cargo tractors. The CNG fleet at DIA is one of the largest in the country. The underground tunnel system connecting the terminal and concourses allows only CNG and electric vehicles. CNG pumping stations are available on site
Detroit Metropolitan Wayne County Airport (DTW)	In 2007, DTW received a $1.4 million VALE grant for gate power and preconditioned air for 26 gates at the new North Terminal
George Bush Intercontinental Airport (Houston) (IAH)	In 2008, IAH was awarded a $25,000 VALE grant for two new electric GSE units. This is only one of several GSE emission reduction programs undertaken by the Houston Airport System (HAS) and airlines that utilize the HAS airports
Indianapolis International Airport (IND)	In 2008, Aircraft Service International Group (ASIG) purchased seven solar-powered hydrant carts for use at IND. (Notably, ASIG also operates these same carts at Seattle-Tacoma International Airport and Fort Lauderdale International Airport)
San Francisco International Airport (SFO)	The airport used an Inherently Low-Emission Airport Vehicle Program grant to purchase alternative-fuel vehicles and infrastructure including 54 electric vehicles such as bag tugs, belt loaders, and push back tractors. The program also included the gasoline-to-propane conversion of 83 vehicles and the purchase of recharging systems for electric vehicles
Example Airline Programs	
Alaska Airlines	Alaska Airlines has converted or replaced a portion of its gas-powered fleet with cleaner-burning propane units or hybrid GSE. Approximately 10 percent of the GSE fleet has been converted to electric
American Airlines	Since 2000, American has converted approximately 30 percent of its GSE bag tractors and belt loaders from gasoline and diesel to electric. American has also installed fast electric chargers at DFW, New York JFK, Chicago O'Hare (ORD), and LAX for its GSE
Continental Airlines	Continental reports that NO_x emissions from GSE have been reduced by approximately 75% at IAH by switching to electric GSE and other emission reduction technologies
Delta Air Lines	In 2010, Delta opened its new GSE facility at ATL where it conducts the majority of the GSE fuel conversions. Delta has also announced plans to purchase approximately 600 new GSE units valued at $50 million including approximately 100 electric GSE units for airports that have the infrastructure to support electric

Continued

Implementer	GSE Program Details
Horizon Air	As of January 1, 2010, over 65 percent of the GSE fleet is electric
Southwest Airlines	As of March 2012, Southwest has purchased or converted more than 850 GSE units to electric including baggage tugs, belt loaders, lavatory trucks, carts, and push back tractors. In doing so, the carrier reduced its GSE fuel consumption by approximately 700,000 gallons annually. Additionally, Southwest has converted to gate service electricity in 61 of the 64 airports it serves, reducing APU fuel consumption by more than 15 million gallons in 2007
United Airlines	United Airlines operates about 325 electric vehicles at DIA ranging from baggage tractors and forklifts to golf carts and also operates approximately 200 natural gas vehicles including tugs, vans, and light-duty pickup trucks
Example GSE Provider Programs	
Aviapartner	Based in Brussels, Aviapartner operates 31 GSE units throughout Europe and is using a new "Visualizer" airport system in order to facilitate more efficient use of its vehicles, thereby reducing fuel use and emissions. It is also assessing the concept of "pooling" GSE for common use among airlines
Elite Line Services	Elite Line Services has converted all of Alaska Airlines' cargo operation to electric forklifts and fast charge. In addition, it is in the middle of a project to upgrade all the Anchorage International Airport (ANC) cargo forklift fleet to electric
Fraport	The operator of the Frankfurt Airport is conducting trials on hydrogenpowered GSE as well as conducting GSE use studies to improve the efficiency of GSE utilization with the objective of reducing fuel use and emissions
Menzies Aviation	Menzies Aviation has implemented electric baggage tugs at its locations and recently added 11 eTugs to its GSE fleet. As of May 2009, 30 tugs out of 110 in its fleet were electric
Rentech Inc.	Rentech has entered into agreements with several airlines in the Los Angeles Basin to use its alternative fuel in GSE. Major airline partners include American, Southwest, Delta, United, and Continental
SwissPort	A leading international ground services and cargo handling provider, SwissPort follows a strict renewal and replacement strategy for its GSE all over the world. Electric bag and cargo tractors are employed in many locations and, where not, they will be introduced over the next few years. SwissPort also works closely with major GSE manufacturers in developing modern vehicles with low fuel consumption and low emissions. All its diesel vehicles have been outfitted with filters for soot particles

This information pertains only to airside GSE and does not include the large number and wide array of other emission reduction and alternative-fuel programs at these airports.

New Words

1. add-on [ˈæd ɒn] *adj.* 附加的
2. ethanol [ˈeθənɒl] *n.* 乙醇
3. biodiesel [ˈbaiəudiːzl] *n.* 生物柴油
4. downtime [ˈdaʊntaim] *n.* （工厂等由于检修、待料等的）停工期
5. mitigation [ˈmitiˈɡeiʃn] *n.* 缓解，减轻，平静

Phrases and Expressions

1. in-ground hydrant fueling systems 地井管线加油系统
2. oxidation catalysts 氧化催化剂
3. three-way catalytic converters 三元触媒转换器
4. particulate traps 微粒过滤器

Notes to the Text

1. In some cases, the primary functions of select types of GSE can be replaced by incorporating fixed point-of-use support equipment into airport terminal gate design.

在某些情况下，一些特定航空地面设备的首要功能可由机场候机楼大门设计中整合固定使用点的支持设备来取代。

2. Importantly, such infrastructure projects are less costly to install when designed as part of new facilities rather than as retrofits to existing facilities.

重要的是，当这些基础设施项目作为新设施的一部分来设计时，其安装成本比改造现有设施更低。

3. However, the relative energy content and on-airport infrastructure requirements to provide alternative fuel may reduce the overall air quality benefits associated with the use of this equipment. In addition, accounting for off-airport electric power generation impacts associated with charging electric GSE also reduces the air quality benefits of this equipment.

然而，相对能量含量和机场内基础设施要求提供替代燃料可能会降低与使用该设备的使用有关的总体空气质量效益。此外，考虑到与电动航空地面设备充电相关的机场外发电影响也降低了该设备的空气质量效益。

Section 14.3 Economic and Environmental Challenges and Considerations with Alternative - Fuel GSE

This section reports on the primary economic and environmental considerations and challenges associated with owning and operating GSE. Because this topic is multifaceted and comprehensive, the economic elements are discussed first and the discussion of the environmental factors follows.

14.3.1　Economic Considerations and Challenges

Compared to using traditional petroleum-based gasoline and diesel fuels, using alternative fuels (i. e., substitutes for traditional liquid, oil-derived motor vehicle fuels) in airport GSE may reduce energy costs, maintenance costs, and dependence on fossil fuels. The following material compares the costs of fueling GSE with conventional, petroleum-based fuels versus alternative fuels. In addition to fuel costs, characteristics such as performance, energy content, cold weather limitations, maintenance costs, and funding opportunities are also presented.

14.3.1.1　Types of Alternative Fuel

The Energy Policy Act of 1992 (EPAct) defines an alternative fuel as a fuel that is substantially non-petroleum and yields energy security and environmental benefits. Congress passed the EPAct to reduce U. S. reliance on foreign oil by providing tax breaks and requirements for the use of alternative fuels (Sections 501 and 507) to fuel federal fleets. The EPAct considers the following fuels to be alternative options to conventional gasoline or diesel fuel:

(1) Mixtures containing 85% or greater ethanol (E85).

(2) Mixtures containing 20% or greater biodiesel meeting ASTM D 6751.

(3) Natural gas (compressed or liquefied).

(4) Liquefied petroleum gas (propane).

(5) Methanol.

(6) Hydrogen.

(7) Electricity.

To focus the evaluation of transitioning from GSE using traditional petroleum-based gasoline or diesel fuels to GSE using alternative fuels, the following paragraphs describe each alternative fuel previously listed, with the exception of methanol.

However, not all the alternative fuels listed under the EPAct of 1992 are available for widespread use in GSE. Availability, especially for biofuels and hydrogen, is particularly limited based on airport location. Moreover, methanol is not often used in the aviation industry because of its lack of widespread availability. Other limiting characteristics of methanol include its corrosive nature, low energy density (about 50% less than gasoline), and poor performance below 45°F. Therefore, although methanol may be used as a component to produce biofuels and has chemical and physical characteristics similar to ethanol, it is not discussed further.

(1)Ethanol. Ethyl alcohol, or ethanol, is a clear, colorless liquid made from fermenting a biomass in carbohydrates. Starch-based or sugar-based ethanol sources include corn grain and sugar cane; cellulose-based sources include grass, wood, and crop residues.

Low-level blends of ethanol and gasoline (less than 10% ethanol) can be used in any gasoline-powered engine without modification, although blends of less than 85% ethanol and 15% gasoline (E85) do not qualify as an alternative fuel under the 1992 EPAct. Typically, E85 is priced lower than gasoline on a gallon-for-gallon basis but more than gasoline on an energy-equivalency basis. Blends containing more than 10% ethanol are only approved for use in flexible fuel vehicles (FFVs), which are capable of running on both E85 and gasoline.

The characteristics of ethanol are as follows:

1) Availability. There are about 200 ethanol production plants located in the United States, primarily in the Midwest. As of August 2010, approximately 8 million FFVs were on U. S. roads (although only a portion of these vehicles actually use ethanol); there were 53 different 2011 FFV models from domestic and foreign automakers.

2) Energy balance and high octane. Despite some misconceptions, the total amount of energy used to produce ethanol (i. e. , by farming, shipping, and production equipment) is less than the energy released when it is burned (known as the energy balance). The energy balance for corn-based ethanol is approximately 1.24 (for 1 unit of energy produced, 1.24 units of energy are released) and it is expected to increase as technology advances. As a high-octane fuel, ethanol increases horsepower, helps prevent pre-ignition or engine knocking, and enables engines to operate at a higher compression ratio. In the United States, ethanol is often added to gasoline in a low-level blend to oxygenate the fuel and reduce air pollution emissions.

3) Cold temperature fuel gelling. Because E85 may freeze in lower temperatures, fueling stations may need to switch to a lower blend of ethanol during winter months to prevent starting problems. All FFVs can transition to E70 or other lower-level blends without any adjustments.

4) Energy efficiency. Ethanol produces less energy per gallon than gasoline, depending on the blending ratio. As the ratio of ethanol to gasoline increases, the fuel economy decreases. E85 generates 15% - 30% lower gas mileage because E85 has approximately 27% - 36% less energy content per gallon than gasoline.

5) Engine modifications. Ethanol is a strong cleaning agent and has the ability to degrade engine parts manufactured from materials such as natural rubber, plastics, and even metals over time. Therefore, E85 should not be used in existing gasoline or diesel engines without performing modifications. Many existing petroleum-based gasoline-powered vehicles can be converted to use E85 through kits approved by the U. S. EPA. A typical conversion kit mounts in a vehicle's engine compartment and continuously monitors engine and emission controls. The kit supplies supplementary fuel injection to allow for the same ethanol/gasoline compatibility as a FFV. Conversion kit costs vary by the engine type and vehicle model.

6) Storage. Ethanol has a shelf life of about 3 months although it can last for several years if it is properly sealed. The ethanol content in E10 can absorb more water than gasoline, and when the water evaporates valuable fuel components are lost, reducing the

efficiency of the fuel. A vehicle's fuel should be used or replaced within 2 to 3 weeks, or even sooner in humid conditions.

(2) Biodiesel. Biodiesel is produced by vegetable oil, animal fat, or cooking grease reacting with alcohol (typically methanol) in the presence of a catalyst. In the United States, common sources for biodiesel production are soybean oil and recycled cooking oil. B100 consists of 100% pure or "neat" biodiesel and contains no petroleum-based diesel. A blend must be at least 20% biodiesel and 80% petroleum diesel (B20) to be considered an alternative fuel under the EPAct.

To be considered fuel-grade biodiesel, B20 must satisfy the performance requirements and the defined physical and chemical properties of the American Society for Testing and Materials (ASTM) outlined in Specification D 7467. B20 meets the EPAct requirements, minimizes the limitations of high-level blends, and is the most common blend in the U. S.; therefore, this section primarily focuses on the key considerations of B20.

The characteristics of biodiesel are as follows:

1) Availability. A total of 613 biodiesel fueling stations (288 public and 325 private of various blends) were located in the United States as of January 2011. The U. S. DOE estimates that the United States has enough soy oil, feedstock, and recycled restaurant grease to provide 1. 7 billion gallons of biodiesel per year (approximately 5 percent of on-road diesel use).

2) Cold temperature fuel gelling. Low-temperature gelling of biodiesel clogs fuel filters and makes the fuel unusable. B20 may begin to gel when the temperature reaches approximately 8°F, depending on the feedstock used to produce it. For example, biodiesel produced from canola, safflower, and sunflower oils are less likely to gel in cold temperatures while coconut and palm oils (high in saturated fat) are more likely to freeze. Therefore, operators should know what feedstock was used to produce the biodiesel prior to use in cold weather. The National Renewable Energy Laboratory and the U. S. DOE do not recommend the use of high-level blends such as B100 due to concerns about cold temperature gelling (at around 32°F), material compatibility, maintenance requirements, and solvency properties. Appling additives to the fuel such as kerosene, using filter and block heaters, and/or storing vehicles indoors may help reduce the likelihood of cold temperature gelling.

3) Energy content. Similar to ethanol, as the proportion of biodiesel to petroleum-based diesel fuel increases, the energy content decreases. Biodiesel (B100) has a 7%-9% lower energy content than petroleum-based diesel fuel, which reduces an engine's fuel economy, peak horsepower, and peak torque. These changes, especially in blends greater than B20, may offset fuel cost savings.

4) Maintenance. Switching from petroleum-based diesel to biodiesel may clog fuel filters because of biodiesel's solvency characteristics. Existing sediment from petroleum-based diesel could be dislodged with the start of biodiesel use (especially higher blends), reducing fuel flow to the engine and causing a stall. If the sediment causes the filter to rupture,

sediment could travel into the fuel lines, pump, and injectors, resulting in expensive repairs. Therefore, during the initial transition from petroleum-based diesel to biodiesel (especially blends of B20 or higher), routine maintenance should be performed to check for and replace clogged fuel filters. Biodiesel blends higher than B20 have a higher viscosity and density than petroleum-based diesel, which may cause unburned fuel to bypass the piston rings and drain into the oil pan. This may cause the accumulation of engine sludge, shortening the engine's lifespan and requiring more frequent engine oil and filter changes.

5) Shelf life. Compared to petroleum-based diesel, stored biodiesel is more likely to react with oxygen and form a gel-like substance; this is a concern when using GSE that is only operated occasionally or when storing GSE for more than 6 months after the manufacture date. The higher the concentration of biodiesel in the blend, the faster it is likely to degrade. Using storage-enhancing additives and/or a dry, semi-sealed, cool container can also alleviate storage concerns. Biodiesel is an active growing environment for microorganisms because it is a greater attractant of water than petroleum-based diesel. If biodiesel is stored for long periods of time, the denser water will collect at the bottom of the fuel tank and promote microbial growth that may cause engine failure, fuel filter clogging, and corrosion.

6) Solvency. Because biodiesel is a natural solvent, high concentrations of biodiesel will soften and degrade rubber compounds that may be located in fuel hoses, gaskets, and fuel pump seals. B20 can be used in most diesel vehicles and fuel-injection equipment manufactured after 1993 without having an impact on operating performance or requiring engine modifications. The U. S. DOE has not received any reported rubber compound problems due to B20 (i. e., ruptured fuel hoses or fuel pumps) since 2006, even with older engines. However, a thorough search for incompatible rubber compounds in the fueling system should be performed prior to fueling GSE with biodiesel.

(3) Compressed Natural Gas. Commonly used to heat stoves and houses, CNG is pressurized natural gas that remains colorless, odorless, and noncorrosive. CNG primarily consists of methane drawn from gas wells, oil wells, and coal bed methane wells, although it may also consist of synthetic gas, landfill gas, and coal-derived gas in smaller quantities. Although vehicles can use natural gas as either a liquid or a gas, most vehicles use the gaseous form compressed in high pressure fuel cylinders at 3,000 to 3,600 pounds per square inch.

The characteristics of CNG are as follows:

1) Availability. CNG is typically imported through pipelines although it may also be transported as a cryogenic (super-cold) liquid. An extensive network of natural gas pipelines is presently located across the United States, connecting wellheads and electrical generation plants to residential, commercial, and industrial buildings for heating and cooling. Natural gas accounts for about one-fourth of the energy used in the United States, although only one-tenth of 1 percent is currently used for transportation fuel.

2) Performance and operating costs. No noticeable difference in horsepower, acceleration, and cruise speed exists between a CNG vehicle and a similarly sized gasoline or diesel vehicle. The cost of CNG is typically 15%-40% percent less than gasoline or diesel and the CNG market has historically been more stable.

3) Maintenance. Oil changes in a CNG vehicle are less frequent compared to a gasoline or diesel vehicle, because CNG burns cleaner, producing fewer oil deposits.

4) Storage requirements. CNG only contains about a quarter of the energy by volume of gasoline. Therefore, the driving range of a CNG vehicle is less than that of comparable gasoline and diesel vehicles, requiring more frequent fueling. Larger storage tanks can be installed to increase range, but the additional weight displaces payload capacity. Furthermore, the higher cost of the fuel cylinders and CNG tanks means that CNG vehicles cost from $3,500 to $6,000 more than their gasoline-powered counterparts.

Operating temperature during refueling must be kept below $-40℉$ to reduce liner stress. To reduce risk, all CNG tanks should have a residual pressure control system.

(4)Propane or Liquefied Petroleum Gas. LPG is a naturally forming gas composed of both petroleum and natural gas. LPG comes from either petroleum refining (45% of LPG used in the United States) or natural gas processing (55%). Because of its versatility and efficiency, LPG is commonly used for heating and cooking in rural areas of the United States that are not connected to natural gas pipelines.

LPG vehicles operate similarly to gasoline vehicles with SI engines. LPG changes to a liquid state in an LPG vehicle's fuel tank, where it is stored at a pressure of about 300 pounds per square inch. Today, most propane vehicles are conversions from gasoline vehicles.

The characteristics of LPG are as follows:

1) Availability. Propane has been used as a commercial motor fuel for over 80 years. As of 2011, there are more than 270,000 on-road LPG vehicles in the United States and more than 10 million worldwide. Many are used in fleets, including light- and heavy-duty trucks, buses, taxicabs, police cars, and rental and delivery vehicles.

2) Maintenance. LPG has an octane rating from 104 to 112 compared with 87 to 92 for conventional gasoline fuel. The higher octane rating increases power output and fuel efficiency while preventing engine knocking. Propane's low carbon and oil contamination characteristics have resulted in documented engine life of up to two times that of gasoline engines. No cold temperature problems are associated with LPG since the fuel mixture (propane and air) is completely gaseous. Propane operating costs for fleet vehicles range from 5%-30% less than that for conventional or reformulated gasoline vehicles.

3) Performance. Since LPG is less dense than gasoline, power may decrease, but operators rarely notice this loss. LPG fleet operators have reported that horsepower and torque capabilities, as well as vehicle cruising speed, are roughly comparable to those for gasoline vehicles. Fuel economy on new engines is also comparable to that of gasoline

engines.

4) Refueling. LPG vehicles have a refueling rate of approximately 10 to 12 gallons per minute, which is comparable to that of gasoline; and presently approximately 10,000 refueling stations are located across the country.

5) Dedicated LPG vehicles. The availability of dedicated LPG vehicles has declined. No LPG passenger cars or trucks have been produced commercially in the United States since 2004. However, certified installers can retrofit vehicles to run on propane. Since the LPG is stored in high-pressure fuel tanks, separate fuel systems are required for bi-fuel vehicles that run on both LPG and conventional fuels. Propane conversions for light duty vehicles from gasoline to dedicated propane cost roughly between $4,000 and $12,000.

6) Energy content. LPG has about 25 percent less energy than a gallon of gasoline, increasing fuel consumption and reducing range. As with CNG vehicles, larger storage tanks can be installed to increase range, but the additional weight will displace payload capacity.

(5) Electric vehicles. Electric vehicles (EVs) are powered exclusively by an electric motor. Most EVs operate with electricity that is stored in a battery that must be recharged by plugging into a suitable outlet. Batteries are also recharged by regenerative braking, a method of storing the kinetic energy from braking into elastic potential energy that can be redistributed and used to power the car. Whenever an EV is not accelerating, the vehicle's momentum can be used to generate electricity. EVs can run on either alternating current (AC) or direct current (DC) power. Unlike vehicles powered by fossil fuels, EVs can also receive their power from nuclear power, solar power, tidal power, wind power, or other sources.

The characteristics of electric vehicles are as follows:

1) Availability. Presently, approximately 10% of the 72,000 GSE units currently in use in the United States are electric. Thus, more GSE units are electric than any other alternative fuel type.

2) Operational costs. Compared to the volatile cost of fossil fuels, the price of electricity is much more stable. The fuel cost of driving an EV is normally less than that for a gasoline or diesel vehicle, although actual cost depends on the cost of electricity per kilowatt-hour and the energy efficiency of the vehicle. Estimating that electricity costs 13 cents per kilowatt-hour, the fuel for an EV with an energy efficiency of 3 miles per kilowatt-hour costs about 4 cents per mile. This translates to only $1 per gallon if 25 miles per "gallon" is assumed. EV charging rates may also vary by time of use (peak vs. non-peak) and season.

As an example, the Metropolitan Airports Commission purchased a flat-bed two-seater Cushman Motors e-Ride exv2 electric utility vehicle for $22,265 to be used by parking management staff. The utility vehicle contains a 72 V AC motor with a driving range of 45 to 55 miles per charge. At Minneapolis-St. Paul International Airport (MSP), the EV can be powered for a cost of approximately $202 per year. Comparatively, MSP pays approximately $818 per year to fuel a Ford Escape Hybrid and $1,653 to fuel a Ford F-150 pickup truck.

Some EV manufacturers include warranties to cover batteries for approximately 80,000 to 100,000 miles. Since the battery is expensive to replace, operators should consult with the dealer prior to purchasing an EV to come to a clear consensus on the expected battery life and warranty.

1) Energy efficiency. An EV can convert approximately 75% of the chemical energy stored in the batteries to power the wheels while an internal combustion engine (ICE) can only convert about 20% of the energy stored in gasoline. In stop-and-go operations, EVs are even more efficient, since electricity is not consumed while the vehicle is stopped (no idling).

2) Performance. The acceleration, speed, and handling of an EV can equal or exceed that of conventional ICE vehicles. EV operation is also much quieter than ICE vehicles. However, EVs have limited towing ability over longer distances and thus cannot be used for some operations (e. g. , towing an aircraft from a gate to a maintenance hangar). EVs may also have difficulty hauling larger loads up inclined ramps.

3) Maintenance. EVs require less maintenance than ICE vehicles. No oil changes, belts, spark plugs, fuel injectors, or emissions tests are involved. The completely sealed cooling systems do not require refilling, replacement, or flushing. EVs also have fewer moving parts, which results in reduced inventories, lower operating capital, and fewer spare parts. Regenerative braking also reduces wear and tear on brake pads.

4) Conversions. The costs of converting a gasoline-powered vehicle to an EV can be high although it could potentially be offset by lower operational and maintenance costs. Converting a GSE powered by an ICE to electric power requires completely removing the engine and adding a battery pack, cabling, electric motor, and metering equipment. Therefore, converting to electric power is most cost effective when the vehicle's engine has reached the end of its life cycle or needs to undergo expensive repairs. Instead of purchasing a new ICE, converting to electric power could be considered.

Converting to electric power does not require certification from the U. S. EPA. However, vehicles that have a gross vehicle weight rating of less than 10,000 pounds, use more than 48 V of electricity, and have a maximum speed greater than 25 mile/h must meet Federal Motor Vehicle Safety Standard 305: Electrolyte Spillage and Electrical Shock Prevention.

5) Range anxiety. EVs have a limited battery storage capacity that must be replenished by plugging the EV into an electrical power source. Neighborhood electric vehicles (NEVs), commonly found at airports, are limited to operating on roads with speed limits of 35 mile/h. However, since NEVs are limited to speeds of 35 mile/h, NEVs are not considered light-duty vehicles and are not eligible for fleet credit under the EPAct of 1992. Battery packs are also heavy, take up considerable vehicle space, are expensive, and may

need to be replaced over the life cycle of the EV.

6) Charging stations. The National Electrical Code (NEC) has established three distinct plug-in electric vehicle (PEV) charging station levels. Each NEC level describes the amount of power that can be supplied to the vehicle to be charged (the more power delivered, the faster the charge). The three NEC levels are defined in Table 14. 2.

Table 14. 2　National electrical code plug-in electric vehicle charging levels

Charging Level	Voltage(AC) V	Current A	Power/ kV · A	Input Phase	Standard Outlet	Estimated Full Charge Time
Level Ⅰ	120	12	1. 44	Single	NEMA 5-15R (Standard 110 V outlet for U. S.)	8 – 14 h
Level Ⅱ	208/240	32	6. 7/7. 7	Single	SAE J1772/3	4 – 8 h
Level Ⅲ	480	400	192	Three	No standard. Some adopting Tokyo Electric Power Company	< 1 h

NEC Level Ⅱ charging is the EV industry standard. The Society of Automotive Engineers (SAE) has approved a standard plug known as SAE J1772.

The cost to provide recharging outlets at existing parking sites can be expensive. The cost for a Level Ⅱ station, which includes engineering, permitting, hardware, weatherproofing, and service costs, is approximately $10,000 per outlet for the first two new outlets; for more than two outlets, the costs would drop to approximately $2,000 per outlet. Installation of recharging stations at surface parking lots is typically more expensive, because trenching is typically required.

At some airports, GSE may be able to share power with the electric motor used to power the jetway for passenger boarding since it is only used a few minutes per hour. The electrical circuit may be able to support charging stations when the jetway motor is not being used. The circuit can also reduce installation costs since wire and conduit runs are shorter.

Other factors to consider prior to installing an EV charging station include airport layout, regulations, and traffic patterns to and from charging stations.

7) Recharge time. Fully recharging an EV may take from 4 – 8 h, although fast-charging stations can be purchased to limit recharging times. However, even a "quick charge" to 80% capacity can take over 15 min. If conveniently located, GSE can be plugged into recharging stations overnight or during break periods/downtime (by recharging EVs overnight operators may be able to take advantage of off-peak rates to decrease the cost of powering EVs).

(6) Hydrogen. The simplest and most abundant element in the universe, hydrogen can be produced from fossil fuels, biomass and other renewable energy sources, or by electrolyzing water.

Hydrogen vehicles either convert the chemical energy of hydrogen into torque by combustion or electrochemical conversion in a fuel cell. Similar to ICE vehicles that combust gasoline or diesel fuel, hydrogen vehicles with ICEs burn hydrogen in the engine to produce energy that powers the vehicle. In a fuel cell, hydrogen reacts with oxygen to produce electricity that powers an electric traction motor.

The characteristics of hydrogen are as follows:

1) Energy content. At 52,000 BTU per pound, hydrogen has the highest energy content per unit weight of any known fuel; this is approximately three times the energy of a pound of gasoline. Therefore, the process of converting hydrogen to energy using engines or fuel cells is much more efficient than the comparable gasoline counterparts.

2) Availability. No fuel-cell vehicles powered with hydrogen are available yet for sale. Hydrogen is available only as an industrial or scientific chemical product, not as a bulk fuel. No bulk hydrogen distribution infrastructure exists near the scale of that for fossil fuels. Transporting hydrogen is also difficult since it must be refrigerated to maintain a liquid state.

3) Distribution. Generating hydrogen, transporting it via truck or pipeline, and storing it aboard the vehicle may be an inefficient and expensive process. Similar to CNG, hydrogen typically requires heavy tanks or insulating bottles if stored as a super-cold liquid.

4) Lifespan. Hydrogen fuel cells have less than half of the lifespan of a traditional ICE vehicle (about 1,900 h or 57,000 miles).

5) Storage requirements. The amount of energy contained in 1 gallon of gasoline is equivalent to the amount of energy stored in 2.2 pounds of hydrogen gas; thus, a light-duty fuel-cell vehicle must store from 11 – 29 pounds of hydrogen to drive 300 miles or more. Storing this much hydrogen on a vehicle would require more space than the trunk size of a typical car. Since hydrogen technology is still in its infancy, expensive, and not readily available, using hydrogen in GSE was not evaluated further.

(7) Hybrid Electric Vehicles. In a hybrid electric vehicle (HEV), a small ICE is connected to an electric generator. Electric power is combined with gasoline, diesel, or an alternative fuel to power the electric traction motors, which in turn power the wheels of the vehicle. The drive is electric (battery powered) at low speeds and powered by the main ICE at high speeds.

The EPAct of 1992 did not originally consider HEVs as alternative-fuel vehicles. However, the National Defense Authorization Act for Fiscal Year 2008 amended the 1992 EPAct to include four new categories of vehicles as "alternative-fueled vehicles" under Section 30B of the Internal Revenue Service Code, including "a new qualified hybrid motor vehicle".

The characteristics of HEVs are as follows:

1) Range/gas mileage. Unlike in EVs, the batteries in HEVs do not need to be plugged in to recharge. HEVs also avoid the inconvenience of long charging times and the cost of charging infrastructure. Some HEVs can be driven up to 70 miles on a single gallon of

gasoline. The electric motor provides additional power to assist the engine in accelerating, passing, or hill climbing, allowing for a smaller, more efficient ICE to be used. In some HEVs, the electric motor solely provides power for low-speed driving conditions where ICEs are least efficient. HEVs usually cost 5 - 7 cents per mile to operate while conventional ICE vehicles cost 10 - 15 cents per mile. To prevent wasted energy from idling, HEVs automatically shut off the engine when the vehicle comes to a stop and restart it as soon as the accelerator is pressed. Like in EVs, regenerative braking systems in HEVs capture deceleration energy and convert it to electricity to propel the vehicle and increase overall efficiency.

2) Maintenance. HEVs must undergo the same maintenance procedures as conventional vehicles although spare parts may be more difficult to find and have a higher cost.

14.3.1.2 Cost of Conventional Fuels vs. Alternative Fuels

This subsection compares the cost of conventional petroleum-based fuels with the cost of alternative fuels, including the historical, current, and forecast costs. Alternative fuels are typically not subject to dramatic price fluctuations because they are less dependent on the price of crude oil (unlike petroleum fuel prices). However, depending on the type of alternative fuel and/or the percentage blend, some alternative fuels still fluctuate based on crude oil prices, national security, spikes in the cost of agricultural products, and other factors.

(1) Historical cost. Figure 14.1 depicts the 11-year average cost of gasoline and diesel fuels compared to alternative fuels from 2000 - 2011 per gallon of gasoline equivalent (a discussion of the gallon of gasoline equivalent values is provided in the subsection Fuel Operating Cost Considerations).

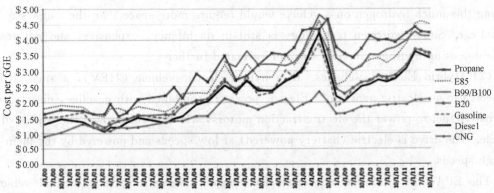

GGE = gallon of gasoline equivalent

Source: U.S. Department of Energy (2012b).

Some figures in this report have been converted from color to grayscale for printing. The electronic version of the report (posted on the web at www.trb.org) retains the color versions.

Figure 14.1 U.S. 11-year average fuel prices in cost per gallon of gasoline equivalent

(2) Current cost. The national average cost per gallon for gasoline, diesel, and alternative fuels in January 2012 is provided in Table 14.3. As shown, CNG had the lowest cost per gallon at $1.24 less than gasoline (on an energy-equivalent basis); E85 was 23

cents less per gallon than gasoline; and propane cost 29 cents less per gallon than gasoline. Compared to the cost of diesel, B20 prices were 9 cents higher and pure biodiesel (B100) prices were 34 cents higher per gallon.

Table 14.3　Average cost per gallon of fuel

National Average Cost Per Gallon	January 2012
Gasoline	$ 3.37
Diesel	$ 3.86
Compressed natural gasa	$ 2.13
Ethanol (E85)	$ 3.14
Propane	$ 3.08
Biodiesel (B20)	$ 3.95
Biodiesel (B100)	$ 4.20

Compressed natural gas is measured on an energy-equivalent basis (gallon of gasoline equivalent).

Source: U. S. Department of Energy (2012)

According to the U. S. Energy Information Administration, the world average gasoline and diesel fuel prices are predicted to increase from $ 2. 35 and $ 2. 44 per gallon, respectively, in 2009 to $ 3. 69 and $ 3. 89 per gallon, respectively, in 2035 (in 2009 dollars). Annual average diesel prices are anticipated to be higher than gasoline prices because of increased demand for diesel. With the estimated increases in the cost of gasoline and diesel fuels, alternative fuels are expected to become more affordable. For example, in 2022, the retail price of gasoline is anticipated to be $ 3. 43 per gallon while the price of E85 is anticipated to be $ 2. 68 on a gallon of gasoline equivalent (GGE) basis (The following paragraphs discuss the gasoline equivalent basis).

14.3.1.3　Fuel Operating Cost Considerations

When viewed separately from other operational cost factors, the cost per gallon of a fuel may be misleading. The energy content and location/availability of an alternative fuel, which are described below, should also be factored in to provide a more accurate estimate of fuel cost.

(1) Energy Content. Because of differing energy content per gallon for fuels, the price paid per unit of energy content differs from the price paid per gallon. Prices for the alternative fuels in terms of cost per GGE are generally higher than their cost per gallon because of their lower energy content. For example, 1. 41 gallons of E85 are required to do the same work as 1 gallon of diesel fuel. Therefore, although E85 was priced at $ 3. 14 per gallon compared to that of gasoline at $ 3. 37 in January 2012, the cost for E85 is actually more expensive than gasoline on a GGE basis ($ 4. 44 per gallon). Table 14. 4 lists conversion factors that should be used to achieve a level playing field as either GGE or gallon

of diesel equivalent (GDE).

Table 14. 4 Energy content equivalency factors

Fuel	Lower Heating Value	Conversion Factor to Dollars per Gallon of Gasoline Equivalent	Conversion Factor to Dollars per Gallon of Diesel Equivalent
Gasoline	115,400 BTU/gal	1.00	—
Diesel	128,700 BTU/gal	—	1.00
Compressed natural gas	960 BTU/ft3	1.00	1.12
Ethanol (E85)	75,670 BTU/gal	1.41	1.58
Propane	83,500 BTU/gal	1.38	1.54
Biodiesel (B20)	—	0.91	1.02
Biodiesel (B100)	117,093 BTU/gal	0.99	1.10

(2) Location/Availability. The price of an alternative fuel is dependent upon where the fuel is manufactured and blended and where fueling infrastructure is located. For example, while gasoline and diesel consumption is highest along America's coasts, most ethanol plants are concentrated in the Midwest where it is absorbed in local and regional markets.

Price also varies depending on whether the purchaser of alternative fuel buys in bulk supply from the producer via rail, pipeline, or barge (spot price); a limited supply from a refueling truck (rack price); or at a traditional pump (retail price). Furthermore, the retail price is also influenced by whether the fueling station is branded or unbranded and the degree of competition in the vicinity of the station.

Biofuels are not often shipped via pipeline so they are generally blended at the local wholesale terminal. Not all fueling stations sell high percentage biofuel (ethanol and biodiesel) blends such as E85 or B100. Biofuel prices are contingent upon seasonal availability; factors involved in growing, processing, and distributing biofuels can contribute to price fluctuations. The use of low-level biofuel blends such as E10 and B5 can be influenced by local air quality regulations or federal and state renewable fuel standards. Additionally, as more alternative-fuel producers and suppliers enter the market, competition will likely increase the available supply of biofuels, potentially lowering the price of biofuels.

(3) Other Fuel Cost Considerations. Federal, state, and local tax provisions may be applicable for certain fuels used for off-highway business use. Fuel cost adjustments may include taxes or tax credits such as excise taxes, alcohol fuel credits, biofuel tax credits, gasoline tax refunds, etc.

Additionally, bulk fuel purchase discounts or in the case of electric vehicles, off-peak electrical charging usage should be considered in the overall fuel costs for each GSE fuel type.

14.3.1.4　Non-Fuel Cost Considerations

Beyond the costs for purchasing the fuel, there are indirect costs that should also be considered when evaluating alternative-fuel GSE.

(1) Labor. Labor costs represent the single largest expense of the total cost of owning and operating GSE. As shown in Table 14.5, ramp labor represents over 80 percent of the total cost to own and operate baggage tractors. Alternative-fuel equipment should be evaluated to determine if their operation could reduce labor costs and/or free up labor resources for other non-fuel emissions-related ground handling operations. Alternative-fuel equipment should be evaluated for the potential to reduce the time to adequately train the operator and/or improve operational learning curves and efficiencies while reducing safety-related incidents and accidents.

Table 14.5　Example of baggage tractor (gasoline/diesel) maintenance cost considerations

Cost Type	Per Tractor	Total (25 Tractors)	Costs per Tractor			
			Annual	Percentage of Total Annual	Annual Non-labor	Percentage of Total Annual Non-labor
Ownership costs						
Initial cost	$ 25,000	$ 625,000				
Average life (years)	20		$ 1,250	1.2%	$ 1,250	10.8%
GSE storage facility (capital costs)a ($ per 20-year period)		$ 100,000				
Average storage costs ($ per tractor)			$ 200	0.2%	$ 200	1.7%
Residual resale value	$ 2,500	$ 62,500	− $ 125	−0.1%	− $ 125	−1.1%
Total ownership costs per year			$ 1,325		$ 1,325	
Operating Costs						
Utilization/day (hours)	8	200				
Utilization/year (hours)	2,920	73,000				
Lifetime utilization (hours)	58,400	1,460,000				
Maintenance (annual hours)	100	2,500				
Maintenance (annual hours)	100	2,500				
Maintenance labor rate ($ 40 per hour)			$ 4,000	3.9%		
Maintenance parts			$ 2,000	1.9%	$ 2,000	17.2%
Annual training costs			$ 1,000	1.0%	$ 1,000	8.6%

Continued

Cost Type	Per Tractor	Total (25 Tractors)	Costs per Tractor			
			Annual	Percentage of Total Annual	Annual Non-labor	Percentage of Total Annual Non-labor
Ramp labor ($30 per hour)			$87,600	84.9%		
Fuel burn per hour ($2.50 per gallon @ 1 gallon per hour)			$7,300	7.1%	$7,300	62.8%
Total operating costs per year			$101,900		$10,300	
Total annual costs			$103,225	100.0%	$11,625	100.0%
Average cost per hour of GSE utilization			$35.35			

Other labor cost reduction strategies include the adjustment to the work schedule. The labor schedule for the ground servicing of aircraft is derived by the aircraft schedule. Where it may not be possible for an airline to support point-to-point passenger service, the hub-and-spoke schedule enables the airlines, especially large "legacy" carriers, to support a vast system network. Aircraft arrive from the spoke stations to the hub station in a scheduled arrival bank. Passengers arriving at the hub station then connect to a closely timed departure bank resulting in the shortest overall travel time for the connecting passenger. While this schedule is preferable to the passenger, a "peaked" hub-and-spoke schedule places the greatest demand on GSE and associated labor resources and increases the potential for airline arrival and departure delays and resultant aircraft fuel expenditures and emissions.

Alternatively, "de-peaking" the schedule places less demand on labor and equipment resources. In a de-peaked schedule, arriving and departing aircraft are scheduled more uniformly throughout the day; thus, fewer resources are required for any given hour in the schedule compared to that of a peaked schedule.

For example, consider a simplified 16-aircraft operation at an airport (eight arrivals and eight departures per day): a peaked schedule could consist of four departures at 8:00 a.m., four arrivals at 12:00 noon, four departures at 4:00 p.m., and four arrivals at 8:00 p.m.; a uniformly distributed schedule could have one departure at 8:00 a.m., an arrival at 9:00 a.m., and alternating arrivals and departures each hour throughout the day. While the GSE fuel cost and GSE emissions may be identical in each of these cases for any given GSE type, the variable labor and GSE inventory requirements for the peaked schedule could be as much as four times that of the de-peaked schedule in this example. It should be noted, however, that in the de-peaked schedule, the average passenger connect times may be expected to increase, which may result in the loss of market share to airline competitors, depending on the alternatives that were available to the passenger. It is the airline passenger that creates the demand for air travel, the demand that the travel occurs during certain times of the day, and the demand that layovers between connecting flights to be limited. To de-peak air travel may require re-regulation and subsequent restructuring of the airline industry.

(2) Other non-fuel cost considerations. Federal, state, and local tax provisions may be applicable for certain vehicles used in off-highway-related businesses. In addition to the purchase price of equipment, net adjustments should include tax credits such as credits for the purchase of alternative-fuel vehicles. Other tax-related considerations would include the applicability of business-related Section 179 depreciation expense for GSE. Other cost considerations include GSE insurance coverage for damage, liability, and business interruption loss, cost of capital (funding costs to purchase GSE), and administration overhead.

14.3.1.5　Cost of Alternative-Fuel GSE

The cost of an alternative-fuel GSE vehicle varies heavily based on several airport-specific factors such as the type of GSE, quantity of GSE purchased (i. e., bulk rates), existing contracts with the airport, the manufacturer, performance capabilities, custom features, lighting and signage, etc. Alternative-fuel GSE, particularly electric, LPG, and CNG vehicles, normally have a higher up-front cost than gasoline or diesel GSE. In some cases, low-level blends of ethanol and biodiesel may be useable in existing vehicles without modifications (although additional maintenance is required and caution should be taken during the transitioning process). On average, the initial cost of electric GSE can be 30%-35% more expensive than gasoline GSE. Similarly, higher cost of the fuel cylinders and tanks means that light-duty CNG and LPG vehicles cost from $3,500 to $6,000 more than their gasoline-powered counterparts.

(1) Life-cycle costs, which incorporate fuel cost savings, maintenance costs, vehicle lifespan, and infrastructure, must also be considered or else it would not make financial sense to convert to electric, LPG, or CNG with existing technology. Cost savings are usually realized when considering life-cycle cost benefits. Additionally, non-cost factors, such as the benefits from improved air quality, GSE performance, and airport marketing and public image, should also be considered.

(2) Maintenance costs. When considering total operating costs, the GSE airport administrator must also consider maintenance costs, which include not only maintenance materials and supplies, but also the hourly rates for mechanics' wages. The GSE administrator should ensure that all GSE propulsion systems are warranted by the original equipment manufacturer to operate on alternative fuels; however, converted propulsion systems are typically not included under the vehicle warranty. As a representative example (for consideration purposes only), Table 14.5 shows how the maintenance cost (labor and parts) of a typical gasoline or diesel baggage tractor can amount to a large percentage of the

total life-cycle cost.

1) Biofuel GSE. Because biofuels (ethanol and biodiesel) are natural solvents, they may degrade rubber compounds found in fuel hoses, gaskets, and fuel pump seals (especially higher blends); this degradation could result in clogged filters, increasing maintenance costs compared to conventional fuels (although engines manufactured after 1993 typically do not experience problems). If the filter ruptures, sediment could travel into the fuel lines, pump, and injectors, causing expensive repair needs. Also, since biofuels are greater attractants of water than petroleum-based fuels, they promote microbial growth in fuel tanks. Microbial growth may cause engine failure, fuel filter clogging, and corrosion. Therefore, if GSE uses ethanol or biodiesel, routine maintenance should be performed to check for and replace clogged fuel filters. The GSE administrator should prepare for increased engine fuel filter and fuel storage filter replacements and maintain equipment inventories accordingly. Prior to fueling GSE with high blends of biofuels, precautions should be taken to verify that no incompatible rubber compounds are in the fueling system.

Maintenance personnel should change the fuel filter following the use of the first tank of biofuels, and fuel filters at dispensing units should be changed when operators notice that the flow of fuel slows. Periodic fuel testing may also be required to ensure fuel quality. Similarly, maintenance personnel should periodically check for free water at the bottom of fuel storage tanks. If biofuels must be stored for over 6 months, additional maintenance and labor may be required to prevent and/or mitigate fuel contaminated by water (e. g. , seasonal fuel tank draining).

To reduce the potential for cold temperature fuel gelling of biofuels, the GSE administrator may need to purchase additives such as kerosene, filter and block heaters, and/or indoor storage space, adding to the maintenance cost.

2) Electric GSE. EVs require no oil changes, belts, spark plugs, fuel injectors, or emissions testing; do not require refilling, replacement, or flushing of cooling systems; and have smaller engine part inventories. EVs (as well as HEVs) also reduce wear and tear on brake pads through regenerative braking, a process that converts kinetic energy from braking to electricity that is stored in the battery. Therefore, maintenance costs (parts and labor) are less than for GSE fueled with gasoline, diesel, or biofuels.

Comparing maintenance costs per hour of conventional fuel GSE to electric GSE is inaccurate since there is no idling time in an EV. Thus, when considering the maintenance cost per hour of a gasoline or diesel GSE to be equal to the cost per hour of an electric GSE, the electric GSE can accomplish 65%-70% more work for the same amount of maintenance; if maintenance is scheduled by hours, a gas unit is maintained almost 2. 5 times more often than an electric.

3) CNG and LPG GSE. The oil in a CNG vehicle does not need to be changed as frequently as a gasoline or diesel vehicle because CNG burns cleaner, producing fewer oil deposits. LPG has an octane rating from 104 to 112 (compared with 87 to 92 for

conventional gasoline fuel), which helps prevent engine knocking. Because of LPG's low carbon and oil contamination characteristics, the engine life of a LPG vehicle can be up to two times that of gasoline engines. Unlike with biofuels, no cold temperature problems are associated with LPG since the fuel mixture is completely gaseous.

14.3.1.6　Training

Compared to conventional-fuel GSE, training costs for alternative-fuel GSE may be higher. Training may help GSE operators identify when GSE charging or alternative-fuel infrastructure is malfunctioning and when potential safety hazards exist. For example, since LPG and CNG are clear and odorless, GSE operators may need to be briefed on adding an odorant to the fuel mixture and identifying signs of leaks in fuel tanks.

Operators of electric GSE should also be informed of the charging time required, when the GSE needs to be recharged to ensure demand is met, and the best time to charge the vehicle if peak electrical usage rates apply. To reduce fuel consumption and maintenance costs, the GSE administrator may consider providing fuel-efficient driving and vehicle-operating training annually to GSE drivers, regardless of the fuel type. The training can help ensure that GSE are used as intended and that driving techniques are used that reduce fuel consumption, greenhouse gas emissions, and accident rates.

14.3.1.7　Cost of Infrastructure

New fueling infrastructure may be necessary to support a fleet of alternative-fuel GSE. In addition to costs, the space available to accommodate new fueling infrastructure must also be considered.

For example, electric charging infrastructure, LPG, or CNG fuel tanks may be required if no existing infrastructure nearby the airport is available. Although ethanol and biodiesel could be stored in existing gasoline and diesel infrastructure (after appropriate cleaning), supplementary fuel tanks would still be required unless the entire fleet is transitioned to run on biofuels.

CNG and LPG fueling stations have high installation costs; for example, since a CNG fueling facility requires dedicated supply lines, compression apparatus, storage cylinders, and special dispensers, the construction cost ranges from $400,000 to $600,000. The high cost also factors in the need for CNG and LPG fuel tanks to be designed to withstand high internal pressures and be resistant to accidental punctures.

Electric charging infrastructure can be expensive at an airport without sufficient existing electric power available. Although some electric GSE could be plugged into a traditional 120 V outlet, the time to fully charge the vehicle could take over 8 h. A Level II or Level III "quick charge" station is likely required to satisfy fleet demand during peak air travel periods. The cost of a charging station can be anywhere from $10,000 (Level II) to $60,000 (Level III) depending on existing electrical outlets, wiring, power demand, capacity, and the quantity purchased.

However, bridge electric power sharing or other opportunities may be available to extract power for charging without the need for additional infrastructure (or could reduce installation costs of new infrastructure). For instance, a jet bridge only uses the power that is supplied to it for about 5% of the day; the remaining 95% could be used for electric GSE charging.

14.3.1.8 Life-Cycle Cost Considerations

Since alternative-fuel GSE and supporting infrastructure typically have a higher initial cost than conventional-fuel GSE, airports with higher annual fuel consumption rates may have a quicker return on investment when purchasing alternative-fuel and/or electric GSE compared to lower fuel-use airports. The break-even fuel cost varies based on the type of GSE, the purchase price, available funding, required maintenance, type of fuel used, infrastructure costs, and other factors. For instance, using electric bag tractors, belt loaders, cargo loaders, lavatory service trucks, and narrow-body aircraft tractors reduces fuel, maintenance, and high spare-part and equipment costs.

As an example of life-cycle cost considerations, the cost-benefit analysis of electric GSE performed by Idaho National Laboratories is described in the following paragraphs.

Idaho National Laboratories GSE cost-benefit analysis study. In February 2007, Idaho National Laboratories performed a study to evaluate the costs associated with operating baggage tractors, belt loaders, and pushback tractors. A cost model was developed to assist airlines and other stakeholders in future evaluations of deploying GSE. The approach included visiting four airports and working with two airlines to obtain data on GSE capital, operating, maintenance, and infrastructure costs.

The study found that electric GSE has lower operating costs than ICE GSE for the baggage tractor, belt loader, and pushback tractor. Capital costs for new ICE GSE are significantly lower than for new electric GSE. The payback time for electric GSE ranges from 3 to 7 years when no cost-sharing is provided. With cost-sharing and/or grants, the payback time for electric GSE can be reduced to 3 years or less, with life-cycle cost savings accruing over the life of the GSE.

The study also showed that converting old ICE vehicles to electric or implementing group purchases can help lower the cost of electric GSE. Techniques such as utilizing existing bridge supply power and utilizing smart power-sharing charge systems to reduce supply requirements can be used to help lower infrastructure costs.

New Words

1. methanol ['meθənɒl] *n.* 甲醇
2. octane ['ɒktein] *n.* 辛烷值
3. gelling ['dʒeliŋ] *n.* 胶凝(作用),凝胶化(作用)
4. fahrenheit ['færənhait] *adj.* 华氏的,华氏温度计的
5. feedstock ['fi:dstɒk] *n.* 进料,给料(指供送入机器或加工厂的原料)